Composite Materials: Properties, Design and Applications

Composite Materials: Properties, Design and Applications

Editor: Gerald Brooks

MURPHY & MOORE
www.murphy-moorepublishing.com

Published by Murphy & Moore Publishing
1 Rockefeller Plaza,
New York City, NY 10020, USA
www.murphy-moorepublishing.com

Composite Materials: Properties, Design and Applications
Edited by Gerald Brooks

International Standard Book Number: 978-1-63987-121-6 (Hardback)

Cataloging-in-Publication Data

Composite materials : properties, design and applications /
edited by Gerald Brooks.
p. cm.
Includes bibliographical references and index.
ISBN 978-1-63987-121-6
1. Composite materials. 2. Engineering design.
3. Materials. I. Brooks, Gerald.
TA418.9.C6 C66 2022
620.118--dc23

Contents

Preface

Materials formed by joining two or more materials which exhibit the desired physical and chemical properties are called composite materials. Plywood, fiberglass and reinforced concrete are some of the examples of composite materials. Being lighter and stronger than conventional materials, they find extensive use in the fields of aerospace, automobile, construction and orthopedics. Composite materials are classified into metal matrix composite, ceramic matrix composite and reinforced-plastics matrix composite. On the basis of structure, they are further bifurcated into sandwich structured composites and monolithic structured composites. Examples of fibres used for reinforcements include metal fibres, boron fibres and carbon fibres. Composite materials are manufactured using the techniques of casting, braiding, winding, moulding and forming. This book traces the progress of this field and highlights some of its key concepts and applications. It includes some of the vital pieces of work being conducted across the world, on various topics related to composite materials. As this field is emerging at a rapid pace, the contents of this book will help the readers understand the modern concepts and applications of the subject.

Various studies have approached the subject by analyzing it with a single perspective, but the present book provides diverse methodologies and techniques to address this field. This book contains theories and applications needed for understanding the subject from different perspectives. The aim is to keep the readers informed about the progresses in the field; therefore, the contributions were carefully examined to compile novel researches by specialists from across the globe.

Indeed, the job of the editor is the most crucial and challenging in compiling all chapters into a single book. In the end, I would extend my sincere thanks to the chapter authors for their profound work. I am also thankful for the support provided by my family and colleagues during the compilation of this book.

Editor

1

On the Behavior of FRP-to-concrete Adhesive Interface: Theoretical Models and Experimental Results

Enzo Martinelli, Antonio Bilotta, Ciro Faella and Emidio Nigro
Department of Civil Engineering, University of Salerno, Fisciano (SA),
Department of Structural Engineering, University of Naples "Federico II", Naples
Italy

1. Introduction

Fiber-reinforced polymers (FRP) are more and more commonly employed for strengthening existing structures of both reinforced concrete (RC) and masonry. Since FRP sheets (cured in situ) or plates (preformed) are externally bonded on a concrete or masonry substrate, the issue of adhesion on those materials generally controls the effectiveness of strengthening in members stressed either in bending or shear (Motavalli & Czaderski, 2007).

The use of composite materials for structural strengthening of civil structures and infrastructures began with some pioneering application at the middle of the '80s (Meier, 1987) of the past century.

Plenty of experimental work and theoretical investigations have been carried out in the following years with the aim of demonstrating the feasibility of strengthening civil structures by means of composite materials (Swamy et al., 1987; Meier, 1995). However, composite materials were already widely used in other fields of structural engineering, such as aerospace (Hart-Smith, 1973), aeronautics and, later, automotive. Thus, the initial research activities about the possible use of composites in civil structures were not mainly focused on the behavior of composites themselves. They were rather intended at addressing two main issues regarding, on the one hand, the different behavior of composites with respect to more traditional materials (basically, steel) commonly used as a reinforcement in civil structures (Arduini & Nanni, 1997; Naaman et al., 2001; Triantafillou et al., 2001) and, on the other hand, the aspects related to the adhesive connection of the FRP laminates to the concrete (or masonry) substrate (Täljsten, 1997; Neubauer & Rostasy, 1997).

The main findings of the research activities carried out in the '90s contributed to guidelines (fib, 2001; CNR-DT200, 2004; ACI 440-2R-08) for designing FRP-based strengthening intervention of RC and masonry members.

Bonding between FRP laminates (sheets or plates) and concrete emerged as a cutting-edge issue from the first decade of research activities on composite materials for civil structures. In particular, several failure modes due to loss of adhesion between the externally bonded FRP element and the concrete substrate have been observed experimentally and recognized as specific features of this kind of members (Meier, 1995; Bonacci, 1996).

As a matter of principle, the following seven failure modes have been defined in the scientific literature (Teng et al., 2002):

a. flexural failure by FRP rupture;
b. flexural failure by crushing of compressive concrete;
c. shear failure;
d. concrete cover separation;
e. plate end interfacial debonding;
f. intermediate flexural crack induced interfacial debonding;
g. intermediate flexural-shear crack induced interfacial debonding.

The last three failure modes are actually related to debonding failure of the FRP laminate from the concrete substrate. Local failure possibly induced by irregularities in the substrate surface can be also observed. Fig. 1 represents those failure modes pointing their typical position throughout the FRP-to-composite adhesive interface.

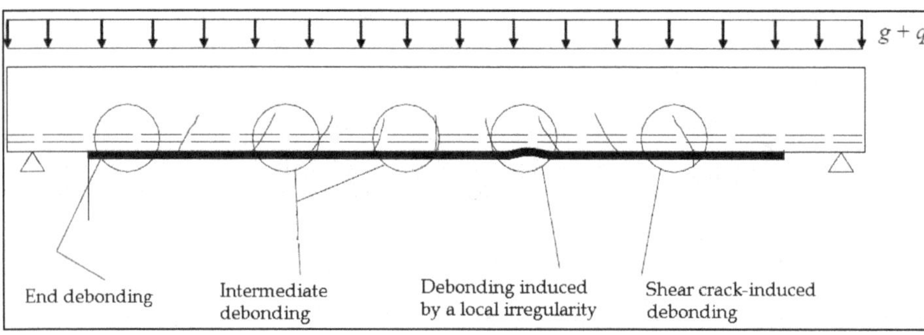

Fig. 1. Possible failure modes due to debonding in RC beams externally strengthened by FRP

Analytical studies have been carried out for determining the actual distribution of stresses throughout the adhesive interface. In particular, well-established elastic models, already used within the framework of structural mechanics, have been proposed as simplified practical methods for determining both shear and normal (peeling) stresses at the FRP-to-concrete interface (Roberts, 1989; Malek et al., 1998; Rabinovich & Frostig, 2000). However, those simplified methods were generally based on the assumption of an elastic behaviour of the above mentioned interface. Although this assumption can be considered under service loads, it cannot be generally accepted for the load values close to the onset of debonding. Micro-cracking phenomena develop as the levels of the interface stresses cannot be resisted by concrete, resulting in a highly nonlinear behaviour for the FRP-to-concrete interface which can be modelled by means of several possible relationships between interface stresses and displacements (Lu et al, 2005). Consequently, more advanced numerical models have been also proposed for simulating the actual distribution of stresses looking after the various fracture phenomena developing in concrete beneath the adhesive interface (Rabinovich & Frostig, 2000; Faella et al., 2008).

More recently, several models have been proposed in the scientific literature for predicting the strength of beams against both plate end-debonding and intermediate-debonding (Smith & Teng, 2001; Teng et al., 2003). They generally derive by mechanical observations carried out on the behaviour of FRP laminates glued on concrete blocks (Täljsten, 1996) and tested in pullout with the aim of measuring some relevant quantities, such as the ultimate strength at debonding and the axial strain distribution throughout the FRP bonded length (Chajes et al., 1996).

Understanding the behaviour of FRP-to-concrete joints tested under pull-out actions is of paramount importance for describing the key mechanical properties of the adhesive interface between FRP and concrete, which plays a key role in the possible debonding failure of externally strengthened beams. Thus, advanced testing and monitoring techniques have been also used for a deeper investigation of the behaviour of those joints (Czaderski et al., 2010). The experimental results obtained by the mentioned pull-out tests can be considered for identifying the non-linear relationships connecting interface stresses and displacements (Faella et al., 2009). In particular, a bilinear elastic-softening relationship between the interface shear stress and the corresponding displacements is often used for the FRP-to-concrete interface. The elastic branch of that stress-strain relationship results from the elastic behaviour of both the adhesive layer and the concrete substrate, mainly stressed in shear. The slope of such an elastic branch (namely, the "slip modulus" according to Lee et al, 1999) is generally much smaller than the value corresponding to the ration between the shear modulus of the resin and its thickness, as it would be determined by assuming a fully stiff behaviour of the concrete block (Faella et al., 2002).

A closed-form analytical solution has been derived in Faella et al. (2002) for the distribution of both shear stresses and relative slips in FRP-to-concrete joints with a bilinear adhesive interface. Further advances have been proposed by Lu et al. (2005), while the influence of different assumptions on the shape of the stress-slip relationship (i.e. bilinear, linearly softening, rigid-plastic, and so on) is discussed in both Chen & Teng (2001) and Wu et al. (2002). However, the fracture energy G_F (Täljsten, 1996) is the key parameter characterizing any various shear-stress-interface-slip relationship. The ultimate pull-out strength of the FRP-to-concrete joints is controlled by that parameter as well as the axial stiffness of the laminate and the bonding length. A limit value of the bonding length beyond which no further increases of the ultimate pull-out force can be observed; it is generally referred as "transfer length" (Bizindavyi & Neale, 1999). Some concepts of Fracture Mechanics are more and more employed in modelling the overall behaviour of both FRP laminates connected to concrete blocks (Yuan et al., 2007) and RC beams externally strengthened by FRP (Rabinovitch & Frostig, 2001; Achintha & Burgoygne, 2008).

The present chapter deals with the mechanical behaviour of the adhesive interface between FRP and concrete, proposing a wide overview of some theoretical models and experimental results which can be useful for understanding such a behaviour.

In the second section, an analytical model will be firstly presented for determining both shear and normal stresses throughout the adhesive interface in the linear range. The nonlinear behaviour of the FRP-to-concrete interface will be also addressed by discussing the ultimate bearing capacity of FRP laminates bonded on concrete members.

The third section will present the features of the most well-established testing techniques for investigating the mechanical behaviour of FRP-to-concrete interface, from the early loading stages up to failure. A series of experimental results of pull-out tests on FRP laminates will be also presented for pointing out the most important behavioural characteristics of the FRP-to-concrete interface in the case of both sheets and plates.

Finally, the fourth section will present some of the most recent design formulae which can be derived by the above mentioned theoretical models and calibrated on the available experimental results, partly reported in the present chapter.

2. Theoretical models

An analytical formulation for modelling the behaviour of the FRP-to-concrete adhesive interfaces is presented and discussed in the present section. The behaviour of FRP-to-concrete joints in the linear range is examined in the first subsection. Then, the aspects more directly related at the ultimate behaviour are addressed in the second one.

2.1 A general analytical model: formulation in the linear range

A simplified model is formulated in the present section with the aim of simulating the behaviour of FRP laminates bonded on concrete and tested in pull-out, simulating how both the in-plane (namely, "slip") and out-of-plane displacement components develop throughout the FRP length. It is based on the following assumptions:

- the FRP strip is simulated as a Bernoulli beam;
- the adhesive layer is modelled as a bi-dimensional elastic domain in plane deformations.

The generalized forces N, V, M applied on the transverse section of the strip at the abscissa z and the interface stresses (shear τ and normal σ, respectively) are represented in Fig. 2. The following equilibrium equations can be stated between those force and stress components:

$$\frac{dN}{dz} + \tau_f b_f = 0 \ , \tag{1}$$

$$\frac{dV}{dz} = -b_f \sigma_y \ , \tag{2}$$

$$\frac{dM}{dz} + \tau_f \frac{b_f t_f}{2} - V = 0 \ . \tag{3}$$

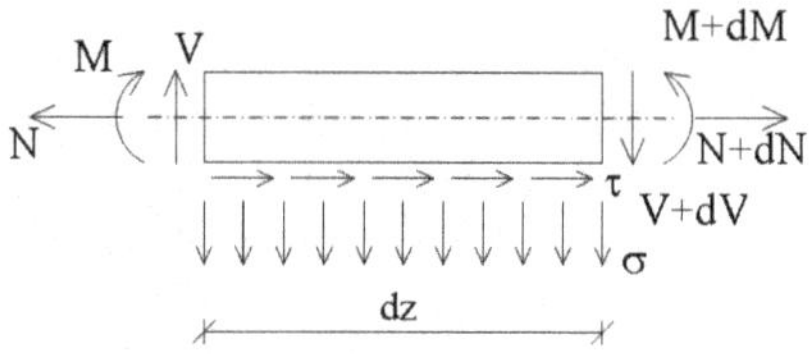

Fig. 2. Forces on the segmental FRP strip element

The parameters v, w and φ completely describe the displacement field of the strip and the following compatibility equations relate them to the axial strain ε and the curvature χ (Fig. 3):

$$\varepsilon = \frac{dw}{dz} \ , \tag{4}$$

$$\chi = \frac{d\varphi}{dz} \ , \tag{5}$$

$$\varphi = -\frac{dv}{dz} \ . \tag{6}$$

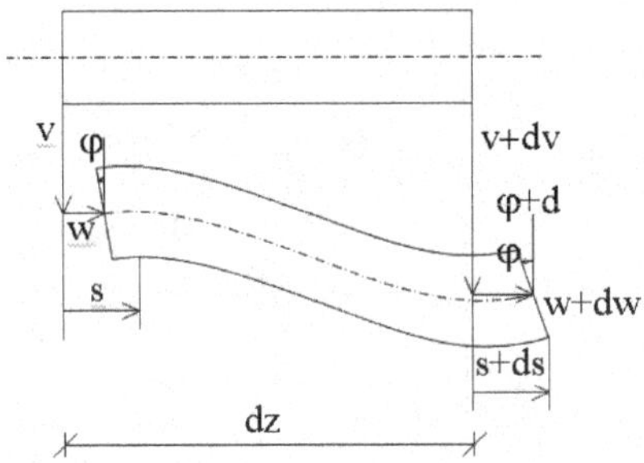

Fig. 3. Displacements components of the segmental element

The interface slip s, namely the axial displacement at the bottom chord of the strip transverse section, can be also related to the displacement components represented in Fig. 3:

$$s = w + \frac{t_f}{2} \cdot \varphi \ . \tag{7}$$

Finally, the usual (generalized) stress-strain relationships can be introduced for the strip modelled as a Bernoulli beam:

$$N = EA_f \varepsilon = E_f b_f t_f \varepsilon \ , \tag{8}$$

$$M = EI_f \chi = E_f \frac{b_f t_f^{\ 3}}{12} \chi \ , \tag{9}$$

where EA_f and EI_f are respectively the axial and flexural stiffnesses which can be assumed for the strip transverse section. Based on the second assumptions reported at the beginning of section 2.1, The following equilibrium equations can be written for the infinitesimal 2D element of resin within the adhesive layer (Fig. 4):

$$\begin{cases} \dfrac{\partial \sigma_{yy}}{\partial y_a} + \dfrac{\partial \tau_{yz}}{\partial z_a} = 0 \\ \dfrac{\partial \tau_{yz}}{\partial y_a} + \dfrac{\partial \sigma_{zz}}{\partial z_a} = 0 \end{cases} , \tag{10}$$

where y_a and z_a are the Cartesian coordinates of the infinitesimal areal element of resin.

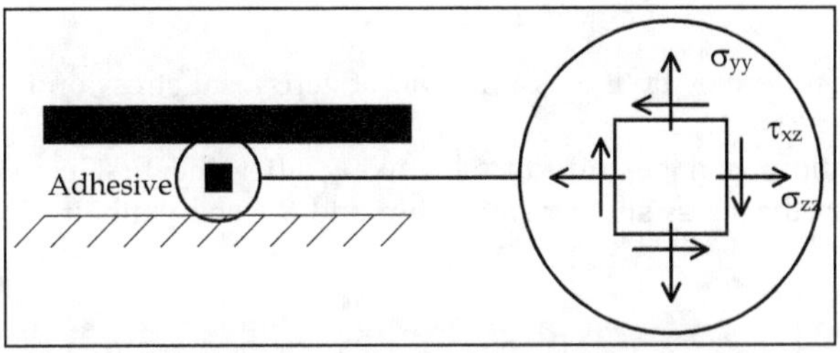

Fig. 4. General plane stress-state for an infinitesimal element of the adhesive layer

The corresponding strain measures can be expressed as functions of the displacement field $v_a(y_a,z_a)$ e $w_a(y_a,z_a)$ within the adhesive layer:

$$\begin{cases} \varepsilon_{yy} = \dfrac{\partial v_a}{\partial y_a} \\ \varepsilon_{zz} = \dfrac{\partial w_a}{\partial z_a} \\ \gamma_{yz} = \dfrac{\partial v_a}{\partial z_a} + \dfrac{\partial w_a}{\partial y_a} \end{cases} , \tag{11}$$

where y_a and z_a are the Cartesian coordinates of the infinitesimal areal element of resin. Finally, the following elastic relationships can be stated between stresses and strains:

$$\begin{cases} \varepsilon_{yy} = \dfrac{1}{E_a} \cdot \left[\sigma_{yy} - \nu_a \cdot \sigma_{zz}\right] \\ \varepsilon_{zz} = \dfrac{1}{E_a} \cdot \left[\sigma_{zz} - \nu_a \cdot \sigma_{yy}\right] \\ \gamma_{yz} = \dfrac{\tau_{xy}}{G_a} = 2 \cdot (1+\nu_a) \cdot \dfrac{\tau_{xy}}{E_a} \end{cases} , \tag{12}$$

where E_a and G_a are the elastic properties of the adhesive layer. In this formulation, it connects the FRP laminates to a stiff substrate representing the concrete element. The upper bound of this layer is connected to that strip and some compatibility equations should be written for introducing this physical constraint into the mathematical model. Thus, the three equations (10)-(12) can be worked out for deriving a relationship between the displacement components $v_a(0,z_a)$ and $w_a(0,z_a)$ and the corresponding stress components $\sigma_{yy}(0,z_a)$ and $\tau_{xy}(0,z_a)$. A key assumption can be introduced for simplifying the analytical expressions of the interface stresses, considering a constant value of the shear stress throughout the adhesive thickness (Rabinovich & Frostig, 2000):

$$\tau_{yz} = \tau_a(z_a) = \tau_a(z) . \tag{13}$$

Consequently, the second one of the two equations in (10) leads to the following simplification in terms of the normal stress in longitudinal direction:

$$\frac{\partial \sigma_{zz}}{\partial z_a} = 0 \Rightarrow \sigma_{zz} = f_1(y_a) \Rightarrow \sigma_{zz} = 0 . \tag{14}$$

The last implication derives by the assumption of zero axial stress on the initial section of the layer adhesive.
Further simplifications can be introduced looking after the first equations in (10) and introducing therein the stress-strain relationship and the compatibility equation within the adhesive layer:

$$\sigma_{yy}(y_a, z_a) = -\int \frac{d\tau_a}{dz} dy + g_1(z_a) = -\frac{d\tau_a}{dz} y_a + g_1(z_a) , \tag{15}$$

and

$$\varepsilon_{yy}(y_a, z_a) = \frac{\sigma_{yy}(y_a, z_a)}{E_a} = -\frac{1}{E_a} \cdot \frac{d\tau_a}{dz} y_a + \frac{g_1(z_a)}{E_a}. \tag{16}$$

The transverse component of the displacement field can be derived by further integration:

$$\begin{aligned} v_a(y_a, z_a) &= v(z_a) + \int \varepsilon_{yy} dy = v(z_a) + \int \left(-\frac{1}{E_a} \cdot \frac{d\tau_a}{dz} y_a + \frac{g_1(z_a)}{E_a} \right) dy = \\ &= v(z_a) - \frac{1}{E_a} \cdot \frac{d\tau_a}{dz} \frac{y_a^2}{2} + \frac{g_1(z_a)}{E_a} y_a \end{aligned}, \tag{17}$$

and the value of the unknown function g_1 can be finally derived by imposing zero value to the displacement at the bottom interface ($y_a = t_a$):

$$v(z_a) - \frac{1}{E_a} \cdot \frac{d\tau_a}{dz} \frac{t_a^2}{2} + \frac{g_1(z_a)}{E_a} t_a = 0 , \tag{18}$$

$$g_1(z_a) = -\frac{E_a}{t_a} \cdot v(z_a) + \frac{d\tau_a}{dz} \frac{t_a}{2} . \tag{19}$$

Consequently, the following explicit expressions can be written for σ_{yy}, ε_{yy} and v_a described in (15), (16) and (17), by introducing the expression of g_1 given by equation (19):

$$\varepsilon_{yy}(y_a, z_a) = \frac{d\tau_a}{dz} \cdot \left(\frac{t_a - 2y_a}{2 \cdot E_a} \right) - \frac{v(z_a)}{t_a} , \tag{20}$$

$$\sigma_{yy}(y_a, z_a) = E_a \cdot \varepsilon_{yy}(y_a, z_a) = \frac{d\tau_a}{dz} \cdot \left(\frac{t_a - 2y_a}{2} \right) - \frac{E_a}{t_a} v(z_a) , \tag{21}$$

$$v_a(y_a, z_a) = v(z_a) \cdot \left(1 - \frac{y_a}{t_a} \right) + \frac{d\tau_a}{dz} \cdot \left(\frac{t_a - y_a}{2E_a} \right) y_a . \tag{22}$$

Shear strain and stress can be related according to the following relationship:

$$\gamma_{yz}(z_a) = \frac{\tau_a(z_a)}{G_a} , \tag{23}$$

and, considering the compatibility equation involving shear strains, further transformations can be carried out by introducing the compatibility equation in (11):

$$\gamma_{yz}(z_a) = \frac{\partial v_a}{\partial z_a} + \frac{\partial w_a}{\partial y_a} . \tag{24}$$

The first derivative of the longitudinal displacement w_a inside the adhesive layer can be written as follows:

$$\frac{\partial w_a}{\partial y_a} = \gamma_{yz}(z_a) - \frac{\partial v_a}{\partial z_a} = \frac{\tau_a(z_a)}{G_a} - \frac{dv}{dz}\cdot\left(1-\frac{y_a}{t_a}\right) - \frac{d^2\tau_a}{dz^2}\cdot\left(\frac{t_a - y_a}{2E_a}\right)y_a \ , \tag{25}$$

and the corresponding function w_a can be derived by integrating eq. (25) and introducing a zero-displacement condition for $y_a=t_a$ (namely, at the bottom of the adhesive layer):

$$w_a(y_a,z_a) = -\frac{\tau_a(z_a)}{G_a}(t_a - y_a) + \frac{dv}{dz}\cdot\left(\frac{t_a}{2} - y_a + \frac{y_a^{\ 2}}{2t_a}\right) + \frac{d^2\tau_a}{dz^2}\cdot\left(\frac{t_a^{\ 3}}{12E_a} - \frac{t_a y_a^{\ 2}}{4E_a} + \frac{y_a^{\ 3}}{6E_a}\right) , \tag{26}$$

The above relationship can be utilized for deriving the expression of the axial displacements beneath the laminate (y_a=0), corresponding to the slip at the interface s(z):

$$w_a(0,z_a) = s(z) = -\frac{t_a}{G_a}\cdot\tau_a + \frac{t_a^{\ 3}}{12E_a}\cdot\frac{d^2\tau_a}{dz^2} + \frac{t_a}{2}\cdot\frac{dv}{dz} \ , \tag{27}$$

as well as equation (22) can be utilized for recognizing that the corresponding transverse component $v_a(0,z_a=z)$ is equal to v(z):

$$\sigma_{yy}(y_a,z_a) = E_a\cdot\varepsilon_{yy}(y_a,z_a) = E_a\cdot\left[\frac{d\tau_a}{dz}\cdot\left(\frac{t_a - 2y_a}{2\cdot E_a}\right) - \frac{v(z_a)}{t_a}\right] , \tag{28}$$

$$\sigma_y(z) = \sigma_{yy}(0,z_a) = -v(z_a)\cdot\frac{E_a}{t_a} + \frac{t_a}{2}\cdot\frac{d\tau_a}{dz} \ . \tag{29}$$

Thus, the rotation field φ(z) of the laminate element can be easily defined through equation (6) and the generalized strain fields can be determined by means of equation (4) and (5).
The above equations can be finally combined for deriving a unique differential equation in terms of interface shear stresses. A first equation can be obtained by differentiating eq. (7) and introducing (4), (6) and (8). After a further differentiation and introducing the definition of interface slip provided by eq. (27), the following differential relationship between the shear stress and the transverse displacements can be obtained:

$$\frac{G_a}{E_f t_f t_a}\cdot\tau_a - \frac{d^2\tau_a}{dz^2} + \frac{G_a}{2}\cdot\frac{d^3 v}{dz^3} + \frac{G_a}{12E_a}\cdot t_a^{\ 2}\cdot\frac{d^4\tau_a}{dz^4} = 0 \ . \tag{30}$$

Another relationship is obtained by differentiating equation (3) and introducing equation (2), (5), (6) and (9) for expressing both the bending moment and the shear force in terms of transverse displacements and interface (shear and normal) stresses. The final expression of an equation in terms of t_f and v (and their derivatives) can be obtained introducing equation (29):

$$\frac{b_f(t_f + t_a)}{2}\cdot\frac{d\tau_a}{dz} - \frac{E_f b_f t_f^{\ 3}}{12}\cdot\frac{d^4 v}{dz^4} - \frac{E_a b_f}{t_a}\cdot v = 0 \ . \tag{31}$$

The two equations (30) and (31) can be easily combined for deriving the following eighth-order differential equation in terms of interface shear stresses:

$$\frac{G_a}{E_f t_f t_a} \cdot \tau_a - \frac{d^2\tau_a}{dz^2} + \left(\frac{t_a^{\ 2}}{3} + \frac{t_a t_f}{6} + \frac{t_f^{\ 2}}{12}\right) \cdot \frac{G_a}{E_a} \cdot \frac{d^4\tau_a}{dz^4} - \frac{E_f t_a t_f^{\ 3}}{12 \cdot E_a} \cdot \frac{d^6\tau_a}{dz^6} + \frac{E_f G_a t_a^{\ 3} t_f^{\ 3}}{144 \cdot E_a^{\ 2}} \cdot \frac{d^8\tau_a}{dz^8} = 0 . \quad (32)$$

Finally, an expression of the displacement function v as a function of the interface shear stresses can be derived by solving equation (31) and introducing the (30):

$$v = \frac{3 \cdot t_a^{\ 2} + 3 \cdot t_a t_f + t_f^{\ 2}}{6E_a} \cdot \frac{d\tau_a}{dz} - \frac{E_f t_a t_f^{\ 3}}{6E_a G_a} \cdot \frac{d^3\tau_a}{dz^3} + \frac{E_f t_a^{\ 3} t_f^{\ 3}}{72 \cdot E_a^{\ 2}} \cdot \frac{d^5\tau_a}{dz^5} . \quad (33)$$

Eight boundary conditions are needed for the problem at hand and they can be symbolically written as follows:

$$\tau_a\big|_{z=0} = 0 \ , \ \tau_a\big|_{z=L} = 0 \ , \quad (34)$$

$$\varepsilon\big|_{z=0} = -\frac{N(0)}{E_f A_f} \ , \ \varepsilon\big|_{z=L} = \frac{N(L)}{E_f A_f} \ , \quad (35)$$

$$M\big|_{z=0} = 0 \ , \ M\big|_{z=L} = 0 \ , \quad (36)$$

$$V\big|_{z=0} = 0 \ , \ V\big|_{z=L} = 0 \ . \quad (37)$$

Typical solutions of the problem described by equation (32) and the boundary conditions (34)-(37) are represented in figures 5 and 6, in terms of shear and normal stresses respectively. In both graphs, the interface stress components are divided by the average shear stress $\tau_m = F/b_f L$ and the abscissa are reported in non-dimensional form as $\bar{z} = \omega z$.

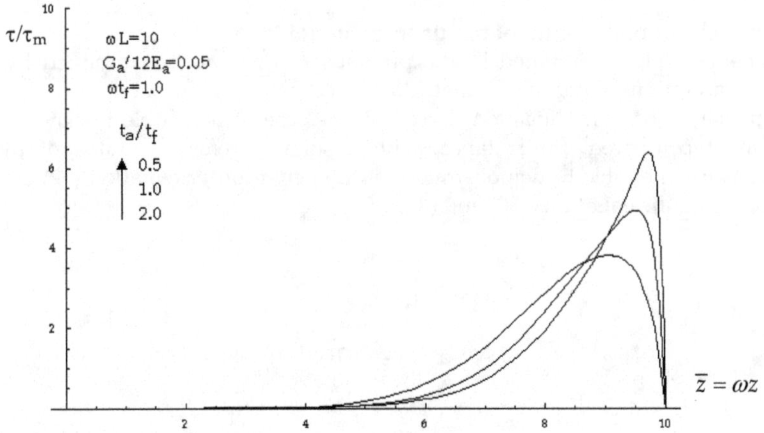

Fig. 5. Possible distributions of interface shear stresses throughout the adhesive interface

The results of the analysis carried out by means of this model are completely controlled by the following non-dimensional parameters:

$$\omega L \ , \quad (38)$$

$$\omega t_f = \sqrt{\frac{G_a t_f}{E_f t_a}} = \sqrt{\frac{k_{s,a} t_f}{E_a}} \ , \tag{39}$$

$$\frac{G_a}{E_a} = \frac{k_{s,a}}{k_{v,a}} \ , \tag{40}$$

$$\frac{t_a}{t_f} . \tag{41}$$

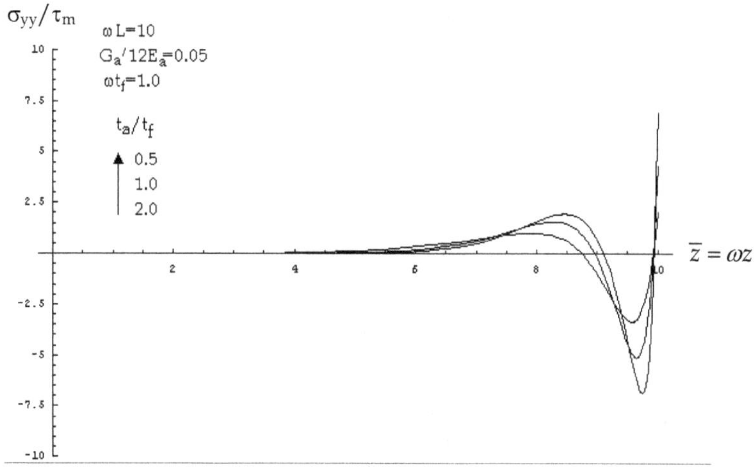

Fig. 6. Possible distributions of interface normal stresses throughout the adhesive interface

2.2 Possible simplifications of the general model

The general model formulated in the previous section can be simplified by introducing further assumptions about the geometric and mechanical assumptions.

In particular, a key simplification derives in the case of negligible thickness t_f of the FRP laminate, if compared, for instance, with a suitable reference value of the length. In particular, it is possible to demonstrate that the height-order equation (32) can be reduced by uncoupling the equations (30) and (31):

$$\frac{d^2\tau_a}{dz^2} + \omega^2 \tau_a = 0 \ , \tag{42}$$

$$\frac{d^4 v}{dz^4} + 12 \cdot \frac{E_a}{t_a t_f^3} v = 0 \ , \tag{43}$$

Under the mathematical point of view, the above simplification derives from the condition $\omega t_f \rightarrow 0$ for whichever value of the non dimensional parameters listed at the end of section 2.1. It means that the thickness of the FRP laminate is much smaller than the characteristic length of the problem ω, which is the main argument of the solutions of the general equation (32). The uncoupled boundary conditions for eq. (42) simplify as a consequence of the condition $\omega t_f \rightarrow 0$, turning in a completely uncoupled expressions:

$$\varepsilon\big|_{z=0} = -\frac{N(0)}{E_f b_f t_f} = \frac{ds}{dz}\bigg|_{z=0} \qquad \varepsilon\big|_{z=L} = \frac{N(L)}{E_f b_f t_f} = \frac{ds}{dz}\bigg|_{z=L} . \tag{44}$$

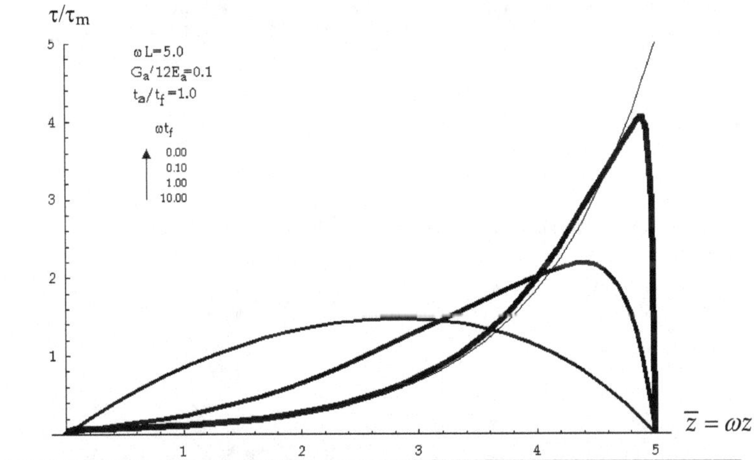

Fig. 7. Interface shear stress for a thin FRP strip

A sound demonstration about how equation (32) reduces to (42)-(43) in the case of $\omega t_f \to 0$, can be obtained by means of some mathematical transformations of both the equation (32) and the boundary conditions (34)-(37) in dimensionless form. These transformations, as well as the discussion about the slight modifications in the out-of-plane stress regime, are omitted herein for the sake of brevity.

2.3 Solutions in the nonlinear range

The assumption introduced in the subsection 2.2 for simplifying the general eighth-order equation formulated in subsection 2.1 can be generally assumed as a reasonable trade-off between the (good) accuracy of the obtained model and the (higher) simplicity of its equations. The approximation introduced by the supplementary assumption is generally accepted especially in the nonlinear range, as further uncertainties are introduced by the fracture behaviour of concrete which cannot be covered by the complex assumptions leading to equation (32).

Thus, the nonlinear response of FRP-to-concrete joints under pull-out actions can be analyzed by assuming the problem described by equation (42) with the boundary conditions in (44). Since the nonlinear response is now of interest, equation (42) can be slightly modified for considering the possibility of a nonlinear relationship $\tau_a=\tau(s)$. Moreover, eq. (1) can be written in terms of axial strains ε in the FRP strip by introducing equation (8):

$$\frac{d\varepsilon}{dz} + \frac{\tau(s)}{E_f t_f} = 0 . \tag{45}$$

and, considering the relationship between axial strains and interface slip derived at the end of the subsection 2.2, the following relationship can be written for generalizing eq. (42) to the nonlinear range:

$$\frac{d^2s}{dz^2}+\frac{\tau(s)}{E_f t_f}=0\ . \tag{46}$$

The bilinear shear-stress-interface-slip relationship is considered herein as a general relationship for the nonlinear response of the interface between FRP and concrete:

$$\tau(s)=\begin{cases}-k_{el}\cdot s & |s|\le s_{el}\\ -k_u\cdot(s_u-|s|)\cdot \mathrm{sign}(s) & s_{el}<|s|\le s_u\\ 0 & |s|>s_u\end{cases}\ , \tag{47}$$

where all symbols are represented in Fig. 8. For the sake of simplicity, only the monotonic behaviour with s≥0 will be considered in the following elaborations.

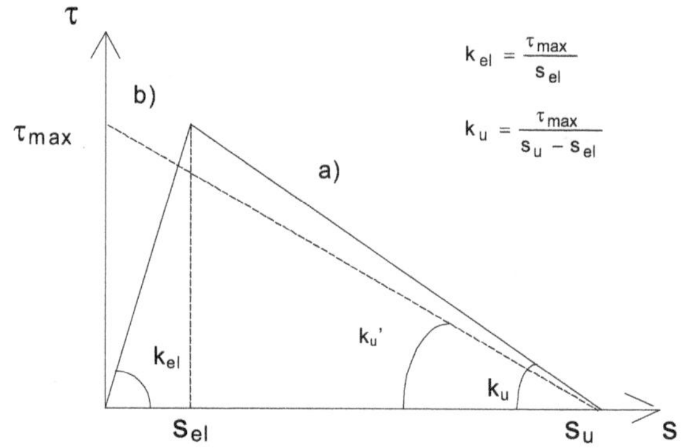

Fig. 8. Bilinear shear-stress-interface-slip relationship (fib - bulletin 14, 2001)

Consequently, two second order differential equations can be obtained by introducing the two nonzero branches of the relationship (47) within the general equation (46). In particular, under low load levels, the condition $0\le s\le s_{el}$ yields throughout all the bonding length and the equation (46) can be written in the following form in terms of interface slips s:

$$\frac{d^2s}{dz^2}-\omega^2\cdot s=0\ . \tag{48}$$

Considering a pull-out force P applied at the end (z=L) of the bonding length, the following boundary conditions can be written after eq. (44):

$$\varepsilon\big|_{z=0}=\frac{ds}{dz}\bigg|_{z=0}=0 \qquad \varepsilon\big|_{z=L}=\frac{ds}{dz}\bigg|_{z=L}=\frac{P}{E_f b_f t_f}\ . \tag{49}$$

Simple mathematical transformations (whose details are omitted herein) lead to the following solution in terms of interface shear stresses

$$\tau=-\omega\cdot\frac{P}{b_f}\cdot\frac{\cosh(\omega\cdot z)}{\sinh(\omega L)}\ , \tag{50}$$

which can be easily turned in terms of interface slips s considering the expression describing the linear elastic branch in eq. (47). This solution can be accepted for values of the shear strength $|\tau| \leq \tau_{max}$ or, in order words, for pull-out forces lower than P_{el}:

$$F_{el} = \frac{\tau_{max}}{\omega} \cdot b_f \ . \tag{51}$$

leading to an interface slip $s=s_{el}$ in $z=L$. As $P>P_{el}$, two different parts of the adhesive interface can be recognized. In the first part, namely for $0\leq z\leq z_{el}<L$, the values of the interface slips s keep smaller than the elastic threshold s_{el}. In the second part, for $z_{el}<z\leq L$, the interface slips s are in the range (s_{el}, s_u), and the following solution can be derived by integrating the two differential equations:

$$\frac{d^2 s_1}{dz^2} - \omega \cdot s_1 = 0 \ \text{for} \ 0<z<z_{el}<L, \tag{52}$$

$$\frac{d^2 s_2}{dz^2} + \frac{k_{in}}{E_f t_f} \cdot s_2 = \frac{k_{in}}{E_f t_f} \cdot s_u \ \text{for} \quad z_{el}<z\leq L. \tag{53}$$

with the following boundary conditions:

$$\varepsilon\big|_{z=0} = \frac{ds_1}{dz}\bigg|_{z=0} = 0 \qquad s_1\big|_{z=z_{el}} = s_{el} \tag{54}$$

$$s_1\big|_{z=z_{el}} = s_1\big|_{z=z_{el}} \qquad \frac{ds_1}{dz}\bigg|_{z=z_{el}} = \frac{ds_2}{dz}\bigg|_{z=z_{el}} \tag{55}$$

The following solution in terms of interface shear stress τ involves the parameter $z_{el}<L$, describing the length of the elastic part of the bonding length:

$$|\tau| = \begin{cases} \tau_{max} \dfrac{\cosh(\omega \cdot z)}{\cosh(\omega \cdot z_{el})} & 0 \leq z \leq z_{el} \\ \tau_{max}\left\{\cos\left[\omega_{in}(z-z_{el})\right] - \dfrac{\omega_{in}}{\omega} \cdot \sin\left[\omega_{in}(z-z_{el})\right] \cdot \tanh(\omega z_{el})\right\} & z_{el} < z \leq L \end{cases} \tag{56}$$

where

$$\omega_{in} = \sqrt{\frac{\tau_{max}/(s_u - s_{el})}{E_f t_f}} = \sqrt{\frac{k_{in}}{E_f t_f}} \ . \tag{57}$$

Fig. 9 shows two typical distributions of shear stresses in the a) linear and b) nonlinear range of the behaviour of the adhesive interface described by the bilinear relationship in eq. (47). It is worth noting that z_{el} can actually range in the interval $[z_{el,min}, L)$:

$$z_{el,min} = \max\{0; z_{el,deb}\} \tag{58}$$

in which $z_{el,deb}$ is the value of z_{el} leading to $\tau(L)=0$ (or $s(L)=s_u$):

$$\cos\left[\omega_{in}\left(L - z_{el,deb}\right)\right] - \sqrt{\frac{k_{in}}{k}} \cdot \sin\left[\omega_{in}\left(L - z_{el,deb}\right)\right] \cdot \tanh\left(\omega \cdot z_{el,deb}\right) = 0 . \tag{59}$$

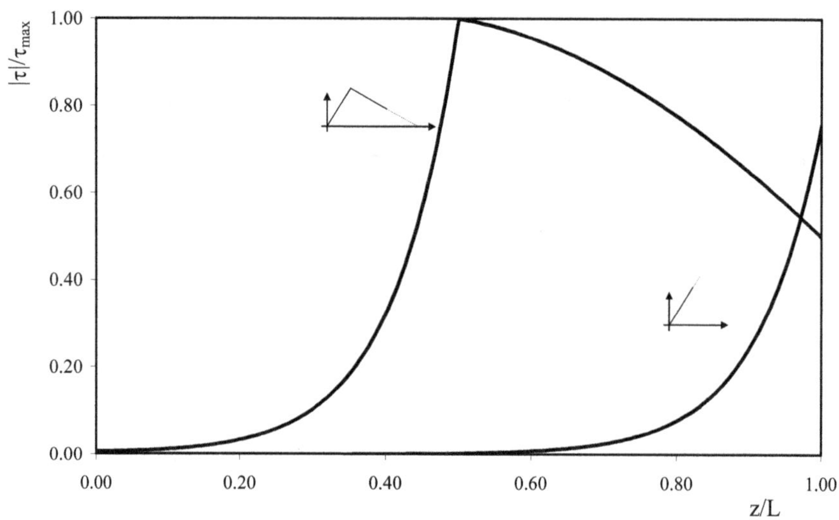

Fig. 9. Typical distribution of interface shear-strength in the linear and nonlinear range

2.4 Key parameters related to the ultimate strength

The case of $z_{el,deb}=0$ is of particular relevance, as it represents the condition of debonding initiation (namely, $s(L)=s_u$) with $\tau(0)=\tau_{max}$. A characteristic length L_{eff} can be defined in that case by means of equation (59):

$$\cos\left[\omega_{in} L_{eff}\right] = 0 \Rightarrow L_{eff} = \frac{\pi}{2} \cdot \sqrt{\frac{E_f t_f}{k_{in}}} . \tag{60}$$

Under the mechanical standpoint, this characteristic value has a paramount conceptual meaning. It represents the border between "short" and "long" anchorage length. In particular, the general expression of the force $F>F_{el}$ corresponding to a given position of z_{el} can be derived by integrating the shear stresses in equation (56):

$$F = \tau_{max} b_f \cdot \left\{ \frac{\sin\omega_{in}\left(L - z_{el}\right)}{\omega_{in}} + \frac{\cos\omega_{in}\left(L - z_{el}\right) \cdot \tanh \omega z_{el}}{\omega} \right\} . \tag{61}$$

It can be applied if and only if $z_{el} \in [z_{el,min}, L)$ (namely, the interface slip s is $s_{el} \le s \le s_u$). In the case of $z_{el}=L$ equation (61) reduces to (51). On the contrary, when $z_{el}=z_{el,deb}=0$ the following expression can be derived for the pull-out force:

$$F_{deb} = \tau_{max} b_f \cdot \frac{\sin\omega_{in} L_{eff}}{\omega_{in}} = \frac{\tau_{max} b_f}{\omega_{in}} = b_f \cdot \sqrt{\tau_{max} \cdot \left(s_u - s_{el}\right) \cdot E_f t_f} . \tag{62}$$

Since generally s_u>>s_e the value F_{deb} is only slightly lower than the maximum strength F_{max} which can be evaluated by applying the theory of Fracture Mechanics (Täljsten, 1996) (or either as a maximum on z_{el} of the for F in eq. (61)):

$$F_{max} = b_f \cdot \sqrt{2G_F \cdot E_f t_f} = b_f \cdot \sqrt{\tau_{max} \cdot s_u \cdot E_f t_f} \ . \tag{63}$$

They can be assumed as coincident in the approximation of rigid-softening behaviour in considered Fig. 9.
Note that equation (63) can be applied only if $L_b \geq L_{eff}$ being L_{eff}, the effective transfer length. If $L_b < L_{eff}$ a smaller value of the ultimate force has to be expected according to the following relationship:

$$F_{max} = \sqrt{2G_F E_f t_f} \cdot b_f \cdot \left[\frac{L_b}{L_{eff}} \cdot \left(2 - \frac{L_b}{L_{eff}}\right)\right]. \tag{64}$$

3. Experimental testing techniques and results

As stated above, the effectiveness of FRP externally bonded on RC members is strongly related to the interface behaviour that depends on the mechanical and geometrical properties of materials. Thus an FRP-to-concrete bond strength model is the key to the accurate prediction of debonding failures in FRP-strengthened RC beams, including plate end debonding failures and shear crack-induced debonding failures, as well as intermediate flexural or flexural-shear crack-induced debonding failures (Teng et al., 2002).
Plenty of pull-out tests have been carried out by researchers during last years in order to experimentally investigate the influence of FRP and concrete mechanical properties on the bonding behaviour.
An overview of most common bond test techniques is reported in the first subsection. Then, the most significant results achieved by such bond tests are showed in the second one.

3.1 Overview of testing techniques

Yao et al. (2005) classified the existing test setups into the following types: (a) double-shear pull tests; (b) double-shear push tests; (c) single-shear pull tests; (d) single-shear push tests; and (e) beam (or bending) tests. These arrangements are based on the definition of the loading condition of the element and on the symmetry of the system (a double or single test refers to the contemporaneous loading of two or one FRP reinforcement applied on the specimen sides). Collectively, the first four test methods, may also be referred to as pull tests, as the plate is always directly pulled by a tensile force. Pull-pull (a) and push-pull (d) test setup are the most popular test methods. In end debonding failures in FRP flexural- strengthened RC beams with longitudinal laminates as well as debonding failures in FRP shear-strengthened RC beams with transverse laminates, the bond strength model developed from pull tests is directly applicable. Furthermore, in intermediate crack-induced debonding failures in FRP flexural- strengthened RC beams with longitudinal laminates, the stress state in the critical region of the beam is also closely similar to that of the concrete prism in a single-shear pull test. In order to extend the results of bond tests to various types of strengthening (flexural, shear, torsion), the pull-pull test setup (a and c) are probably the configuration giving the loading condition more similar to the actual one in RC elements, but it is also the most difficult to realize with a reliable setup. Furthermore the test setup (a), as well as (b) could lead to underestimate the bond strength due to the influence of detailing (Blontrock et al., 2002; Ceroni et al., 2008). On the contrary, the push-pull test (d), where the concrete is partially in compression, is easier to be realized but it gives comparable predictions of the bond strength, if the pushing force is applied on the concrete block sufficiently far from the external reinforcement (Yao et al., 2005). Indeed, if the compressed area of concrete is too extended, the volume of material involved in the debonding failure can be lower and, thus, the related fracture energy decreases.

Asymmetrical schemes (c and d) are, in general, preferable to the symmetrical ones (a and b) mainly because the latter are more influenced by the alignment detailing of the two strengthened sides. Moreover the specimen symmetry is however lost when the debonding starts on one side and prevents from following correctly the post-peak behaviour.

As concerns the asymmetrical push-pull test setups (d), they are commonly realized by positioning a single concrete block in a stiff steel frame with an upper plate compressing the specimen, while the end of the FRP reinforcement, glued on one side, is clamped in the grips of a tensile machine (Nigro et al., 2008; Savoia et al. 2009; Ceroni and Pecce, 2010). An alternative configuration can be also realized by fixing the end of the concrete block, placed horizontally, and applying tension to the FRP reinforcement with an hydraulic jack (Yao et al., 2005; Mazzotti et al., 2009). As concerns the asymmetrical pull-pull test setups (c), they can be realizes by installing metallic threaded rods inside the concrete specimen which can be clamped in the testing machine grips (Bilotta et al. 2011).

Asymmetrical single-shear test setup (c or d) appear to be promising candidate as a standard setup for determining the FRP-to-concrete bond strength; hence they were adopted to perform the tests showed in the following section.

3.2 Experimental results

A total of 58 SST tests in four sets (i.e. I-12 tests, II-16 tests, III-12 tests, IV-18 tests) were performed on CFRP plates (38) and sheets (20) applied on two opposite longitudinal faces of 29 concrete prisms. Concrete mix was specifically designed to obtain low compressive concrete strength to simulate the FRP application on existing concrete. Compressive tests were performed at 28 days after casting: the cylinder mean strength, f_{cm}, was equal to 23.82 MPa, 21.46 MPa, 26.00 MPa and 19.00 MPa for set I, II, III and IV respectively. CFRP Young's moduli, E_f, were obtained by means of tensile flat coupon tests: their mean values were between 110-220 GPa for plates and 170-240 GPa for sheets. Prior to FRP installation, the concrete surface was treated by sand paper (sets I, II, III) or by bush hammering (set IV), in order to eliminate the mortar till the aggregate became clear. Primer was always used to consolidate the concrete surface except for specimens of set VI. Further details related to experimental tests of each set can be founded in (Nigro et al. 2008, Savoia et al. 2009, Bilotta et al. 2009, Bilotta et al. 2011).

Note that several strain gauges were applied along the FRP laminates in order to measure axial strains during the bond test (see Fig. 10b,c). Experimental readings allow theoretical models of the adhesion laws to be to verified or calibrated.

The behaviour of the FRP-to-concrete interface was analyzed by plotting the experimental axial strains $\varepsilon_i(z)$ measured by means of strain gauges during the test along the FRP reinforcement. Strains recorded along FRP plates at different load level are plotted in Fig. 12a,b for sheets and plates respectively. The first debonding load, P_{fd}, identified the beginning of debonding.

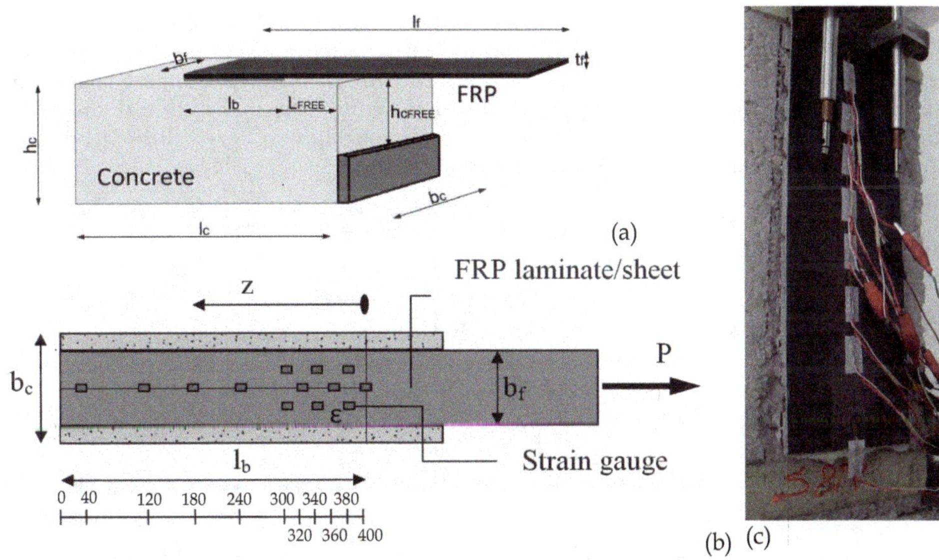

Fig. 10. Test setup scheme (a) – strain gauges distributions: scheme (b) and photo (c)

Fig. 11. Specimen after test and typical failure mode

Note that the maximum strain recorded at the loaded ends of FRP sheets was $\varepsilon_{sheets} \approx 0.6\%$; for plates it was significantly lower ($\varepsilon_{plates} \approx 0.2\%$), even if, for both types of reinforcement, the bond length was higher than the theoretical effective one, $L_{eff} = \sqrt{(E_f \cdot t_f)/(2 \cdot f_{ctm})}$, evaluated according to *fib* bulletin 14 (2001) and CNR-DT200/2004 (2004), where f_{ctm} is the mean tensile strength. Such result is mainly due to the higher stiffness of plates; indeed, as showed above by (63), the FRP stiffness, $E_f t_f$, is the main parameter affecting the debonding strain, being maximum strain at debonding phenomenon being the maximum strain due to debonding proportional to the factor $\sqrt{1/(E_f t_f)}$. In particular, the ratio $\left(\sqrt{1/(E_f t_f)}\right)_{sheets} / \left(\sqrt{1/(E_f t_f)}\right)_{plates}$ of FRP reinforcements investigated in Fig. 12 is 2.5

similar to $\varepsilon_{sheets} / \varepsilon_{plates} \approx 3$. Moreover, since plates are about eight times thicker than sheets while the ratio between Young moduli is very close to unity (i.e. $E_{f,S}/E_{f,P}=1.35$), this result seems to confirm that the reinforcement thickness particularly affects debonding behaviour. Indeed, the greater the thickness, the higher the increase in the normal and shear stresses at FRP to concrete interface and consequently the probability of premature debonding occurrence (Oehlers and Moran, 1990; Tounsi et al., 2009).

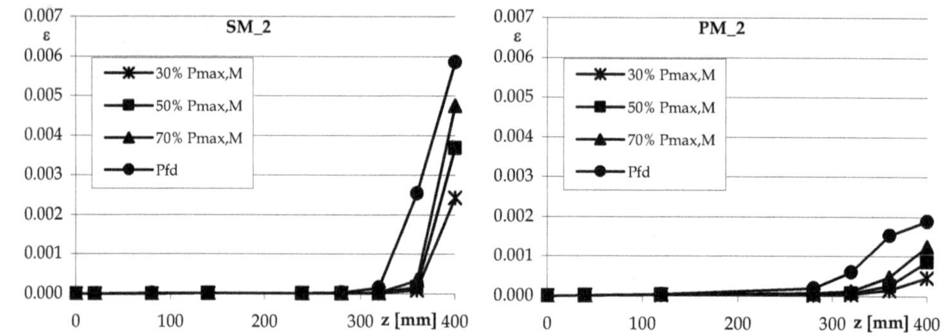

(a) (b)

Fig. 12. Strain recorded on sheet (a) and plate (b) reinforced specimens

The interface shear stresses $\tau_i(z)$ were obtained by the variation of axial stress and, hence, strain throughout the FRP by the following relationship between two strain gauges at distance $\Delta z_{i,i+1}$

$$\tau_{i,i+1} = \frac{\varepsilon_{i+1} - \varepsilon_i}{\Delta z_{i,i+1}} \cdot E_f \cdot t_f \tag{65}$$

where E_f and t_f are FRP Young's modulus and thickness, respectively. Typical shear stress profiles assessed for sheets and plates, respectively, are reported in Fig. 13a,b.
Note that, at the loaded end of the reinforcement, the shear stresses assessed for loads close to the debonding of the reinforcement are lower than those assessed for lower loads. This indicates that in this zone of the reinforcement the shear-stress-slip law is in the softening stage typical of a post-elastic behaviour.

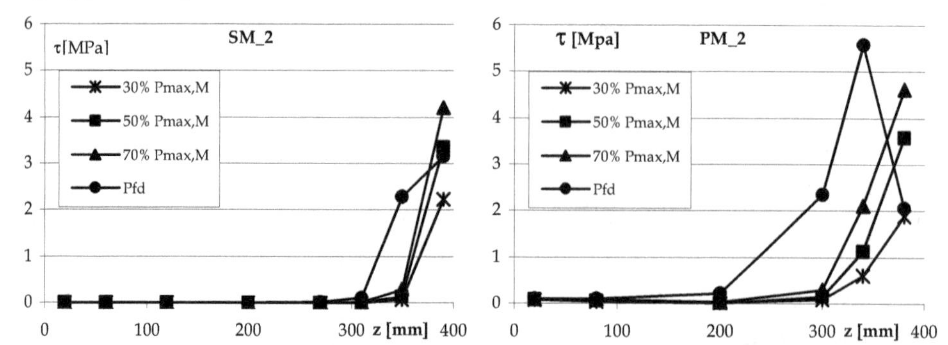

(a) (b)

Fig. 13. Shear stresses assessed on sheet (a) and plate (b) reinforced specimens

The shear-stress-slip relationship, describing the FRP-to-concrete interface law, can be identified starting from the experimental results of pull-out tests; the distribution of shear stresses could be, in fact, obtained by Eq. (65) and the corresponding slips values by integrating the axial strains measured during the test by the following relationship

$$s_{i,i+1} = \sum_{k=0}^{i} \frac{(\varepsilon_{k+1} + \varepsilon_k)}{2} \cdot (x_{k+1} - x_k) \tag{66}$$

In particular the bond law at the FRP-to-concrete interface (see Fig. 14) can be obtained by calculating the shear stresses using Eq. (65) (considering the strains recorded by two gauges nearest the loaded end of FRP, see Fig. 10) and the corresponding slips using Eq. (66) (considering all the strain gauges applied on the FRP reinforcement).

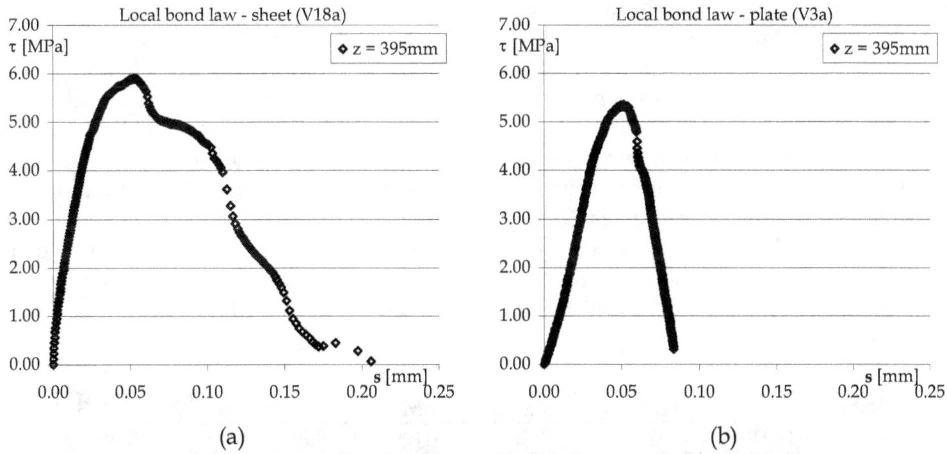

Fig. 14. Experimental bond law at the FRP-to-concrete interface: sheet (a) and plate (b)

In this way a direct identification method (DirIM) is put in place since the interface law is obtained directly with respect to experimental values of shear stresses and relative slips.

As stated above, according to the study made by Taljsten (1996), derived by linear fracture mechanics, the maximum value F_{max} of the force applied to the FRP-to-concrete joint can be obtained by (63) where G_F is the specific fracture energy of the interface depending on the mechanical properties of the adhesive FRP-to-concrete interface.

Although plenty of alternative models are available in the literature (i.e. Savoia et al., 2003) to describe FRP-to-concrete interface interaction, a simplified bi-linear relationship can reproduce the key aspects of the interface behaviour as pointed out by the comparative study carried out by Lu et al. (2005). Elasto-softening bilinear relationship is one of the most widely accepted expressions for the interface relationship and fracture energy; three mechanical parameters completely identifies such relationship (maximum shear stress τ_{max}, ultimate slip s_u, and elastic stiffness k_e), but obviously the knowledge of G_F by solving equations (63) and (64), is not sufficient for determining such parameters.

As widely discussed in Faella et al. (2009), two approaches are mainly possible to calibrate an interface relationship: the first approach (Direct Identification Method - DirIM) starts

from the experimental strains measured during the test: the distribution of shear stresses can be roughly evaluated through equilibrium of the axial stresses depending on the axial strain by (65); the corresponding slip values can be obtained by integrating throughout the FRP the axial strains by (66). The couples of values $\left(\bar{s}_j, \bar{\tau}_j\right)$ can be "directly" used to calibrate the τ-s relationship through a numerical regression, such as the least square method. Although the DirIM is very easy under the analytical standpoint, the drawback of the method is that the distribution of shear stresses cannot be directly compared with data provided by the pull out tests, because both interface shear stresses and local displacements cannot be directly measured during the usual pull-out tests. In order to overcome such problem, the indirect identification procedure (IndIM) can be pursued to calibrate an interface relationship starting from the availability of axial strain evolution in FRP plate at different load levels up to debonding failure: once a bi-linear interface relationship has been assigned, the corresponding theoretical strain distributions can be evaluated and compared with experimental data; the procedure is iterative and ends when the difference between theoretical and experimental strains is less than a prefixed tolerance (Faella et al., 2003).
With reference to some experimental data (Nigro et al., 2008) both DirIM and IndIM have been applied in order to plot the τ-s relationship and derive the corresponding specific fracture energy G_F, computed by totalling up the area under the bi-linear curve.
Table 1 shows that IndIM leads to specific fracture energy G_F larger than those computed by DirIM in almost all analyzed cases. Moreover, in the case of experimental tests carried out by the authors, values of G_F obtained with IndIM are as stable as the measured failure loads P_{test}, confirming the superior accuracy of IndIM with respect to DirIM. Some of the most representative bi-linear interface relationships obtained by applying both DirIM and IndIM are reported in Fig. 15.

Label		PM_1	PM_2	PM_3	SM_1	SM_2	SM_3
G_F	(DirIM)	0.45	1.20	0.86	0.66	0.86	0.48
G_F	(IndIM)	0.73	1.75	1.03	0.66	0.67	0.72
$G_{F,(IndIM)}/G_{F,(DirIM)}$		1.61	1.46	1.20	0.99	0.78	1.50

Table 1. Specific fracture energy GF: (DirIM) Vs (IndIM)

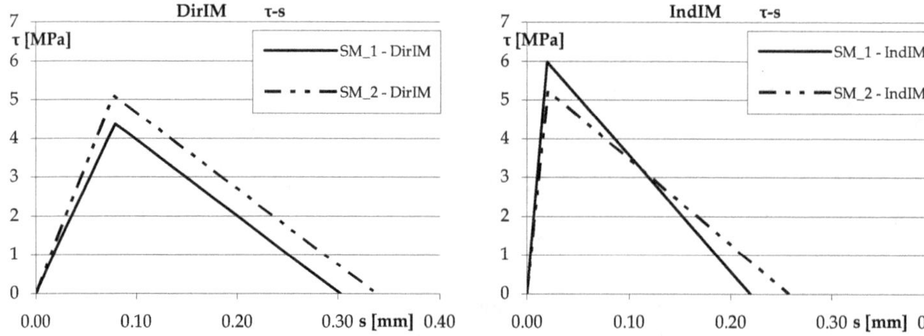

Fig. 15. Bi-linear interface relationship: (DirIM) Vs (IndIM)

Two methods lead to results that could be even significantly different, especially in terms of ultimate slips. Furthermore, the ascending branches of DirIM curves are characterized by a lower slope with respect to IndIM.

Comparisons between the theoretical and experimental values of the axial strains throughout the FRP plate are reported, for different values of load test, in Fig. 16. The results are referred to test SM_3. The theoretical strain values have been obtained by closed-form solutions reported in Faella et al. (2003).

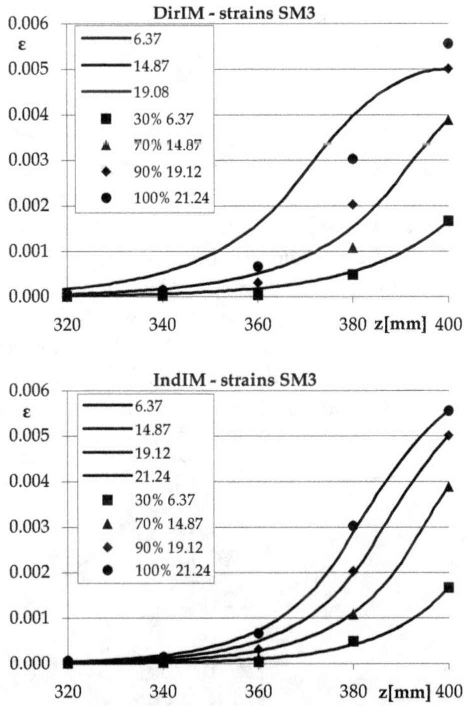

Fig. 16. Theoretical and experimental strains: (DirIM) Vs (IndIM)

The theoretical values of strains obtained assuming the interface relationship identified through IndIM are much closer to the observed ones than the values which can be derived assuming the τ-s relationship calibrated by means of DirIM. Finally it is worth observing that the experimental data can be simulated with good accuracy even if a simplified adhesion law (i.e. a bilinear adhesion law) is assumed.

4. Design formulae

The fracture energy, Γ_F, depends on both the strength properties of adherents, concrete and adhesive, and the properties of the concrete surface. When the reinforcement is correctly applied, the detachment of the reinforcement occurs with debonding in the concrete layer and the specific fracture energy of the interface law can be written in a form similar to that used for the shear fracture (mode I). Therefore, in order to obtain design formulae, the fracture energy can be expressed as a function of the shear strength in the concrete: $\Gamma_F(\tau_{b,max})$, where $\tau_{b,max}$ depends on the tensile and compressive strength of concrete.

4.1 Theoretical formulations of debonding load

In most formulations the fracture energy is directly expressed as function of the tensile and/or compressive strength of concrete and it is also function of a shape factor depending on the FRP to concrete width (b_f/b_c).

Based on experimental tests, many theoretical formulations have been proposed in the past (Taljsten, 1994; Neubauer and Rostasy, 1997; Brosens and Van Gemert, 1997; fib bulletin, 2001; Chen and Teng, 2001; Smith and Teng, 2002; CNR-DT200, 2004; Lu et al., 2005; Seracino et al., 2007) to evaluate bond strength at the end of R.C. beams externally strengthened with FRP.

The lay out of the existing formulations is often similar; the differences are essentially related to the numerical coefficients calibrated on experimental results and to the eventual presence of safety factors. This last point is an important issue in using these formulations for design when the safety level (mean, characteristic or design provisions) has to be chosen. Some of these formulas are used to predict the bond strength at the intermediate crack too, through the exchange of some factors (Teng et al., 2003; Chen et al., 2006; CNR-DT200, 2004). For illustrative purposes an example of a calibration procedure of a simplified design relationship based on experimental data is reported in the following section. The example was made with reference to the theoretical relationship (67) suggested in CNR DT200-2004 but the procedure can be used to calibrate also different formulations.

$$N_{f,max} = \frac{1}{\gamma_{f,d}\sqrt{\gamma_c}} \cdot \beta_L \cdot b_f \cdot \sqrt{k_G \cdot k_b}\sqrt{2 \cdot E_f \cdot t_f \cdot \sqrt{f_{ck} \cdot f_{ctm}}} \quad (67)$$

$$L_e = \sqrt{\frac{E_f \cdot t_f}{2 \cdot f_{ctm}}} \quad (68)$$

$$k_b = \sqrt{\frac{2 - b_f/b_c}{1 + b_f/400}} \quad (69)$$

$$\beta_l = \frac{L_b}{L_e} \cdot \left(2 - \frac{L_b}{L_e}\right) \quad \text{if} \quad L_b \leq L_e; \quad \beta_l = 1 \quad \text{otherwise} \quad (70)$$

f_{ck} being the characteristic value of cylindrical compressive strength of concrete, k_G a coefficient regarding the experimentally calibrated fracture energy and equal to 0.064 for mean value provision or 0.03 for design value. The safety factor for debonding failure, $\gamma_{f,d}$, is usually assumed equal to 1.2 or 1.5 (non-controlled or controlled gluing application); γ_c is the safety factor for concrete (equal to 1.5, EC2, 2004).

4.2 Simplified calibrating procedure

The mean value of the maximum axial strain in FRP corresponding to debonding failure can be expressed by means of a relationship obtained by a deterministic model and fine-tuned on experimental data through a probabilistic approach, as suggested in (EN1990 – Annex D) and applied by Monti et Al. (2009).

By assuming a simplified formulation similar to the design relationship (67), taking into account no safety partial factors, the debonding strain can be expressed as:

$$\varepsilon_{fd}\left(k_{bf}, f_c, f_{ct}, E_f, t_f, b_f, k_b\right) = k_{bf} \cdot \sqrt{\frac{2 \cdot k_b \cdot \sqrt{f_c \cdot f_{ct}}}{E_f t_f}} \tag{71}$$

where the coefficient k_{bf} can be calibrated based on experimental results using a least-square procedure consisting in the resolution of the following minimum problem (Bilotta et al. 2011):

$$k_{bf} = \arg\min_{k_{bf}} \sum_{i=1}^{n} \left[\varepsilon_{fd}\left(k_{bf}, f^{(i)}, f_{ct}^{(i)}, E_f^{(i)}, t_f^{(i)}, b_f^{(i)}, k_b^{(i)}\right) - \varepsilon_{exp}^{(i)} \right]^2 . \tag{72}$$

Furthermore, a random variable δ can be defined, for each ith test, as the ratio of the experimental debonding strain, $\varepsilon_{exp,i}$, to the theoretical one, $\varepsilon_{fd,i}$, evaluated by considering the geometric and mechanical data characterizing that test:

$$\delta_i = \frac{\varepsilon_{exp,i}}{\varepsilon_{fd,i}} . \tag{73}$$

The error function, δ, can cover the uncertainties of the simplified model considered as follows:

$$\varepsilon_{fd} = \varepsilon_{fd}\left(k_{bf}, f_c, f_{ct}, E_f, t_f, b_f, k_b\right) \cdot \delta , \tag{74}$$

therefore, the mean value of the intermediate debonding strain can be obtained by a coefficient k_m, starting from k_{bf} adjusted by means of the mean value of the error parameter $\overline{\delta}$, being in general $\overline{\delta} \neq 1$ because the regression line is imposed to intercept the origin:

$$k_m = k_{bf} \cdot \overline{\delta} . \tag{75}$$

Obviously, this strain is proportional to the debonding stress, being linear elastic the FRP constitutive law.

If the random variable represents a strength, its characteristic value is often defined for design purposes as the 0.05 percentile of the frequency distribution associated to the examined variable. Gauss distribution is the most generally considered for describing the errors. Assuming that the Young modulus of the FRP reinforcement, E_f, and the concrete tensile and compressive strengths, f_{ctm} and f_{cm}, are the only mechanical parameters influencing the value of the maximum axial strain developed in FRP at debonding, the expressions for the both general model and the calibrated one involving the coefficient k_{bf} as well as $\overline{\delta}$ are:

$$\varepsilon_{th} = \varepsilon_{th}\left(E_f, f_{cm}, f_{ctm}\right) , \tag{76}$$

$$\varepsilon_{th,m} = \varepsilon_{th,m}\left(E_f, f_{cm}, f_{ctm}, \bar{\delta}, k_{bf}\right). \tag{77}$$

E_f, f_c and f_{ct} are assumed normally and independently distributed random variables, with a priori values of the coefficients of variation according to the design relationships provided by codes and literature information (Bilotta et al. 2011). Also the variable δ can be assumed as normally distributed. However, the hypothesis of normal distribution is required to be checked at least by comparing the experimental curve of the cumulative frequency to the theoretical one corresponding to a Gaussian distribution having the same mean value and standard deviation. Hence, under the hypothesis of normal distribution for the variable δ, the strain provision corresponding to the 0.05 percentile of the Gaussian distribution is:

$$\varepsilon_{th,k} = \varepsilon_{th,m} - 1.64 \cdot \left[\mathrm{Var}(\varepsilon_{th,m})\right]^{0.5} \tag{78}$$

where the variance of $\varepsilon_{th,m}$ can be expressed as:

$$\mathrm{Var}\left(\varepsilon_{th,m}\right) = C_{Efm}^2 \cdot \mathrm{Var}\left(E_f\right) + C_{fcm}^2 \cdot \mathrm{Var}\left(f_{cm}\right) + C_{fctm}^2 \cdot \mathrm{Var}\left(f_{ctm}\right) + C_{\delta m}^2 \cdot \mathrm{Var}\left(\delta_m\right) \tag{79}$$

$$C_{Efm} = \left|\frac{\partial \varepsilon_{th,m}}{\partial E_f}\right|_{\overline{E_f}}, \quad C_{f_{cm}} = \left|\frac{\partial \varepsilon_{th,m}}{\partial f_{cm}}\right|_{\overline{f_{cm}}}, \quad C_{f_{ctm}} = \left|\frac{\partial \varepsilon_{th,m}}{\partial f_{ctm}}\right|_{\overline{f_{ctm}}}, \quad C_{f_{ctm}} = \left|\frac{\partial \varepsilon_{th,m}}{\partial \delta_m}\right|_{\overline{\delta_m}} \tag{80}$$

If the Eqs. (80) and (79) are substituted in the Eq. (78), the following general expression is obtained, providing the 0.05 percentile of the debonding load:

$$\varepsilon_{th,k} = \varepsilon_{th,m} - 1.64 \cdot \varepsilon_{th,m} \cdot \left[a \cdot \mathrm{CoV}_{E_f}{}^2 + b \cdot \mathrm{CoV}_{f_{cm}}{}^2 + c \cdot \mathrm{CoV}_{f_{ctm}}{}^2 + \mathrm{CoV}_{\delta_m}{}^2\right]^{0.5} \tag{81}$$

where the coefficient a, b, c depend on the functional relationship of E_f, f_{cm} and f_{ctm} in the expression of ε_{th}; moreover the coefficients of variation are defined for each parameter as the ratio of the mean value to its standard deviation:

$$\mathrm{CoV}_{E_f} = \frac{\overline{E_f}}{s_{E_f}}, \quad \mathrm{CoV}_{f_{ctm}} = \frac{\overline{f_{ctm}}}{s_{f_{ctm}}}, \quad \mathrm{CoV}_{f_{cm}} = \frac{\overline{f_{cm}}}{s_{f_{cm}}}, \quad \mathrm{CoV}_{\delta_m} = \frac{\overline{\delta_m}}{s_{\delta_m}} \tag{82}$$

Clearly the coefficient of variation of the variable δ_m, $CoV_{\delta m}$, depends on the data distribution. The Eq. (81) can be written as:

$$\varepsilon_{th,k} = k_k \cdot \varepsilon_{th} \tag{83}$$

assuming:

$$k_k = k_m \cdot \left(1 - 1.64 \cdot \left[a \cdot \mathrm{CoV}_{E_f}{}^2 + b \cdot \mathrm{CoV}_{f_{cm}}{}^2 + c \cdot \mathrm{CoV}_{f_{ctm}}{}^2 + \mathrm{CoV}_{\delta_m}{}^2\right]^{0.5}\right). \tag{84}$$

Lower percentiles can be obtained by substituting in the Eq. (78) the coefficient 1.64, related to the 0.05 percentile of the frequency distribution, with the coefficients 2.58 and 3.08 corresponding to the 0.005 and 0.001 percentiles, respectively.

The use of percentiles lower than 0.05 can be alternative to the use of safety factors that usually have to be additionally applied to characteristic provision to take into account the model uncertainness (EN1990 - Annex D).
In Table 2 the coefficient k_m and the R^2, that is a measure of the reliability of the regression, are reported: two different experimental samples were considered for sheets and plates respectively. Note that for the cured-in-situ FRP systems (sheet) the R^2 value is quite high (0.855); on the contrary for the preformed systems (plate) the R^2 value is quite low (0.349), mainly due to a higher scattering of the experimental data.
In addition the mean value of the variable δ_m and its CoV are reported. In both cases the CoV, that represents a measure of the model significance, is lower than the threshold value of 40% (Monti et al., 2009) so that all the models can be considered reliable.
Finally, Table 2 shows that different scattering of experimental data leads to very different 0.05, 0.005 and 0.001 percentile of the calibrating coefficient k, for sheet and plate respectively.

FRP type	k_m	R^2	$\overline{\delta_m}$	CoV_{δ_m}	$k_m \cdot \overline{\delta_m}$	$k_{k,5\%}$	$k_{k,0.5\%}$	$k_{k,0.1\%}$
Sheet	0.270	0.855	1.027	17.7%	0.278	0.192	0.143	0.117
Plate	0.236	0.349	1.064	23.2%	0.251	0.152	0.095	0.064

Table 2. Statistical data

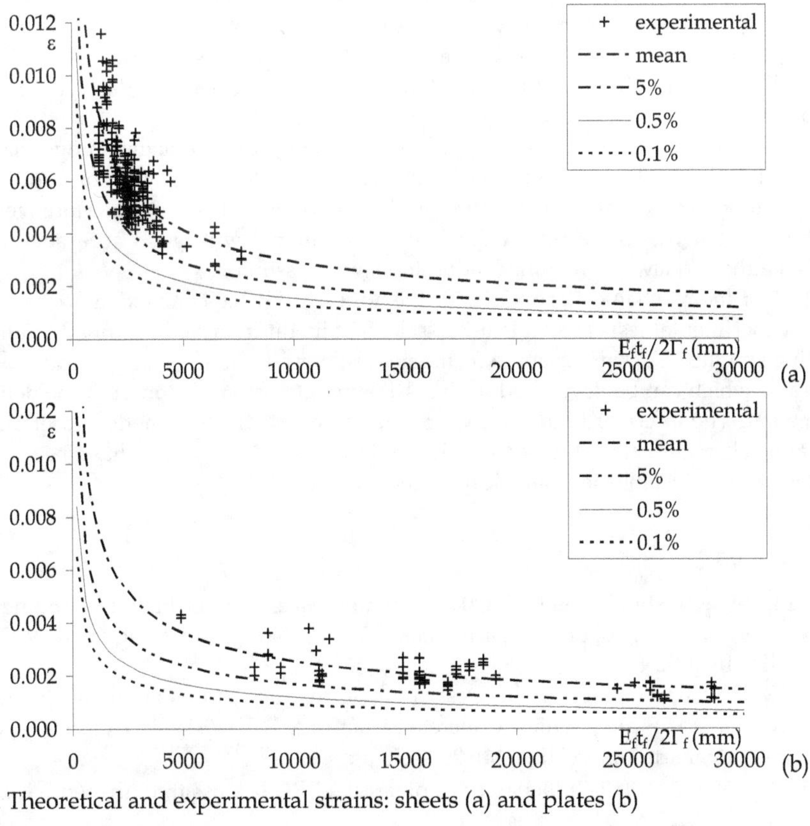

Fig. 17. Theoretical and experimental strains: sheets (a) and plates (b)

Theoretical debonding strain (71) are plotted in Fig. 17a,b versus the parameter $E_f t_f / k_b \sqrt{f_{cm} f_{ctm}}$, for sheets and plates respectively. Each curve is obtained with reference to each percentile of the k coefficient reported in Table 2. Theoretical debonding strains, obtained by means the calibrating procedure showed above, can be easily compared with experimental ones (singular point).

5. Conclusions

A wide overview of the key contributions in understanding the mechanical behavior of FRP laminates bonded on a concrete substrate has been reported in this chapter. Several original contributions have been also proposed.

First of all, a general analytical model has been formulated for describing the distribution of shear and normal stresses within the adhesive layer connecting the FRP laminate to the substrate. The key dimensionless parameters governing such distributions have been clearly pointed out and their influence has been showed in numerical application. The possible simplification of that general analytical model in a simpler one based on the assumption of a layer of elastic springs connecting the FRP laminate to the substrate has been also proposed. This model leads to much simpler differential equations with a reasonable reduction of accuracy. Moreover, it can be easily extended to the nonlinear branch, especially in the case of a supposed bi-linear (namely, elastic-softening) behaviour of the interface in terms of relationship between shear stresses and interface slip. Closed-form solutions of the stress and strain distribution throughout the FRP laminate can be derived under this assumption.

Those closed-form solutions can be employed for identifying the actual properties of the interface relationship. The numerical stability of an indirect-identification procedure based on those solutions has been demonstrated in the section 3 of this chapter. Moreover, it has been showed that the experimental data can be simulated with good accuracy even if a simplified adhesion law (i.e. a bilinear adhesion law) is assumed.

The values of the mechanical properties of the adhesive FRP-to-concrete interface obtained by some experimental tests available in the scientific literature have been finally utilized for calibrating simplified design-oriented formulae for determining the values of the maximum axial strains which can be developed in the FRP laminate before the onset of the debonding phenomenon. The effect of both the material randomness and the model uncertainties in defining the characteristic value of the axial strain in FRP at debonding have been handled by means of a well-established statistical procedure.

6. References

Achintha P. M. M. & Burgoyne C. J. (2008). Fracture Mechanics of Plate Debonding, *ASCE Journal for Composites for Construction*, Vol. 12, No. 4, July/August 2009, 396-404, ISSN 1090-0268

ACI 440.2R-08 (2008). Guide for the Design and Construction of Externally Bonded FRP Systems for Strengthening Concrete Structures. *ACI440.2R-08* American Concrete Institute, Farmington Hills, MI, 2008, 76 pp.

Arduini M. & Nanni A. (1997). Behavior of precracked RC beams strengthened with carbon

FRP sheets, *ASCE Journal for Composites for Construction*, Vol. 1, No. 2, 63–70, ISSN 1090-0268

Bilotta A., Ceroni F., Di Ludovico M., Nigro E. & Pecce M. (2011). Bond Tests on concrete elements strengthened with EBR and NSM FRP systems, *Proceedings of FRP RCS 10*, Tampa, Florida, April 2011, accepted.

Bilotta A., Ceroni F., Nigro E. & Pecce M. (2011). Design by testing of debonding load in RC element strengthened with EBR FRP materials, *Proceedings of FRP RCS 10*, Tampa, Florida, April 2011, accepted.

Bilotta A, Di Ludovico M. & Nigro E. (2009). Influence of effective bond length on FRP-concrete debonding under monotonic and cyclic actions. *Proceedings of 9th International Symposium on Fiber Reinforced Polymer Reinforcement for Concrete Structures*, Sydney (Australia), July 2009 (CD ROM).

Bizindavyi L. & Neale K. W. (1999). Transfer lengths and bond strengths for composites bonded to concrete, *ASCE Journal for Composites for Construction*, Vol. 3, No. 4, July/August 1999, 153-160, ISSN 1090-0268

Blontrock, H., Taerwe, L. & Vanwalleghem, H. (2002). Bond testing of externally glued FRP laminates at elevated temperature, *Proceedings of the International Symposium "Bond in Concrete: from research to standards"*, 648-654, ISBN 963-420-714-6, Budapest, November 2002, Publisher, City

Bonacci J.F. (1996). Strength, Failure Mode and Deformability of Concrete Beams Strengthened Externally with Advanced Composites, *Proceedings of the 2nd International Symposium on Advanced Composite Materials in Bridges and Structures*, ISBN, Montreal (Canada), 419-426, Publisher, City

Ceroni, F., Pecce, M., Matthys, S. & Taerwe, L. (2008). Bond tests on concrete elements with CFRP and anchorage systems, *Composites: Part B* - Elsevier, Vol. 39, No.3, April 2008, 429–441, ISSN 1359-8368

Ceroni F. & Pecce M. (2010). Evaluation of bond strength and anchorage systems in concrete elements strengthened with CFRP sheets, *ASCE Journal of Composites in Construction*, in press, ISSN 1090-0268

Chajes M.J., Finch W.W., Januszka T.F. & Thomson T.A. (1996). Bond and force transfer of composite-material plates adhered to concrete, *ACI Structural Journal*, Vol. 93, No. 2, February 1996, 208–217, ISSN 0889-3241

Chen J.F. & Teng J.G. (2001): Anchorage Strength Models for FRP and Plates Bonded to Concrete, *ASCE Journal of Structural Engineering*, Vol. 127, No. 7, July 2001, 784-791, ISSN 0733-9445

CNR-National Council for Research (2004), *Guide for the Design and Construction of Externally Bonded FRP Systems for Strengthening Existing Structures-Materials, RC and PC structures, masonry structures*

Czaderski C., Soudki K. & Motavalli M. (2010). Front and Side View Image Correlation Measurements on FRP to Concrete Pull-Off Bond Tests, *ASCE Journal for Composites for Construction*, Vol. 14, No. 4, July/August 2010, 451-463, ISSN 1090-0268

European Committee for Standardization. *EN 1990 – Eurocode - Basic of Structural Design*. 2002.

Faella C., Martinelli E. & Nigro E. (2002). Aderenza tra calcestruzzo e Lamine di FRP utilizzate come placcaggio di elementi inflessi. Parte II: modelli teorici ed elaborazioni numeriche, Proceedings of the XIV C.T.E. Conference, Mantua (Italy), 7-8 November 2002, 237-245, (in Italian);

Faella C., Martinelli E. & Nigro E. (2008). Formulation and Validation of a Theoretical Model for Intermediate Debonding in FRP Strengthened RC Beams, *Composites Part B*, Vol. 39, No. 4, 645-655, June 2008, ISSN 1359-8368

Faella C., Martinelli E. & Nigro E. (2009). Direct versus Indirect Method for Identifying FRP-to-Concrete Interface Relationships, *ASCE Journal for Composites for Construction*, Vol. 13, No. 3, May/June 2009, 226-233, ISSN 1090-0268

fib Bulletin No. 14 (2001). *Externally bonded FRP reinforcement for RC structures*, Technical report, ISBN 978-2-88394-054-3, Lausanne (CH)

Hart-Smith, L. J. (1973). Adhesive-bonded double-lap joints, NASA Technical Report, CR-112235, Hampton, Va. (USA)

Lee Y. J., Boothby T. E., Bakis C. E. & Nanni A. (1999). Slip Modulus of FRP Sheets Bonded to Concrete, *ASCE Journal of Composites for Construction*, Vol. 3, No. 4, November 1999, 161-167, ISSN 1090-0268

Lu X.Z., Teng J.G., Ye L.P. & Jiang J.J. (2005). Bond-slip models for FRP sheets/plates bonded to concrete, *Engineering Structures*, Vol. 27, No. 6, May 2005, 920-937, ISSN 0141-0296

Mazzotti C., Savoia M. & Ferracuti B. (2009). A new single-shear set-up for stable debonding of FRP–concrete joints, *Construction and Building Materials - Elsevier*, Vol. 23, No.4, April 2009, pp. 1529-1537. ISSN 0950-0618

Malek A. M., Saadatmanesh H. & Ehsani M. R. (1998). Prediction of failure load of R/C beams strengthened with FRP plate due to stress concentration at the plate end, *ACI Structural Journal*, Vol. 95, No. 2, March/April 1998, 142-152, ISSN 1090-0268

Meier U. (1987). Brückensanierungen mit Hochleistungs-Faserverbundwerkstoffen, Material und Technik, Vol. 15, No 4, 125-128 ISSN

Meier U. (1995). Strengthening of structures using carbon fibre/epoxy composites, *Construction and Building Materials*, Vol. 9, No. 6, June 1995, 341–351, ISSN 0950-0618

Monti G., Alessandri S. & Santini S. (2009). Design by Testing: A Procedure for the Statistical Determination of Capacity Models, *Journal of Construction and Building Materials, Special Issue on FRP Composites Elsevier*, Vol. 23, 1487-1494. ISSN 0950-0618

Motavalli M. & Czaderski C. (2007). FRP Composites for Retrofitting of Existing Civil Structures in Europe: State-of-the-Art Review, *Proceedings of the American Composites Manufacturers Association 2007*, ISBN, Tampa, FL (USA), October 2007, Publisher, City

Naaman A.E., Park S.Y., Lopez M.M. & Till R.D. (2001). Parameters Influencing the Flexural Response of RC Beams Strengthened Using CFRP Sheets, *Proceedings of the Conference FRPRCS-5*, Cambridge (UK), 117-125, ISBN 0727730290

Neubauer, U. & Rostasy, F.S. (1997). Design Aspects of Concrete Structures Strengthened with Externally Bonded CFRP Plates, Concrete and Composites, *Procedings of 7th International Conference On Structural Faults and Repair*, Vol. 2, 109-118, ISBN, ECS Pub. Edinburgh, Scotland.

Nigro E., Di Ludovico M. & Bilotta A. (2008). FRP- Concrete Debonding: experimental Tests under Cyclic Actions, *Proceedings of 14th World Conference on Earthquake Engineering*, ISBN, Beijing (China), October 2008, Publisher, City

Oehlers D.J., Moran J.P. (1990). Premature failure of externally plated reinforced-concrete beams, *Journal of Structural Engineering - ASCE,* Vol. 116, No.4, April, 1990, 978-995. ISSN 0733-9445

Rabinovich O. & Frostig, Y. (2000). Closed-form High-order Analysis of RC Beams Strengthened with FRP Strips, *ASCE Journal of Composites for Construction,* Vol. 4, No. 2, March-April 2000, 65-74, ISSN 1090-0268

Rabinovich O. & Frostig, Y. (2001). Delamination Failure of RC Beams Strengthened with FRP Strips—A Closed-Form High-Oorder and Fracture Mechanics Approach, *Journal of Engineering Mechanics,* Vol. 127, No. 8, August 2001, ISSN 0733-9399

Roberts T.M. (1989). Approximate analysis of shear and normal stress concentrations in the adhesive layer of plated RC beams, *The Structural Engineer,* Vol. 67, No. 12, December 1989, 229–233, ISSN 0039-2553

Savoia M., Bilotta A., Ceroni F., Di Ludovico M., Fava G., Ferracuti B., Mazzotti C., Nigro E., Olivito R., Pecce M. & Poggi C., (2009). Experimental round robin test on FRP concrete bonding, *Proceedings of FRPRCS9,* Sydney (Australia), July, 2009 (CD ROM).

Smith S.T. & Teng J.G. (2001), Strength Models for Plate End Debonding in FRP-Strengthened RC Beams, *Proceedings of the Conference FRPRCS-5,* ISBN 0727730290, Cambridge (UK), 419-428, 2001, Publisher, City

Smith S. T. & Teng J. G. (2002). FRP-strengthened RC beams. I: review of debonding strength models, *Engineering Structures,* Vol. 24, No. 4, April 2002, 385-395, ISSN 0141-0296

Swamy R.N., Jones R. & Bloxham J.W. (1987). Structural behaviour of reinforced concrete beams strengthened by epoxy-bonded steel plates, *The Structural Engineer,* Vol. 65A, No. 2 February 1987, 59–68, ISSN 0039-2553

Taljsten B. (1996). Strengthening of concrete prisms using the plate bonding technique, *International Journal of Fracture,* Vol. 82, No., 253–266, ISSN 0376-9429

Taljsten B. (1997). Strengthening of beams by plate bonding, *ASCE Journal of Materials in Civil Engineering,* Vol. 9, No. 4, November 1997, 206–211, ISSN 1943-5533

Teng J.G., Chen J.F., Smith S.T. & Lam L. (2002). *FRP Strengthened RC Structures,* John Wiley & Sons Ltd., ISBN 0-471-48706-6, 245pp. Chichester (UK)

Teng J. G., Smith S. T., Yao J. & Chen J. F. (2003). Intermediate crack-induced debonding in RC beams and slabs, *Construction and Building Materials,* Vol. 17, No. 6-7, September-October 2003, 447-462, ISSN 0950-0618

Tounsi A., Hassaine Daouadji T., Benyoucef S. & Addabedia E.A. (2009). Interfacial stresses in FRP-plated RC beams: Effect of adherend shear deformations, *International Journal of Adhesion & Adhesives - Elsevier,* Vol. 29, No. 4, June, 2009, 343-351, ISSN 0143-7496

Triantafillou T., Matthys S. & Taerwe L. (2001). Design of Concrete Members Strengthened with Externally Bonded FRP Reinforcement, *Proceedings of the Conference FRPRCS-5,* Cambridge (UK), 157-166, ISBN 0727730290, Publisher, City

Wu, Z., Yuan, H. & Niu, H. (2002). Stress transfer and fracture propagation in different kinds of adhesive joints, *ASCE Journal of Engineering Mechanics,* Vol. 128, No. 5, May 2002, 562-573, ISSN: 0733-9399

Yao J., Teng J.G. & Chen J.F. (2005). Experimental study on FRP-to-concrete bonded joints, *Composites: Part B,* Elsevier, Vol. 36, No. 2, March, 2005, 99–113, ISSN 1359-8368

2

To Carry Out Tolerance Analysis of an Aeronautic Assembly Involving Free Form Surfaces in Composite Material

Wilma Polini
Dipartimento di Meccanica, Strutture, Ambiente e Territorio, Università degli Studi di Cassino, Cassino Italy

1. Introduction

The increasing competition in aerospace industries has brought to cutting programs in manufacturing, design and assembly of aircraft structural frame. An aircraft is made of several assemblies including 3D complex shapes and the different functional requirements from design and manufacturing that these assemblies should respect are various. Nowadays the most important policy is the "Concurrent Engineering" which rule is to lead in a parallel way, design and manufacturing, making them communicate with the aim of reducing reworking times and discard products; such requirements are strongly felt in the aerospace industries.

Tolerance analysis has a considerable weight in the Concurrent Engineering and represents the best way to solve assembly problems in order to ensure higher quality and lower costs. It is a critical step to design and to build a product and its importance has grown in the past years. In fact, the need to assign dimensional and geometric tolerances to assembly components is due to the standardization of the production and to the correct working of the assembly. The appropriate allocation of tolerances among the different parts of an assembly can result in lower costs per assembly and higher probability of fit, reducing the number of rejects or the amount of rework required on components.

A product is designed and manufactured to perform a task, and its issue depends on one or more parameters of the assembly that are commonly called "project functions". A project function is a dimension or a geometric variable of the assembly whose value depends by the dimensions, the geometry and the tolerances assigned to the components constituting the assembly. The nominal value and the tolerance range of the project function allow us to guarantee the assembly functionality. Practically, the dimensions and the tolerances of the assembly components combine, according to the assembly sequences, and generate the tolerance stack-up functions. Solving a tolerance stack-up function means to determine the nominal value and the tolerance range of a project function by combining the nominal values and the tolerance ranges assigned to the assembly components.

Tolerance analysis may consider alternative assembly cycles in order to identify that one allowing to obtain the assembly functionality with the maximum value of the tolerance range assigned to the components. Huge problems may present during the assembly process if the tolerance study on a sub-component was not carried out or was ineffectual (Whitney, 2004). It is even possible that the product design may have to be subsequently changed because of unforeseen tolerance problems not detected prior to actual assembly took place. In this case costs to the business will be high. It was estimated that 40%-60% of the production cost is due to the assembly process (Delchambre, 1996).

The study of the tolerance stack-up functions, during the design stage, is very critical for aeronautic field whereas the complexity of the structures, to which high performances are required; so advanced material, design techniques and assembly technologies are needed. In fact, the aeronautic structures involve free-form surfaces, which are often made in composite material. They may be considered as non-rigid parts that could be subjected to significant distortion after the removal of manufacturing forces. This condition, know as free-state variation, is principally due to weight and flexibility of the part and the release of internal stresses resulting from fabrication.

Many well-known approaches exist in the literature to tolerance analysis (Hong & Chang, 2002; Shen et al., 2004). However, these methods are not easy to apply, especially for complex aerospace assemblies, since they were born to deal with elementary features, such as plane, hole, pin and so on. So the aid of computer is called for. In the recent years, the development of efficient and robust design tools has allowed to foresee manufacturing or assembly problems during the first steps of product modeling by adopting a concurrent engineering approach. Today Computer Aided Tolerance (CAT) Software is readily available, but even if these tools provide good results they have not been widely used. Commercial CATs are not completely true to the GD&T standards and need improvement after a better mathematical understanding of the geometric variations. The user needs expertise and great experience combined with a through understanding of the packages' theoretical base plus modeling principles to build a valid model and obtain relatively accurate results. Computer Aided Tolerance software efficiently deals with mechanical assemblies where the feature to align are planes, hole-pin, but it hardly treats of free-form surfaces to connect.

The present work deals with the tolerance analysis of freeform surfaces belonging to parts in composite material and that may be non-rigid. The aim of this chapter is to present the steps to carry out the tolerance analysis of an assembly involving free-form surfaces in composite material by using a commercial CAT software. The great effort of the present work is overcome the limits of the CAT software to deal with dimensional and geometric tolerances applied to free form surfaces in composite material. This paper tries to answer to some questions on tolerance analysis without clear answer: What are the functions of the product, how do we flow down these key product functions through into its detail parts? How to model a free-form surface for a tolerance analysis? How to deal with a composite material for a tolerance analysis? How to improve the assembly process in order to reduce the tolerance impacts on these functions?

An aeronautic component is considered as case study. This is an internal frame of a winch arm mounted on an helicopter. The part addressed in this paper is the after internal frame that allows to mount the fairing covers of the head of the winch arm directly on its structural beam. It is made in carbon fiber composite material. The frame is made with five layer of carbon fiber imbued in a matrix of epoxy resin for a total stack thickness of 1.65 mm. After the stratification process, a curing process is carried out by autoclave with a control of temperature and pressure. The weight of the part is about 350 g. Assembly operation among the parts are done by using special grub screws that are inserted in the holes of each component. No adhesives or resins are used to secure the surfaces, since a perfect adhesion is not required.

The paper is organized as follows: In Sec. 2, the mean of tolerance analysis is deeply discussed. In Sec. 3, the main models found in the literature for tolerance analysis are presented. In Sec. 4, the main functions of the commercial CAT softwares are shown. In Sec. 5, the steps of the proposed method for tolerance analysis of an assembly involving free-form surfaces in composite material are shown. In Sec. 6, the application of the proposed steps to an aeronautic case study is deeply discussed.

2. Tolerance analysis

The aim of a tolerance analysis of an assembly is to evaluate the cumulative effect due to the tolerances, that are assigned to the assembly components, on the functional requirements of the whole assembly. Each functional requirement is schematized through an equation, that is usually called stack-up function, whose variables are the model parameters that are function of the dimensions and the tolerances assigned to the assembly components. It looks like

$$FR = f(p_1, p_2, \ldots, p_n) \tag{1}$$

where FR is the considered functional requirement, $p_1, \ldots, p_n$ are the model parameters and $f(p)$ is the stack-up function, that is usually not linear.

A functional requirement is usually a characteristic that relates two features. Its analytical expression is obtained by applying the equations of the *Euclidean geometry* to the features that define the functional requirement or to the points of the features that define the functional requirement.

A stack-up function has to model two possible assembly variations. The first variation is due to the tolerances assigned to the features of the assembly components. The obtained model (that is called *"local model"*) has to be able to schematize all the tolerance kinds, i.e. dimensional, form, and so on, but in the same time it has to be able to represent the Envelope Principle (Rule # 1 of ASME standard) or the Independence Principle (according with ISO 8015 standard) applied to different dimensions of the same part. The *local model* has to define the range of variation of the model's parameters from the assigned tolerances and it has to schematize the interaction among the assigned tolerance zones. The second variation is due to the contact among the assembly components. The variability of the coupled features, by which the link among the parts is made, gives a deviation in the location of the coupled parts. The resulting model (that is called *"global model"*) has to be able to schematize the joints with contact and the joints with clearance between the coupled features.

Once modeled the stack-up functions, they may be solved by means of a worst case or a statistical approach (Creveling, 1997). To carry out a worst case approach, it is needed to define the worst configurations of the assembly (i.e. those configurations due to the cumulative effect of the smallest and the highest values of the tolerances assigned to the assembly components) that satisfy its assigned tolerances. This means to solve a problem of optimization (maximization and/or minimization) under constraints due to the assigned tolerances. Many are the methods developed by the literature to carry out a worst case approach (see Luenberger, 2003). To carry out a statistical approach, it is needed to translate each tolerance assigned to an assembly component into one or more parameters of the stack-up function. Therefore, a Probability Density Function (PDF), that is function of both the manufacturing and the assembly processes, is assigned to each parameter. Being the definition of the relationship among the production and assembly processes and the probability density function of the tolerance of the component strongly hard to estimate, the commonly used assumption is to adopt a Gaussian probability density function. Moreover, a further assumption is to consider independent the parameters used to represent the variability of the features delimiting each dimensional tolerance. The variation of the FR is obtained by means of a Monte Carlo simulation technique (Nigam & Turner, 1995; Nassef & ElMaraghy, 1996) it is usually calculated as ± three times the estimated standard deviation (three sigma paradigm of (Creveling, 1997)).

3. Literature review

The foremost works on the tolerance modeling problem are found in (Requicha, 1983; Requicha, 1993) that introduced the mathematical definition of the tolerance's semantic. He focused on constructing semantically correct tolerance zones and he proposed a solid offset approach for this purpose. Since then, a lot of models are proposed by the literature to perform the tolerance analysis of an assembly whose components may be considered as rigid parts (Hong & Chang, 2002).

The *vector loop* model uses vectors to represent relevant dimensions in an assembly (Chase et al., 1995; Chase et al., 1996; Chase et al., 1997a). Each vector represents either a component dimension or an assembly dimension. Vectors are arranged in chains or loops to reproduce the effects of those dimensions that stack together to determine the resultant assembly dimensions. Three types of variations are modelled in the vector loop model: *dimensional variations*, *kinematic variations* and *geometric variations*. Dimensional variations defined by dimensional tolerances are incorporated as +/- variations in the length of the vector. Kinematic variations describe the relative motions among mating parts, i.e. small adjustments that occur at assembly time in response to the dimensional and geometric variations of the components. Geometric variations capture those variations that are imputable to geometric tolerances.

The *variational* solid modelling approach involves applying variations to a computer model of a part or an assembly of parts (Martino & Gabriele, 1989; Boyer & Stewart, 1991; Gupta & Turner, 1993). To create an assembly, the designer identifies the relevant features of each component and assigns dimensional and geometrical tolerances to them. In real conditions (i.e. manufactured part), the feature has been characterized by a roto-translational displacement with respect to its nominal position. This displacement is modelled to summarize the complete effects of the dimensional and geometric variations affecting the part by means of a differential homogeneous transformation matrix. Once the variabilities of the parts are modelled, they must be assembled together. Another set of differential homogeneous transformation matrices is introduced to handle the roto-translational deviations introduced by each assembly mating relation.

The *matrix* model aims at deriving an explicit mathematical representation of the boundary of the entire spatial region that encloses all possible displacements due to one or more variability sources. In order to do that, homogenous transformation matrices are considered as the foundation of the mathematical representation. A displacement matrix is used to describe any roto-translational variation a feature may be subjected to. The *matrix* model is based on the positional tolerancing and the Technologically and Topologically Related Surfaces (TTRS) criteria (Clément et al., 1998); by classifying the surfaces into several classes, each characterized by some kind of invariance with respect to specific displacement kinds (e.g. a cylinder is invariant to any rotation about its axis) the resulting displacement matrix can be simplified (Clément et al., 1994).

In the terminology adopted by the *jacobian model* approach, any relevant surface involved in the tolerance stack-up is referred to as *functional element* (FE). In the tolerance chain, FEs are considered in pairs: the two paired surfaces may belong to the same part (internal pair), or to two different parts, and paired since they interact as mating elements (kinematic pair, also referred to as external pair). The parts should be in contact to be modelled by this model. Transformation matrices may be used to locate a FE of a pair with respect to the other: these matrices can be used to model the nominal displacement between the two FEs, but also additional small displacements due to the variabilities modelled by the tolerances. The main peculiar aspect of the *jacobian* approach is how such matrices are formulated, i.e. by means of an approach derived from the description of kinematic chains in robotics (Laperrière & Lafond, 1999; Laperrière & Kabore, 2001).

The *torsor* model uses screw parameters to model three dimensional tolerance zones (Chase et al., 1996). Screw parameters are a common approach adopted in kinematics to describe motion; they are used to describe a tolerance zone, since a tolerance zone is the region where a surface is allowed to move. The screw parameters are arranged in a particular mathematical operator called *torsor*, hence the name of the approach. To model the interactions between the parts of an assembly, three types of torsors (or Small Displacement Torsor SDT) are defined (Ballot & Bourdet, 1997): a *part SDT* for each part of the assembly to model the displacement of the part; a *deviation SDT* for each surface of each part to model the geometrical deviations from nominal; a *gap SDT* between two surfaces linking two parts to model the mating relation.

The *Tolerance Maps* model is being developed at the Arizona State University (Davidson et al., 2002; Mujezinovic et al., 2004; Ameta et al., 2007). It is based on a two-levels model: the *local model*, that models part variations in order to consider the interactions of the geometric controls applied to a feature of interest and the *global model* that interrelates all control frames on a part or assembly. A Tolerance-Map (T-Map) is a hypothetical solid of points in n-dimensions which represent all possible variations of a feature or an assembly. Overlaying the coordinates of the T-Map the stack-up equations to perform the tolerance analysis are obtained.

However, these methods are not easy to apply, especially for complex aerospace assemblies, since they were born to deal with elementary features, such as plane, hole, pin and so on. So the aid of computer is called for. In the recent years, the development of efficient and robust design tools has allowed to foresee manufacturing or assembly problems during the first steps of product modeling by adopting a concurrent engineering approach.

Efforts to deal with the tolerance analysis in aeronautic field were carried out. Sellakh proposed an assisted method for tolerance analysis of aircraft structures through assembly graphs and TTRS theory (Sellakh et al., 2003). Marguet presented a methodology to analyse and optimise the assembly sequence of simple shape assemblies (Marguet & Mathieu, 1998; Marguet et Mathieu, 1999). Ody showed a comparison among Error Budgeting techniques and 3D Tolerance Software Packages (Ody et al., 2001). Those papers present solutions of typical mechanical assemblies that involve the alignment of plane, holes and pins, but the aeronautic surfaces have a free form generally.

Some are the works found in the literature to deal with the tolerance analysis of compliant assemblies, i.e. assemblies that contain deformable parts, such as sheet-metal, plastics, composites and glass. Some of the first research related to this area was done by (Gordis & Flannely, 1994). They used frequency domain analysis to predict in-plane loads and displacements from misalignment of fastener holes in flexible components. Liu and Hu have used simple finite element models to predict assembly variation of flexible sheet metal assemblies (Liu & Hu, 1995; Liu & Hu, 1997). Their work was focused on the effect of part fixturing and order of assembly.

Merkley uses the assumptions of Francavilla and Zienkiewicz (Francavilla & Zienkiewicz, 1975) to linearize the elastic contact problem between mating flexible parts (Merkley, 1998). Merkley derived a method for predicting the mean and the variance of assembly forces and deformations due to assembling two flexible parts having surface variations. He describes the need for a covariance matrix representing the interrelation of variations at neighboring nodes in the finite element model. The interrelation is due to both surface continuity, which Merkley calls geometric covariance, and elastic coupling, which he calls material covariance. Merkley used random Bezièr curves to describe surface variations and to calculate geometric covariance. Bihlmaier presents a new method for deriving the covariance matrix using spectral analysis techniques (Bihlmaier, 1999). The new method, called the Flexible Assembly Spectral Tolerance Analysis method, or FASTA, also includes the effect of surface variation wavelength on assemblies.

4. Computer Aided Tolerancing software

Today Computer Aided Tolerance (CAT) software is readily available for tolerance analysis of rigid part assemblies, but even if these tools provide good results they have not been widely used (Turner & Gangoiti, 1991; Chase et al., 1997b; Salomons et al., 1998; Prisco & Giorleo, 2002). The CAT system known as CE/TOL® is based on the *vector loop* model. Many commercial CAT software packages are based on the *variational* model, such as eM-TolMate of UGS®, 3-DCS of Dimensional Control Systems®, VisVSA of UGS®. Commercial CATs are not completely true to the GD&T standards and need improvement after a better mathematical understanding of the geometric variations. The user needs expertise and great experience combined with a through understanding of the packages' theoretical base plus modeling principles to build a valid model and obtain relatively accurate results. Computer Aided Tolerance software efficiently deals with mechanical assemblies where the feature to align are planes, hole-pin, but it hardly treats of free-form surfaces to connect.

A CAT software shares the same user interface and the same database of a CAD package; the CAT information is stored within the CAD model with no need of translation.

The CAT software used in the following of this work is eM-TolMate of UGS® and, therefore, further information about this package has been discussed in the following. EM-TolMate involves the building of the model through the feature definition, the tolerance specification, the assembly and the measurement definitions.

The feature definition process must be performed for each component of the final assembly (there must be a separate CAD model for each component); it uses the existing CAD geometry of the model to create its own features. Therefore, full associativity with CAD entities is assured. The basic features supported in *eM-TolMate* are plane, pin (cylindrical, tapered, threaded), hole, point, tab, slot, constant profile surface, constant cross section, sphere, surface of revolution, general 3D surface. Edge features (for thin-walled parts) are also available. In many cases, the user needs to mathematically derive some feature from existing features (e.g. line of intersection between a plane and a parallel cylinder, centroid of several points or best fit line between several points).

The tolerance specification allows the user to define the dimensional or geometric tolerances supported by the existing international standards of GD&T (ASME and ISO). There is no limit on the number of tolerances applied to a feature. Diameter modifier and material modifier (MMC and LMC) may be added to tolerances applied to a feature of size. Available statistical distributions are normal (default), uniform, triangular, exponential, gamma, Weibull, Laplace or Pearson distribution.

The assembly definition creates an empty CAD model which represents the assembly. The sequence of assembly, represented by a tree structure, has to be defined by using the source components into the correct position of the assembly tree. Afterwards the assembly operations have to be specified by the selection of the mating features that are involved in each operation. Except for the first component of the assembly, the assembly operation has to be defined for each component which will be constrained to the target features of the components inserted before it in the tree. The constraint scheme can be isostatic, but also under or overconstrained.

The measurement definition specifies the geometric relationship that is to be put under control in the assembly or in the single part. EM-TolMate supports various types of measurements: linear distance, angle, clearance, virtual size.

The built model is simulated according to Monte Carlo technique in a statistical approach. EM-TolMate determines simulation order for each feature on each component, the tolerance priority order and the tolerance degree of control (i.e. the degrees of freedom constrained by the tolerance on a feature). The system warns the user about any lack of completeness or ambiguity in tolerancing scheme (e.g. loops, unreferenced datums). Based on this information, eM-TolMate uses random numbers to create several sets of feature variations from nominal geometry, according to the specified tolerances. The simulations (one for each set of variations) proceed based on the mating features specified by the user and components are assembled into position relative to each other. Finally, measurements of the actual assembly are performed and the results are stored in memory. Worst-case limits can be estimated when the sample size of simulations are big enough and setting the estimated limits to "Actual". To assure a high level of accuracy, a large number of simulations has to be carried out.

The user performs the results analysis phase interactively after the simulations are finished. There are two kinds of results: the variation analysis, which computes statistical parameters and reports the overall variation range for each measurement, and the contributors analysis, which determines the sources of variation and presents this information in a sorted list. The user may view the results analysis on the screen or output the information to a file in various file formats.

5. The proposed method

Concurrent Engineering imposes that the design of the manufacturing and the assembly jigs, of the manufacturing and assembly cycles, be in the same time of the product design. In fact, only the concurrent design of the final product, of the single components, of the manufacturing processes and jigs, of the assembly cycle and jigs and of the inspection procedures assures to achieve a high quality with the smallest costs.

The proposed method adapts the steps of the tolerance analysis process suggested by the eM-TolMate® package and described in the previous paragraph to the case of an assembly involving non-rigid freeform surfaces in composite material. It is constituted by the following 7 steps:

a. *Verify the stiffness of the assembly components.* Non-rigid parts may be subjected to significant distortion after the removal of manufacturing forces. This condition, known as free state variation, is principally due to weight and flexibility of the parts and the release of internal stresses resulting from fabrication. This distortion is acceptable if the part surfaces remain within the indicated tolerances, once the part is coupled to another part of the final product by applying reasonable forces. Therefore, the first step of the proposed method is to analyse the internal stress and the relative strain of each assembly components due to the assembly sequence, since those components have freeform surfaces of a small thickness in composite material. This means to simulate the assembly configuration of each assembly components through a fixturing equipment on a coordinate measuring machine and, then to measure the displacements from the nominal configuration represented by the CAD model through an inspection process. This is true if the effect of the inspection process on the measured deviation is indeed non significant and this hypothesis needs to be verified. The displacements between the real part and the nominal one, obtained in the inspection process, may be used as input values for a structural numerical analysis with the objective to obtain the stress and strain fields induced in the material when the part is assembled in the final product. The numerical analysis may be carried out with a Finite Element Method®, but before implementing the real case simulation, an assessment of the software capability on composite material is required through a simple 2D case study. If the results of the numerical analysis demonstrate that the deformations induced by the assembly sequence are at least of an order of magnitude smaller than the applied tolerances, the part may be considered rigid and the models or software of the traditional tolerance analysis literature may be applied. In this way the complexity to consider the compliant of the assembly components may be avoided.
b. *Design the measurements of the assembly.* A first analysis of the assembly working allows to identify all the functional requirements that the product should perform. This means translating the functionality of the design assembly in all the couplings or kinematic constraints among assembly components that are needed for the assembly to work correctly. Then, the pairs of assembly components involved in each functional requirements are defined. The features of each pair of components, that are involved in each functional requirement, are identified and are generally assigned to them a set of tolerances. Therefore, the functional requirements and the assembly constraints are translated into measurement of the assembly, while the coupling parts and their feature are passed to the following step of the method. A widespread measurement in aeronautic field is the distribution of the minimum distances along two faced freeform surfaces, e.g. when the adhesive thickness between the faced parts to connect should be as constant as possible in order to obtain an efficient structural connection and to avoid local compressions at the interface of faced parts due to strong reduction of thickness. In this case it has been demonstrated that the minimum distance among the two patterns of points, belonging to the two faced surfaces, calculated by projecting the point to point distance along the direction perpendicular to the nominal surfaces is very

near to the minimum distance and it is easy to calculate by eM-TolMate® (Polini et al., 2007).

c. *Define the features of the assembly components, specify the tolerances and the related statistical distribution.* The features of the coupling parts of the assembly may be free-form surfaces or basic surfaces. The free-form surfaces of each assembly components have to be schematically represented by a set of control points. The number and distribution of the control points depends by the slope of the free form surface. If two freeform surfaces are faced, the pattern of control points of the second surface may be obtained by intersecting the second surface with the pattern of vectors, that are normal to the first surface in each control point. The tolerances applied to the freeform surface should be applied to each control point defined on the surface. The basic surfaces are plane, pin, hole, point, tab, slot, constant profile surface, constant cross section, sphere, surface of revolution, general 3D surface, edge features and derived features. Dimensional tolerancing is supported as +/- tolerances (symmetric or not, sizes or distance from an origin feature) and for size tolerances of pins, holes, slots and tabs. Tolerance attributes can be freely modified to reflect effectively the parameters of actual features. The most common attribute is the tolerance probability density function for each tolerance applied to each feature of the assembly component. The probability density function depends by the manufacturing process of the feature and it is generally hardly to estimate. Therefore, the default distribution is assumed to be the Gaussian, since it is symmetric, it is simply to use and it is well known by the literature. However, it is always to change it. Finally, it is needed to verify the datum reference frames (DRFs), the assembly should have its DRF, that is generally considered "global"; while each component should have its DRF, that is considered "local". Moreover, the elements (planes, axes and so on) that constitute a local DRF should be easily identified, manufactured with the design specifications and accessible for verification. The elements of the local DRFs should be qualified by tolerances. The elements of the local DRFs should be used to manufacture, to assembly and to control the assembly components. These simple rules on the definition of the DRFs allow to obtain whose components may be manufactured, assembled and controlled. Obviously, they have to be applied by means of a concurrent approach that is implemented by a designer of product, by a planner of manufacturing processes, by a designer of assembly cycles and by a designer of inspection processes.

d. *Design the assembly cycle and jigs.* If required, the assembly jigs should be designed at first. The assembly jigs should have a plane and two pins in order to locate and fix the coupling component. The tooling hole (TH) system is the assembly system that is widespread in aeronautic field and consists in jigs mated with the assembly components by means of pins that are inserted into tooling holes (TH) located on the assembly components. The tolerances of the TH are very critical, since they influence the location and orientation among the different components of the assembly and, therefore, the functional requirements of the assembly. Some suggestions to choice the tolerances of a TH may be found in (Marziale & Polini, 2007).The positioning of a component on its jig is made by the contact of two planar surfaces (one on the part and one on the assembly jig) and the clearance fit between two Tooling Holes (positioned on the part) with the corresponding pins (positioned on the assembly jig), as shown in Fig. 1. The position of the THs has an important role in order to limit the flag waving effect of the component as regards its assembly jig, since a small change in THs position cause a great variation in the location and the orientation of the part on the corresponding jig due to the large dimensions of the parts. Therefore, the distance among the tooling holes must be as larger as the dimensions of the part increases, but this condition is difficult to achieve in some practical cases. Another important consideration is that it is preferred to position the THs out of the part shape, on appendixes that will be trimmed once the part is assembled, in order to have lower production costs. Through these considerations an assembly jig has to be designed for each part. Each assembly jig has a set of tolerances, since it is realised by manufacturing processes.

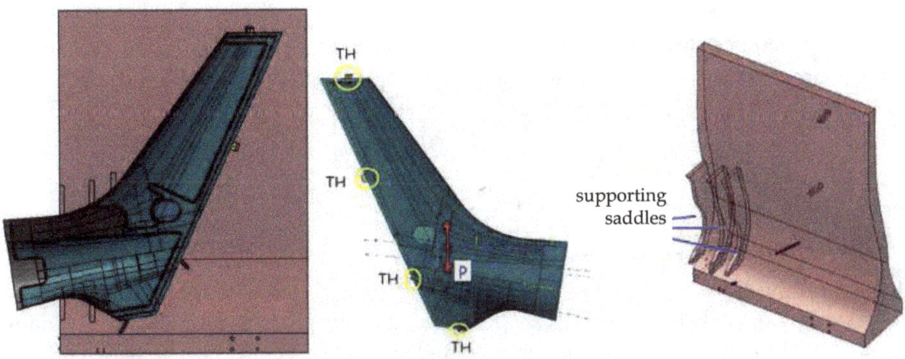

Fig. 1. Aircraft skin's jig with tooling hole system

The design of the assembly cycle foresees as first step the location of each component on its designed jig. Each component has micro-movements respect to its jig due to the tolerances of the jig and of the component applied to the coupling system, i.e. to the TH system. These micro-movements influence the positioning of the part and, therefore, they have to be characterized by means of probability density function, such as done for tolerances of the parts. The probability density function depends by the manufacturing process of the jig and the component and by the assembly constraints. It is generally hardly to estimate, since it should involve many experimental tests. The default probability density function is the Gaussian. Therefore, each couple of components is assembled by moving near the related jigs and, then, by putting the constrain elements, such as adhesive, bolts or screws. In this way a sub-assembly more and more complex is generated up to obtain the final product. The positioning of the components and, therefore, the accuracy of the final product, depends by the stack-up function of dimensions and tolerances belonging to the components and to the jigs.

e. *Choose the simulation runs and carry out simulations.* Monte Carlo simulation has been used to reproduce the combined effect of all the assigned tolerances and the component degrees of freedom respect the assembly jigs on the assembly functional requirements. The assigned tolerances and the component degrees of freedom have been simulated by means of the probability density functions assigned in the previous steps. The number of runs of simulation has been fixed at the value for which the estimated mean and standard deviation of the obtained probability density function of each functional requirement achieve an asymptotic trend. A step by step procedure has been adopted to evaluate the simulation runs, the starting value has been 10000 runs.

f. *Analyse the results.* During the assembly simulation, the results of the measurement calculations are stored in memory. When the assembly simulation is finished, a table containing all the measurement results remains in memory. These results may be exported to make statistical analysis. It is possible to evaluate the distribution of the results of each measurement and, therefore, to calculate the percentage of parts outside the tolerance ranges. Another possibility is to evaluate the reasonable tolerance range to assign to each measuring by starting from its estimated distribution in order to satisfy the tolerance analysis. Finally, a sensitivity analysis may be carried out in order to identify the tolerances mainly affecting each gap trend. The percentage weight of each tolerance has been calculated by evaluating the variance of a measurement σ_{TOT}, when all the tolerances are applied to assembly's components, and the variance of the same measurement σ_{TOT-Ti}, once i-th tolerance is not applied to the assembly's components:

$$C_i = \frac{\sigma_{TOT}^2 - \sigma_{TOT-Ti}^2}{\sigma_{TOT}^2} \, [\%] \tag{2}$$

6. Application example

6.1 Case study

The proposed approach has been carried out considering as case study an aeronautical component. This is a winch arm mounted on an helicopter (see Fig. 2). The winch arm is the external unit that allows the helicopter to rescue people or things where it is impossible to land through a rope lowered by an electrical motor.

The winch arm have a main arm and a head covered by three thin fairings.

On the main arm (the part with number 1 in Fig. 3a) are assembled two frames (the yellow parts 5 and 6 in Fig. 3a) that are two cambered elements connecting and supporting all the structure. The parts with number 5 and 6 in Fig. 3a are the forward and the after internal frames that allow to mount the fairing covers of the head (parts 3 and 4 in Fig. 3a) of the winch arm directly on its structural beam. Finally, the Mid Cowling (part 2 in Fig. 3a) is the central exterior fairing that covers the head of the winch arm The part 7 in Fig. 3a is a cover for the overhaul of the winch arm. All the components of the winch arm assembly are made in carbon fiber composite material. Both of the frames are made with five layers of carbon fibers imbued in a matrix of epoxy resin for a total stack thickness of 1,65 mm. The carbon fibers are oriented in a ± 45° alternated arrangement within a steel cast which reproduces the part. After the stratification process, a curing process is carried out by autoclave with a control of temperature and pressure. The obtained part has size of 300 mm x 400 mm x 500 mm.

The weight of each part is about 350g. Assembly operations among the parts are done using of special grub screws that are inserted in the holes of each component. No adhesives or resins are used to secure the surfaces, since a perfect adhesion is not required.

Even if the form of the frame may appear simple, the shapes of its surfaces aren't regular. The After Frame is coupled to a metallic part, called Box Support, that allows to fix the engine of the winch arm (see Fig. 3b).

Each part has three datum planes as component references. Those planes are external to the part geometry, as shown in Fig. 4 for the Mid Cowling (dimensions and tolerances are not reported in this chapter, since they are reserved). The nominal location and tolerances of coupling holes are defined as regards to this external reference frame. Profile tolerances are defined as general requirements for the whole assembly by the manufacturer. The value declared is ±0.75 mm. In this case form errors may have effect on the airflow around the head of the winch arm.

Furthermore, the parts may have a significant geometric distortion after removal of the forces applied during fabrication. This is a characteristic of composite laminates and occurs as soon as the stratification cast is removed. This is due to its low thickness, weight, flexibility and to the release of internal stresses resulting from the manufacturing processes. When the manufacturing process is completed all the references to the nominal configuration are lost. In order to re-obtain the nominal shape it is necessary to apply forces able to win the elastic recovery of the part. This problem is present during the assembly of the Mid Cowling with its mating components.

Further applications of the proposed method may be found in (Polini et al., 2007; Marziale & Polini, 2008).

a b

Fig. 2. a) winch arm mounted on the helicopter, b) 3D model of winch arm

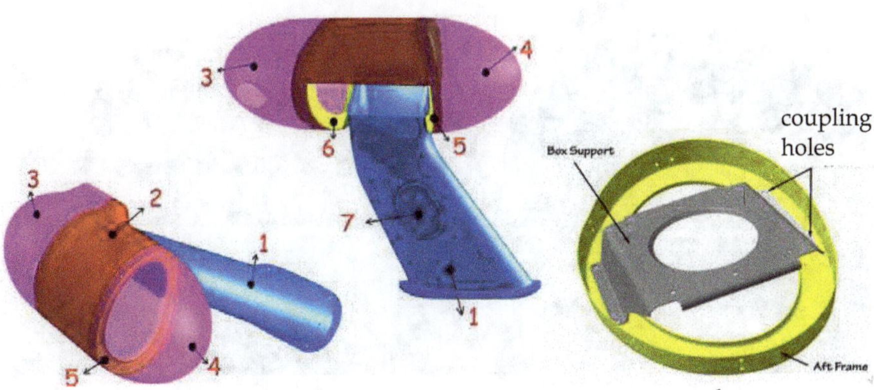

a b

Fig. 3. a) Components of winch arm: 1-Hoist Mount, 2-Mid Cowling, 3-After Cowling, 4-Forward Cowling, 5-Forward Frame, 6-After Frame, 7-Cover; b) Coupling of the After Frame with the Box Support

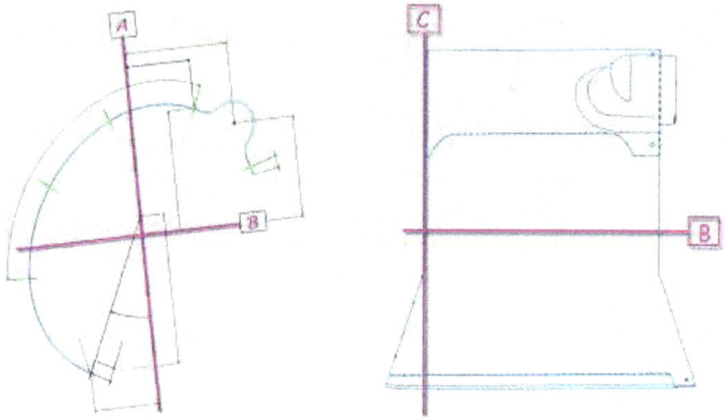

Fig. 4. Datum planes of the Mid Cowling

6.2 Verify the stiffness of the assembly components

For example the procedure has been applied to the Mid Cowling, but the same should be applied to all the components of the winch arm. To be inspected, the Mid Cowling has to be mounted on a fixturing equipment that reproduces its assembly positions, see Fig. 5. Because it is joined to after and forward frames by means of special grub screws that are inserted in each component holes, the fixturing equipment has to reproduce the position of these coupling holes. Therefore, the fixturing equipment has to reproduce the three reference planes A, B and C in Fig. 4. The part may be connected to the equipment through pins with cone-shape heads that couple with the part holes and through blocking elements such as c-shaped clamps and springs. The pins allow to accurately locate the part holes inside the CMM working volume. Finally, to spatially orientate the elements of the equipment, ball-and-socket joints may be used. All the details on the design of the fixturing equipment may be found in (Ascione & Polini, 2010).

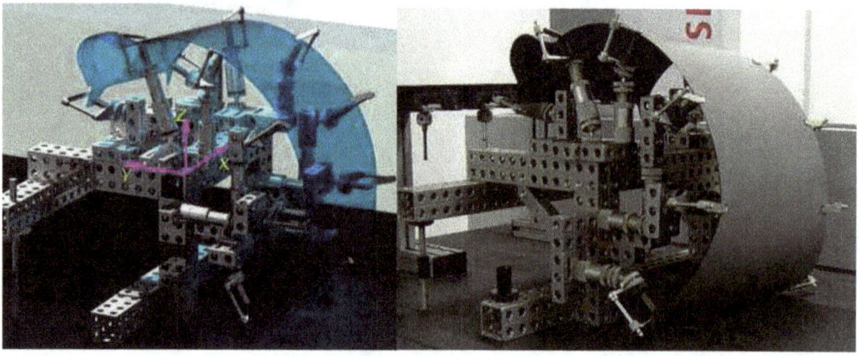

Fig. 5. Fixturing equipment of the Mid Cowling

Five different patches have been defined on the Mid Cowling (see Fig. 6). Each has been measured by continuous scanning along path $\pi/4$ rad tilted. The distance between two following scanning lines was set to 2 mm; thus, involving 10000 points acquired on each patch.

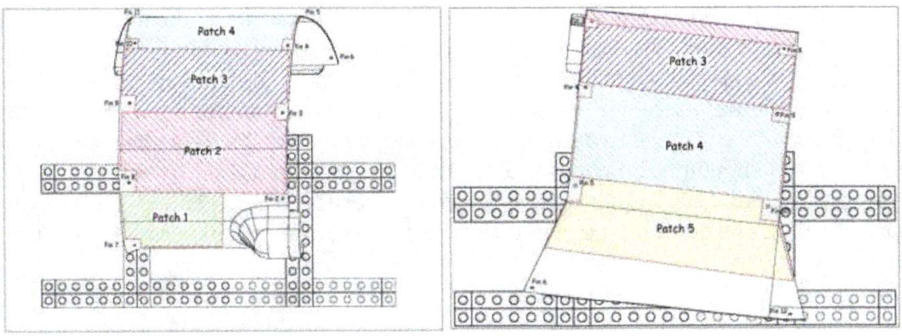

a) b)

Fig. 6. Detected scan patches; a)top view; b) front view

The force of 0.1N applied by CMM probe to the non-rigid freeform surface of the Mid Cowling does not produce a significant deformation. In fact, the maximum deviation is equal to 0.35 μm as shown in (Moroni et al., 2007), while typical value of profile tolerance in this kind of application is around 50 μm.

The deviation along the normal curvature of the surface of the part was calculated for each acquired point. The distribution of the deviations on the surface represents the overall error of shape. Such a value may be seen as the sum of three different contributions; the first due to component production process; the second due to the assemblage of the part and the third due to the measuring procedure. The deviation was calculated for each acquired point with Holos® measuring software.

The data analysis has been performed on each patch extracted from the Mid Cowling. From the analysis of distance colour maps it is possible to observe that no systematic behaviour of the deviations is present (see Fig. 7).

To estimate the repeatability of the measuring procedure, the scanning on the 5 patches was repeated 5 times, maintaining the same Mid Cowling coupling position on the equipment. Therefore, the inspecting procedure may be considered the only responsible of variability of the measuring results. Then, the average deviation was calculated for each of the 50000 measuring points. Finally, the difference between each measurement point and the average deviation was calculated. The values thus calculated were distributed according to a Gaussian probability density function, as verified with the Anderson-Darling statistic test (p-value > 0.1).

The value of the overall repeatability of the measuring process, that has been calculated as the standard deviation of the sample of 250000 values, is equal to $8.5\mu m$. This value is very small; this means to say that the effect of the inspection process on the measured deviation is indeed non significant.

The idea introduced in this work is that the displacement between the real part and the nominal one, obtained in the inspection process, may be used as input values for a structural numerical analysis with the objective to obtain the stress and strain fields induced in the material when the part is assembled in the final product. That is possible because the measured field of displacement represent the real configuration of the component in its assembly state. The inspection process simulate the assembly configuration of the real part; therefore, the measured displacements are the differences from the real part, assembled on the winch arm, and its nominal configuration represented by the CAD model. The measured displacements have been used as loads applied on the components in its nominal configuration for the numerical simulation. In this way the results of the structural analysis are characteristic values of the assembled parts, since the assembly process locks the components of the assembly in the same way the assembly is bounded in the inspection process and in the numerical simulation. The numerical analysis has been carried out with the MSC_Patran/Nastran® software but, before implementing the real case simulation, an assessment of its capability on composite material has been required to obtain a validation of the results for our case. In this sense a matching procedure between experimental results and numerical results on the same material in the same configuration has been necessary.

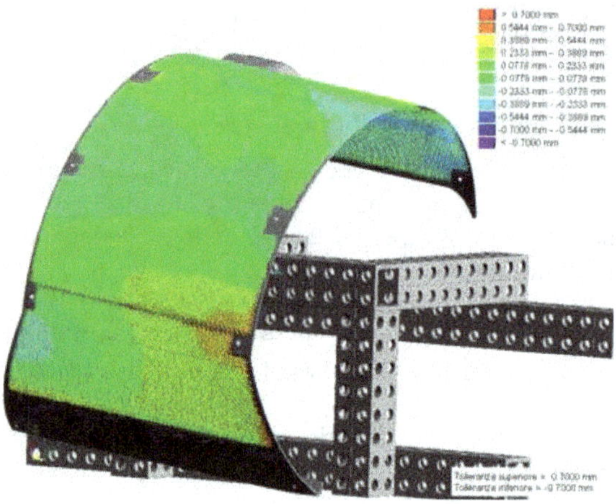

Fig. 7. Chromatic mapping of the deviations between measured points and nominal geometry

Having available experimental results for a tensile test on the composite material constituting the Mid Cowling, the matching procedure has been carried out by simulating the same test mean the numerical software and then comparing the obtained values with the correspondents of experimental test. All the details of the tensile test are reported in (Ascione & Polini, 2010). Fig. 8 shows in blue the stress-strain curve obtained from the analysis for the first node of the mesh (the node at the intersection of the two plane of symmetry) and in green the stress-strain curve for the experimental test. It is possible to see that the results obtainable with the used numerical software on the considered material are reliable.

The same procedure used to simulate the tensile test has been performed to model the Mid Cowling. In Fig. 9 it is possible to see the finite element representation of the Mid Cowling, where triangular shell elements have been used to replace the continuous geometry. In this case triangular elements have been used, since this kind of elements is the best approximation of the distribution of the displacements measured through the inspection process. In Fig. 9 there are green and light blue elements, since the green elements represent the part's geometry, while the light blue elements represent the points where the displacement has been applied. Further details on the setting of the numerical analysis are deeply described in (Ascione & Polini, 2010).

The results of this simulation have been the stress and strain fields induced in the Mid Cowling subjected to the loads explained above. The strain field is showed in Fig. 10. The values of the strain showed in these results are congruent as magnitude with those ones of the tensile test and with the values of the displacements imposed as loads. Therefore, it is possible to consider the obtained values representative of the real state of the component in its real configuration of assembly. The maximum value of the strain field is about 7.28 μm that is very smaller than the applied tolerances. Therefore, the Mid Cowling may be considered as a rigid part for the tolerance analysis.

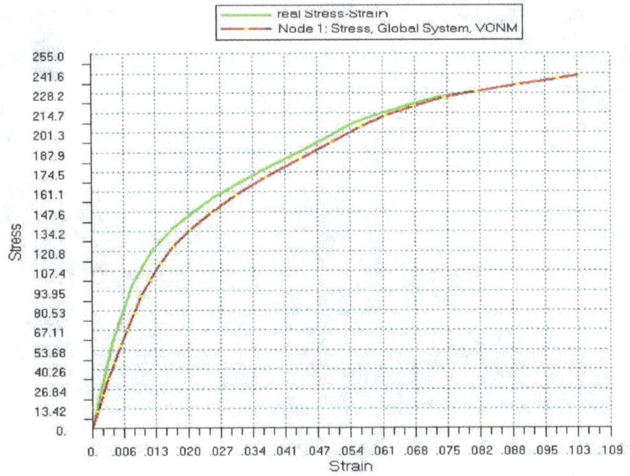

Fig. 8. Stress-strain diagram with green curve for the experimental test and with red curve for the numerical results

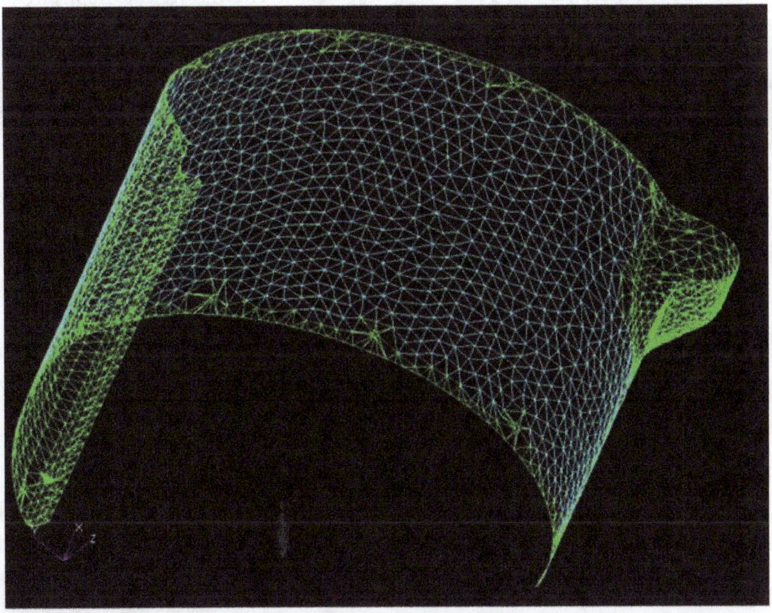

Fig. 9. FEM model used to simulate the Mid Cowling

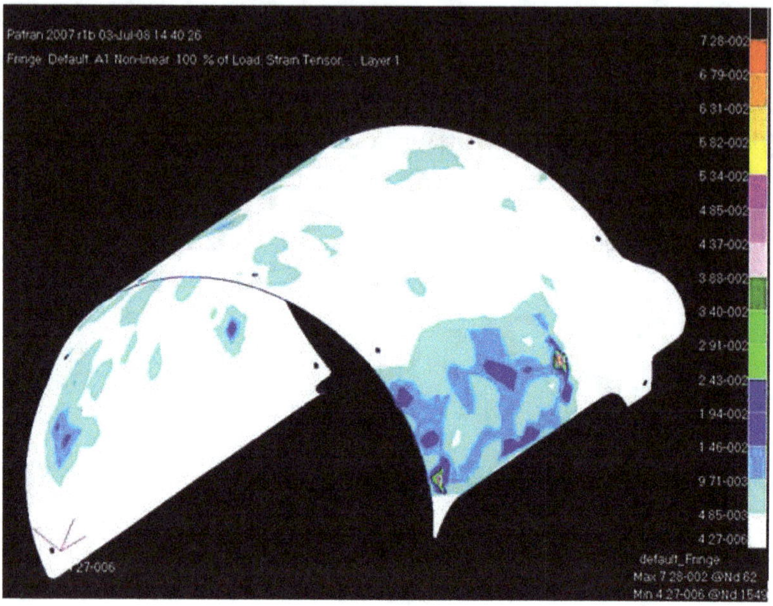

Fig. 10. Strain field resulting from the FEM Analysis of the Mid-Cowling.

6.3 Design the measurements of the assembly

The effect of the deviations when the After Frame is positioned on the Hoist Mount and the Box Support is positioned on the After Frame cause a change in the position of the Box Support as regards the nominal one. Therefore, the axis of the engine of the winch arm, that is positioned at the centre of the hole of the Box Support, may deviate from the nominal position and this deviation may invalidate the right working of the engine. The evaluation of the deviation of the axis of the engine from the nominal position is the first measurement of the assembly. It has been evaluated on the plane of the coupling of the engine with the Box Support along the two axes of the Datum Reference Frame of the Box Support (see Fig. 11).

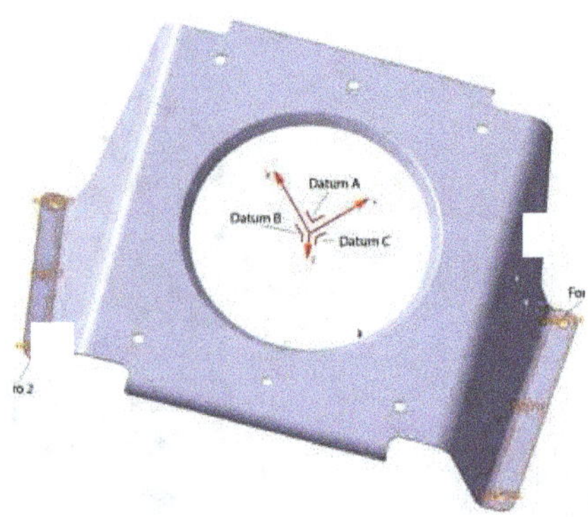

Fig. 11. Datum Reference Frame of the Box Support

The effect of the deviations, when the After Frame is positioned on the Hoist Mount and the After Cowling is positioned on the After Frame, cause a change in the orientation of the After Cowling as regards the nominal one. The aerodynamic drag of the After Cowling depends by the angle of incidence the air rush on the Cowling itself. Therefore, the deviation in the orientation of the After Cowling may invalidate the right working of the winch arm. The evaluation of the deviation of the orientation of the After Cowling from the nominal is the second measurement of the assembly. It has been evaluated as the two translational motions of the centre of the ellipse along the two axis of the Datum Reference Frame of the After Cowling and the three rotations of the axis of the Cowling as regards the nominal position measured in the Datum Reference Frame of the After Cowling.
The same reasoning may be repeated for the Forward Cowling.

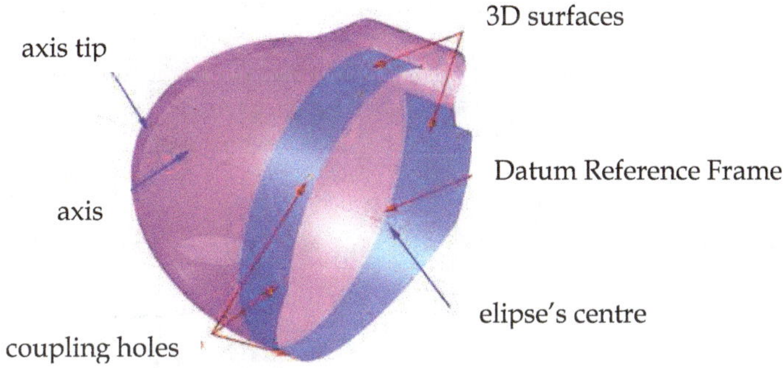

Fig. 12. Features and Datum Reference Frame of the After Cowling

6.4 Define the features of the assembly components, specify the tolerances and the related statistical distribution

The features of the Hoist Mount are the two planes and the two patterns of six holes that allow to couple the part with the After and Forward Frame. The holes have a dimensional tolerance of ±0.25 mm. The Y-axis of the Datum Reference Frame passes from the centre of the holes 1 and 2 of the pattern that allows to couple the After Frame, as shown in Fig. 13 on the left. The Z-axis is perpendicular to the coupling plane with the After Frame and the X-axis is perpendicular to ZY plane. The origin is in the centre of the hole 2 of the pattern projected on the coupling plane.

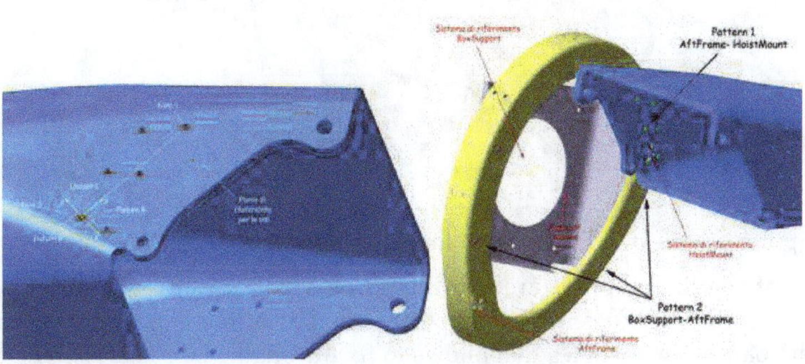

Fig. 13. Features and Datum Reference Frame of the Hoist Mount

The features of the After Frame are the pattern 1 of six holes that couples the part with the Hoist Mount and the pattern 2 of four holes that couples the part with the Box Support, as shown in Fig. 14. The X-axis of the Datum Reference Frames is due to the intersection of the two planes containing the two patterns. The Z-axis is perpendicular to the plane that contains the pattern 1 and that mates the Hoist Mount. The Y-axis is perpendicular to ZY plane. The origin is due to the intersection of the X-axis with the cylindrical surface on the side of the After Frame. Moreover, there are two further patterns of six and eight holes on the lateral cylindrical surface for the coupling with the Mid Cowling and the After Cowling (see Fig. 15). All the holes have a dimensional tolerance of ±0.25 mm. To the lateral cylindrical surface a profile tolerance of 0.076 mm and a dimensional tolerance to the thickness of ±0.10 mm are applied, as shown in Fig. 16.

The same features and the same datum reference frame may be found in the Forward Frame.

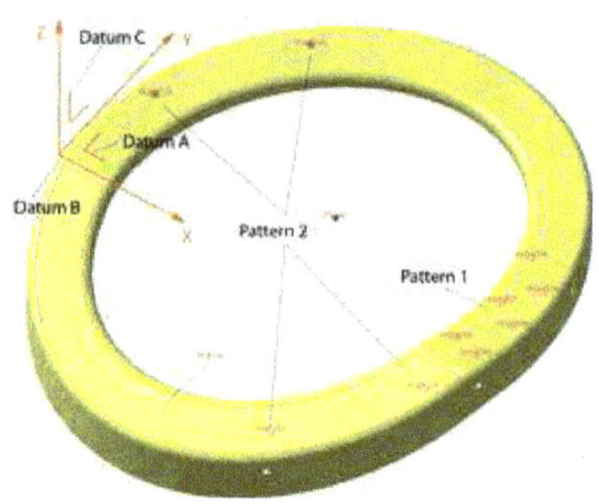

Fig. 14. Features and Datum Reference Frame of the After Frame

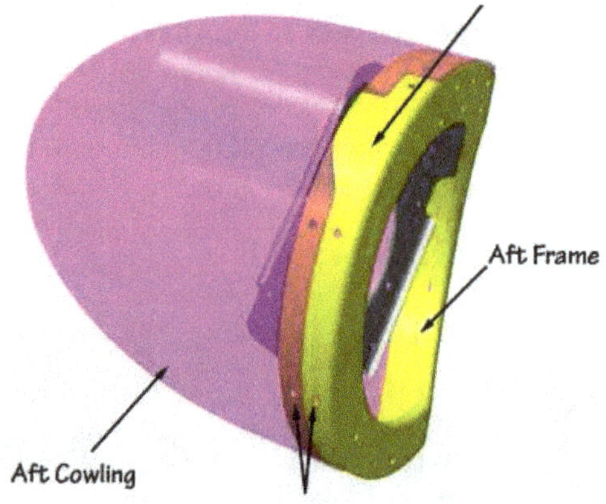

Fig. 15. Coupling of the After Frame with the After Cowling

The features of the Box Support are constituted by a pattern of four holes that allows to couple the After Frame, the central hole where the axis of the engine of the winch arm is fitted and the pattern of six holes that allows to fix the engine (see Fig. 11). The central hole and the pattern of six holes lie on the same plane. The pattern of four holes lies on a plane that mates with the plane of the After Frame. The Z-axis of the Datum Reference Frame is perpendicular to the plane where the central hole lies. The origin is the centre of the hole to fit the engine. The X-axis connects the origin with the centre of one hole of the pattern to fix the engine. The holes of the patterns have a dimensional tolerance of ±0.25 mm.

The After Cowling is constituted by a cylindrical surface, that is signed in blue in Fig. 12, where there is a pattern of eight holes for coupling it to the After Frame. To this surface a profile tolerance of 0.076 mm is applied, while its thickness has a dimensional tolerance of ±0.10 mm (see Fig. 16). Moreover, it has an elliptical surface, that is signed in pink in Fig. 12, with aerodynamic properties. The Datum Reference Frame is located in the centre of the ellipse that generates the cylindrical surface for coupling with the After Frame. The Y-axis is parallel to the largest axis and the X-axis is parallel to the smallest axis of the ellipse. The Z-axis is perpendicular to YX plane with a positive orientation outside the material. The origin is in the centre of the ellipse.

The same features and the same datum reference frame may be found in the Forward Cowling.

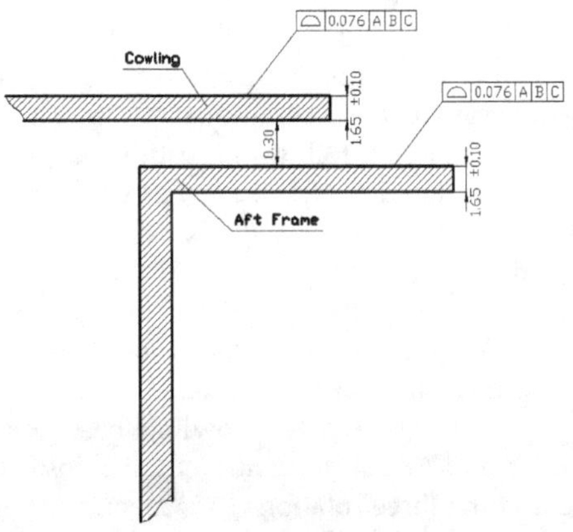

Fig. 16. Tolerances applied to the coupling surfaces of the After Frame and the After Cowling

The applied tolerances have been considered distributed as a Gaussian probability density function, whose parameters (mean and standard deviation) have been estimated as the nominal value of the surface property and one third of the applied tolerance range respectively.

6.5 Design the assembly cycle and jigs

After the curing process, the patterns of holes are manually made by means of a brace and brit with the help of a mask that is placed on the part as guide. Therefore, the After Frame is coupled with the Host Mount by means of special grub screws that are inserted in the holes of each component, as shown in Fig. 17 on the left. No adhesives or resins are used to secure the surfaces, since a perfect adhesion is not required. The Box Support is coupled to the obtained sub-assembly, as shown in Figure 17 in the middle. Finally, the Frames (Forward and Cowlings) are coupled, as shown in Fig. 17 in the middle and on the right. No jigs have been used, since the assembly is small and it is manually caught.

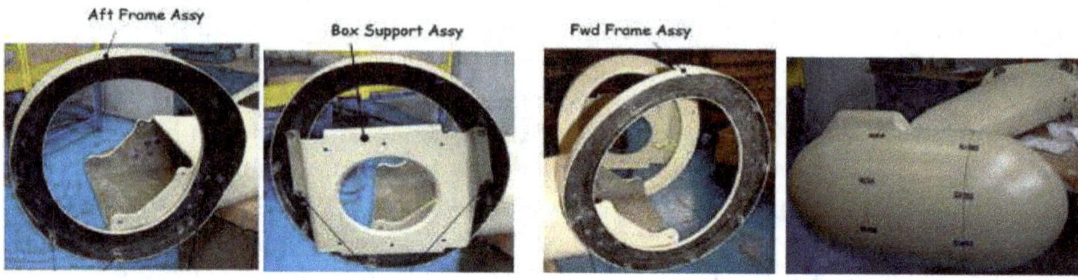

Fig. 17. Assembly cycle of the winch arm

6.6 Choose the simulation runs, carry out simulations and analyze the results

Monte Carlo simulation has allowed to perturb dimensions and geometry of the winch arm components inside the tolerance ranges in order to give the statistical distribution of the deviations of the axis of the engine and of the orientation of the After Cowling. Monte Carlo simulation has been carried out by a well-known Computer Aided Tolerance software, eM-Tolmate of UGS®. The number of runs of simulation has been fixed at 200000, once many tests have been carried out, since this value guarantees a stable estimation of the mean and the standard deviation of the probability distribution characterising the two measurements.

The deviation of the axis of the engine follows an estimated probability density function near to Gaussian along the considered X and Y-axes, as shown in Fig. 18 and Fig. 19. The estimated mean is equal to -0.0005 mm along the X-axis and 0.000 mm along the Y-axis, while the estimated standard deviation is equal to 0.251 mm along the X-axis and 0.294 mm along the Y-axis. Therefore, the axis of the engine may range between -0.88 mm and +0.88 mm (i.e. $\pm 3\sigma$) that include 99.73% of the obtained Gaussian distribution. This range does not significantly influence the right working of the engine.

The deviation of the orientation of the After Cowling has been evaluated as the three translational motions (ΔX, ΔY and ΔZ) of the centre of the ellipse along the three axis of the DRF of the After Cowling and the three rotations ($\Delta\alpha$, $\Delta\beta$ and $\Delta\gamma$) of the axis of the Cowling around the three axes of the DRF of the After Cowling. The three translations and the three rotations follow an estimated probability density function that may be considered Gaussian, as shown in Fig. 20 and in Fig. 21. Table 1 shows the mean and standard deviation that have been estimated for the obtained probability density functions.

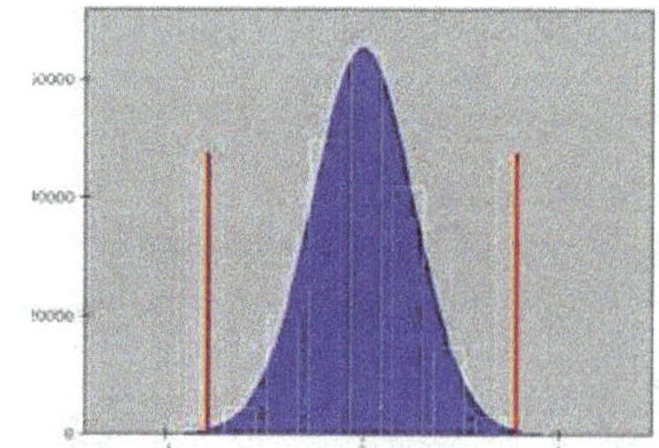

min= -0.784 mm	max= 0.783 mm	range= 1.568 mm

Fig. 18. Probability density function of the deviation of the axis of the engine along the X-axis of the DRF belonging to the Box Support

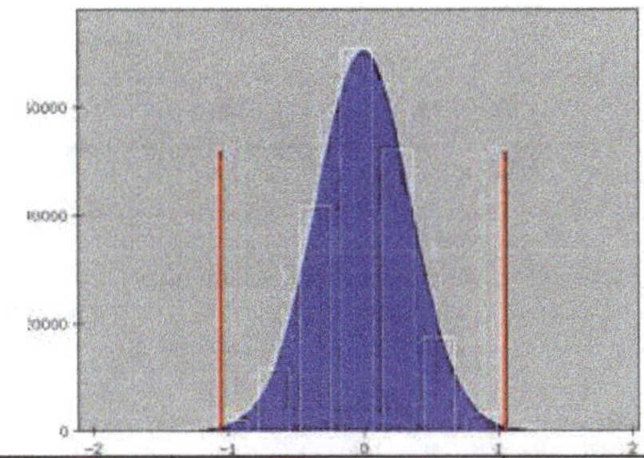

min= -1.049 mm	max= 1.050 mm	range= 2.100 mm

Fig. 19. Probability density function of the deviation of the axis of the engine along the Y-axis of the DRF belonging to the Box Support

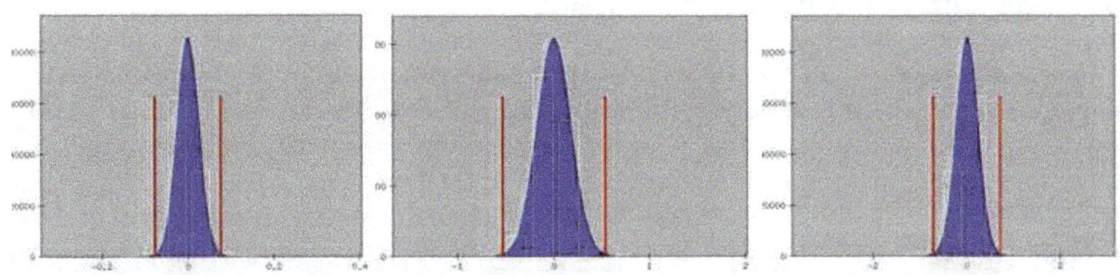

Fig. 20. Probability density function of the three translations motions of the centre of the ellipse of the After Cowling along the three axes of its DRF: X-axis on the left, Y-axis in the middle and z-axis on the right

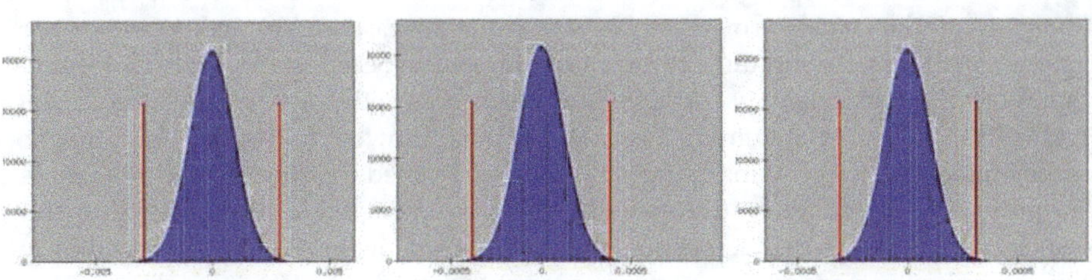

Fig. 21. Probability density function of the three rotations of the axis of the After Cowling around the three axes of its DRF: X-axis on the left, Y-axis in the middle and Z-axis on the right

The obtained results underline that the After Cowling may significantly translate along the largest axis of the ellipse (Y-axis) of about ± 0.531 mm and along the axis perpendicular to the plane of coupling of the After Cowling with the After Frame (Z-axis) of about ± 0.709 mm, while the translation along the X-axis is negligible. Moreover, the After Cowling may significantly rotate around the Z of about ± 0.724 radiants, while the rotations around X and Y axes are negligible. This is due to the coupling of the After Cowling with the After Frame by means of the screws that get into the holes of the parts with a certain clearance.

Measurements of the After Cowling	nominal value	mean	standard deviation (σ)	3σ
ΔX [mm]	0.0000	0.0001	0.0256	0.0768
ΔY [mm]	0.0000	-0.0007	0.1771	0.5313
ΔZ [mm]	0.0000	-0.0005	0.2364	0.7092
Δα [rad]	0.0000	0.0000	0.0010	0.0030
Δβ [rad]	0.0000	0.0000	0.0010	0.0030
Δγ [rad]	0.0000	-0.0005	0.2414	0.7242

Table 1. Estimated parameters of the probability density function of the After Cowling

7. Conclusion

The present work deals with the tolerance analysis of freeform surfaces belonging to parts in composite material and that may be non-rigid. This chapter presents a method to carry out the tolerance analysis of an assembly involving free-form surfaces in composite material by using a commercial CAT software. The great effort of the present work is overcome the limits of the CAT software to deal with dimensional and geometric tolerances applied to free form surfaces in composite material. The proposed method is constituted by seven steps. The first one verifies the stiffness of the assembly components by means of Finite Element Method® in order to evaluate the possibility to apply the tolerance analysis models developed for rigid assemblies. The second step designs the measurements of the assembly; the third defines the features of the assembly components; the fourth specifies the tolerances and the related statistical distribution; the fifth designs the assembly cycle and jigs; the sixth chooses the simulation runs and carry out simulations and the seventh analyses the results.
The method has been applied to the winch arm mounted on an helicopter. All the components of the winch arm assembly are made in carbon fiber composite material. Assembly operations among the parts are done through special screws that are inserted in the holes of each component. No adhesives or resins are used to secure the surfaces, since a perfect adhesion is not required. The method has allowed to verify some functional requirements of the winch arm. In particular, it has verified that the applied tolerances do not compromise significantly the assembly of the engine to rescue people or things. Moreover, it has underlined the possible changes in the Cowlings' orientation and location that may modify the aerodynamic performances of the winch arm.

8. References

Ameta, G., Davidson, J.K. & Shah, J.J. (2007). Influence of form on Frequency Distribution for 1-D clearance which is generated from Tolerance-Maps, *Proceedings of 10th CIRP Conference on Computer Aided Tolerancing*, ISBN/ISSN: 978-3-8322-6036-1, Erlangen, Germany, March 21st-23rd

Ascione, R. & Polini, W. (2010). Measurement of nonrigid freeform surfaces by coordinate measuring machine, *The International Journal of Advanced Manufacturing Technology*, Vol. 51, No. 10, 1055-1067, doi: 10.1007/s00170-010-2684-5

ASME Standard. (1994). *Dimensioning and tolerancing, ASME Y14.5M,* American Society of Mechanical Engineering, New York.

Ballot, E. & Bourdet, P. (1997). A computational method for the consequences of geometric errors in mechanisms, *Proceedings of the 5th CIRP Seminar on Computer Aided Tolerancing,* April 27-29, Toronto, Canada.

Bihlmaier, B.F. (1999). *Tolerance Analysis of Flexible Assemblies using Finite Element and Spectral Analysis,* MS Thesis, Brigham Young University, Provo, Utah

Boyer, M. & Stewart, N.F. (1991). Modeling spaces for toleranced objects, *Int. Journal of Robotics Research,* Vol.10, No.5, 470-582.

Chase, K.W., Gao, J. & Magleby, S.P. (1995). General 2-D tolerance analysis of mechanical assemblies with small kinematic adjustments, *J Des Manuf,* Vol. 5, 263-274

Chase, K.W., Gao, J., Magleby, S.P. & Sorensen, C.P. (1996). Including geometric feature variations in tolerance analysis of mechanical assemblies, *IIE Trans,* Vol. 28, 795-807

Chase, K.W., Gao, J. & Magleby, S.P. (1997a). Tolerance analysis of 2-and 3D mechanical assemblies with small kinematic adjiustments, In: *Advanced Tolerancing Techniques,* Zhang H.C., (Ed.), 103-138, Wiley, ISBN 978-0-471-14594-3, New York

Chase, K.W., Magleby, S.P. & Glancy, C.G. (1997b). A compressive system for Computer-Aided Tolerance Analysis of 2-D and 3-D Mechanical Assemblies, *Proc. of the 5th Int. CIRP Seminar on Computer-Aided Tolerancing,* April 27-29, Toronto, Canada.

Clément, A., Rivière, A. & Temmerman, M. (1994). *Cotation tridimensionelle des systèmes mécaniques, théorie & pratique,* PYS, ISBN 2-85330-132-X, Ivry-Sur-Seine, France.

Clément, A., Riviére, A., Serré, P. & Valade, C. (1998). The TTRSs: 13 Constraints for Dimensioning and Tolerancing, In: *Geometric design tolerancing: theories, standards and applications,* ElMaraghy H.A., (Ed.), 122-131, ISBN 978-0-412-83000-6, Chapman & Hall, London.

Creveling, C.M. (1997). *Tolerance Design,* Addison-Wesley, ISBN 0201634732, Boston.

Davidson, J.K., Mujezinovic, A. & Shah, J.J. (2002). A new mathematical model for geometric tolerances as applied to round faces, *ASME Trans, J Mech Des,* Vol. 124, 609-622

Delchambre, A. (1996). *CAD Method for Industrial Assembly. Concurrent Design of Product, Equipment and Control Systems,* Wiley, ISBN 978-0-471-96261-8, New York

Francavilla, A. & Zienkiewicz, O. (1975). A Note on Numerical Computation of Elastic Contact Problems, *Int. Journal for Numerical Methods in Engineering,* Vol. 9, 913-924

Gordis, J. & Flannelly, W. (1994). Analysis of stress due to fastener tolerance in assembled components, *AIAA Journal,* Vol. 32, No. 12, 2440-2446

Gupta S. & Turner, J.U. (1993). Variational solid modelling for tolerance analysis; *IEEE Computer Graphics and Applications,* Vol.13, 64-74.

Hong, Y.S. & Chang, T.C. (2002). A comprehensive review of tolerancing research. *Int. J. Prod. Res.,* Vol. 40, No. 11, (July 2002) 2425-2459

ISO Standard, (1985). *ISO 8015: Fundamental Tolerancing Principle*

Laperrière, L. & Lafond, P. (1999). Modelling tolerances and dispersions of mechanical assemblies using virtual joints, *Proceedings of ASME Design Engineering Technical Conferences,* ISBN 0791819671, September 12-15, Las Vegas, Nevada, USA.

Laperrière, L. & Kabore, T. (2001). Monte Carlo simulation of tolerance synthesis equations, *Int. J. Prod. Res,* Vol. 39, 2395-2406

Liu, S. & Hu, S. (1995). An Offset Finite Element Model and Its Applications In Predicting Sheet Metal Assembly Variation, *Int. Journal of Machine Tools and Manufacture,* Vol. 35, No. 11, 1545-1557

Liu, S. & Hu, S. (1997). Variation Simulation for Deformable Sheet Metal Assemblies using Finite Element Methods, *Journal of Manufacturing Science and Engineering, Transactions of the ASME*, Vol. 119, No. 3, 368-374

Luenberger, D.G. (2003). *Linear and Nonlinear Programming*, Addison-Wesley, ISBN 020115794, Boston

Marguet, B. & Mathieu, L. (1998). Tolerancing problems for aircraft industries, In: *Geometric Design Tolerancing: Theories, Standards and Applications*, ElMaraghy H.A., (Ed.), 419-427, ISBN 978-0-412-83000-6, Chapman & Hall

Marguet, B. & Mathieu, L. (1999). Aircraft Assembly analysis method taking into account part geometric variations, In: *Global Consistency of Tolerances*, Van Houten F. and Kals H., (Eds.), 365-374, ISBN 0-7923-5654-3, Kluwer Academic Publishers

Martino, P.M. & Gabriele, G.A. (1989). Application of variational geometry to the analysis of mechanical tolerances, Failure Prevention and Reliability, *ASME Paper n.DE* - Vol.16, 19-27.

Marziale, M. & Polini, W. (2007). Tooling Hole Tolerance Analysis in Aeronautic Field, *Proceedings of 10th CIRP International Seminar on CAT*, ISBN 978-3-8322-6036-1, March 21st-23rd, Erlangen, Germany

Marziale, M. & Polini, W. (2008). Tolerance Analysis of an Aircraft Tail Beam, *Proceedings of 18th FAIM International Seminar on Flexible Automation and Intelligent Manufacturing*, 30 June - 2 July, Skovde, Sweden.

Merkley, K. (1998). *Tolerance Analysis of Compliant Assemblies.* Ph.D Thesis, Brigham Young University, Provo, Utah

Moroni, G., Polini, W., & Rasella, M. (2007). Non-Rigid Free Form Surface Inspection: an aeronautical composite component case study, *Proceedings of the 10th CIRP Conference on Computer Aided Tolerancing*, Erlangen, Germany, March 21-23.

Mujezinovic, A., Davidson, J.K. & Shah, J.J. (2004). A new mathematical model for geometric tolerances as applied to polygonal faces, *ASME Trans, J Mech Des*, Vol. 126, 504-518

Nassef, A.O. & ElMaraghy, H.A. (1996). Probabilistic Analysis of geometric tolerances. In: *Computer Aided Tolerancing*. Kimura F., (Ed.), 187-203, Chapman & Hall, London

Nigam, S.D. & Turner, J.U. (1995). Review of statistical approaches to tolerance analysis. *Computer Aided Design*. Vol. 27, 6-15

Ody, R., Burley, G., Naing, S. & Corbett, J. (2001). Geometric and Dimensioning Tolerances in error budgeting for assembly-centric design of aero-structures, *Proceedings of the 7th CIRP Int. Seminar on Computer Aided Tolerancing*, pp. 111-120, ENS de Cachan, France, April 24-25.

Polini, W., Marrocu, M. & D'Ambrosio, L. (2007). Tolerance Analysis of Free-Form Surfaces in Composite Material, *Journal of Computing and Information Science in Engineering*, Vol. 7, (March 2007) 31-43

Prisco, U. & Giorleo, G. (2002). Overview of Current CAT Systems, *Integrated Computer Aided Engineering*, Vol. 9, No. 4, 373-387

Requicha, A.A.A. (1983). Toward a theory of geometric tolerancing, *Int J Rob Res*, Vol. 2, No. 4, 45-60

Requicha, A.A.A. (1993). Mathematical definitions of tolerance specifications, *Manuf Rev*, Vol. 6, No. 4, 269-274

Salomons, O.W., van Houten, F.J.A.M. & Kals, H.J.J. (1998). Current status of CAT Systems, In: *Geometric Design Tolerancing: Theories, Standards and Applications*, ElMaraghy H.A., (Ed.), 438-452, Chapman & Hall, London.

Sellakh, R., Rivière, A., Chevassus, N. & Falgarone H. (2003). An assisted method for tolerancing applied to aircraft structures, *Proc. of 8th CIRP Int. Seminar on Computer Aided Tolerancing*, pp. 187-198, Charlotte, North Carolina

Shen, Z.; Ameta, G.; Shah, J.J. & Davidson, J.K. (2004). A Comparative Study of Tolerance Analysis Methods, *Proceedings of ASME 2004 Design Engineering Technical Conferences and Computers and Information in Engineering Conference,* Paper No. DETC2004-57699, Salt Lake City, Utah, Semptember 28-October 2

Turner, J. & Gangoiti, A.B. (1991). Tolerance Analysis Approaches in commercial software, *Concurrent Engineering,* Vol. 1/2, 11-23

Whitney, D.E. (2004). *Mechanical Assemblies. Their Design, Manufacture and Role in Prnsoduction Development,* Oxford University Press Inc., ISBN 0-19-565782-6, New York

3

Application of Optical Fibre Sensors to Measuring the Mechanical Properties of Composite Materials and Structures

Wojciech Błażejewski, Paweł Gąsior and Jerzy Kaleta
Wrocław University of Technology
Poland

1. Introduction

The role of polymer composites in mechanical structures has been steadily increasing. The key factor here is the coefficient of specific strength, i.e. a ratio of tensile strength to specific gravity. As compared with metals, this coefficient assumes a high value for composites, including the polymer ones. Other major factors determining the choice of materials are: the product price, mass manufacturability, high strength and durability, corrosion resistance, material production and machining energy consumption and a few more. Also the latter criteria increasingly favour composites. Materials engineering as applied to composites offers the possibility of creating entirely new materials and modifying the traditional materials, which in both cases better meet designers' expectations.

The ever increasing popularity of composite structures is first of all due to the considerable reduction in weight in comparison with metal structures, at a sufficiently high mechanical durability. Nevertheless, the safety requirements require that the stress-strain state of composite structure be monitored. Standard NDE methods such as radiography, interferometric holography, ultrasonic scanning or visual inspection are usually not effective in on-line monitoring. Even if they were used for periodic checks, they would not be able to detect promptly defects critical to the condition of the monitored structure. Moreover, conventional measurement devices, such as electrical resistance strain gauges, often get damaged in adverse environmental conditions.

State-of-the-art measuring methods based on optical fibre technology are increasingly often used to monitor the structural health of industrial objects. Optical fibre sensors have many advantages over the conventional devices. The obvious one is that they can work at high levels of electromagnetic interference and in other adverse conditions (high dust concentration, high temperature, high pressure, significant deformations). Moreover, such sensors are characterized by high (deformation and temperature) measuring sensitivity in a wide measuring range. Owing to their small geometric dimensions and a relatively small weight they can be installed inside a structure (e.g. by embedding them in the composite material) or on its surface. Thanks to their high potential for multiplexing it is possible to create a nervous system of the object being monitored, enabling health and damage assessment. In addition, optical fibre based sensors ensure spark-proof safety which is an important consideration for inflammable applications.

Because of the wide range of problems connected with the measurement of mechanical quantities in composites by means of optical fibre sensors it becomes necessary to concentrate on major problems. The authors decided to focus on the structural health monitoring (SHM) aspects for all kinds of composite structures. Solutions being mainly the result of the authors' own research, with reference to the literature on the subject, will be highlighted.

Considering the above, the objective is to present the fundamentals of the measurement method based on optical fibre sensors as applied to:

- monitoring the behaviour of, among others, laminates and pultruded composites used as reinforcement in building structures and aeronautical structures;
- measuring (static and cyclic) strains in Smart composites where strong magnetic fields limit or exclude the use of other measuring techniques;
- monitoring the structural health of composite structures subjected to extreme strain, such as composite pressure vessels for storing compressed hydrogen and methane car fuels.

The focus is on the experimental aspect of measurements, and the numerical modelling of composite materials and structures is deliberately neglected. However, one should note that modelling plays an important role in many measurement situations (e.g. it is used to determine the number and location of sensors, the theoretical state of strain on a meso- and macroscale, the distribution of temperature, etc.).

2. Composite materials and importance of SHM

Currently most of the research and applications are concerned with composites whose matrix is made of polymers, ceramic materials or metals (e.g. aluminium, titanium, nickel and other alloys) and which are reinforced with glass or carbon fibres. Such composites have found wide application in the manufacturing of fuselages and other structural components of planes, gliders, space shuttles and rockets, car body components, fuel tanks, yacht hulls and masts, wind turbine blades and domestic appliances components, in medicine and in many other fields. Moreover, composites are used to reinforce the existing building structures, including the historic ones (e.g. to preserve bridges, churches, etc., or to increase their load-bearing capacity).

The percentage of composite materials used in high-tech applications increases every year, causing a rise in the production of reinforcing fibres. It is estimated that the demand for carbon fibres in 2018 will increase by about 275% (from 41240 to 113500 tons) as compared with that in 2010, (Sloan, 2010). The highest increase in demand is forecasted for applications connected with the wind power industry (700%), the storage of compressed fuels (~410%) and the broadly understood aerospace industry (~220%).

The interest in composite materials is due to their very good mechanical (strength) parameters and low specific gravity. Highly valuable is the possibility of creating a composite with strictly defined parameters for a given structure. Moreover, increasingly often various (mainly optical fibre based) sensors are incorporated into the structure of composite materials. Thanks to this, different parameters (e.g. strain and temperature) can be measured directly in the structure's material. Another major advantage of optical fibre sensors (OFS) is that they can be used to detect defects and their accumulations and to determine the safe life of the material during its service. In recent years most of the research on OFS has been devoted to their use for the continuous monitoring of the structural health of composite elements.

Novel structural health monitoring (SHM) systems for composite structures are currently developed. Thanks to the use of measuring systems integrated with the structure the latter can be optimally managed. Continuous monitoring enables the detection of hazards in the structure, undetected by inspections or modelling. Consequently, the damaged components can be promptly replaced whereby the safe service life can be extended. Moreover, thanks to SHM systems the knowledge about structures and their behaviour in real conditions can be improved whereby the next generation of such structures will be better designed. Ultimately, the costs connected with building new structures will be lower and the safety, reliability and life of the latter will increase (Glisic & Inaudi, 2007).

3. Optical fibre sensors and measuring technique fundamentals

Currently optical fibre sensors (OFS) are increasingly widely used to monitor the mechanical properties of composite materials and structures in their operating conditions where the reliability and safety of the structures (often working under extreme strains) is critical. Typical applications of various OFS to condition monitoring are presented below. Strain (displacement) and temperature are the main parameters which can be measured by such techniques. On the basis of the measured strain and temperature one can estimate other mechanical parameters, such as stress/strain distribution, force, pressure, curvature, leakage and so on.

3.1 Fibre bragg gratings

Fibre Bragg gratings (FBG) are the most popular fibre based sensors. A Bragg grating is a structure made inside the core of a single-mode optical fibre, characterized by periodical changes in the value of the refractive index (Fig. 1). Due to the presence of such a modulated

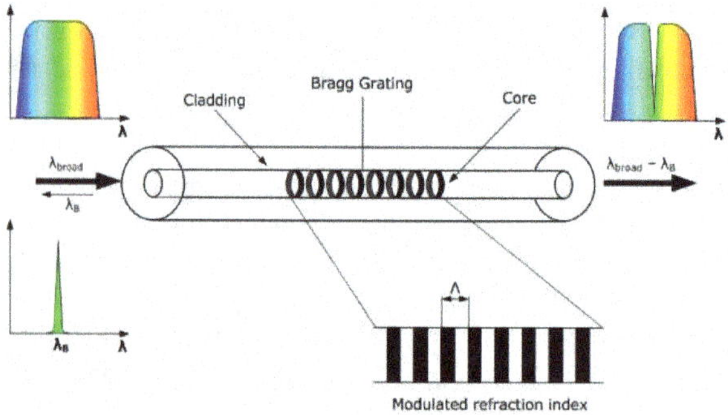

Fig. 1. Scheme of fibre Bragg grating principle of operation

structure inside the core of the optical fibre some of the optical radiation transmitted through the latter is reflected from the grating structure while the remainder is propagated without any loss (Yu & Yin, 2002). The wavelength reflected from the Bragg grating, the so-called Bragg wave (λ_B), is described by the relation:

$$\lambda_B=2\cdot n_{eff}\cdot\Lambda \quad (1)$$

where: n_{eff} - the effective refractive index of the optical fibre core, Λ - a Bragg grating constant.

Sensors bonded to the external surface of the monitored structure or located inside the material are subject to deformations causing changes in the Bragg wavelength, which becomes the touchstone for the measured deformations. It should be noted, however, that a change in λ_B in real measuring systems is the result of the simultaneous influence of temperature and deformations, which can be measured in ranges: -270°C ÷ 800°C and -3% ÷ 3%, respectively.

3.2 Interferometric SOFO® sensors

Another type of OFS applicable to composite structures are interferometric sensors (i.e. SOFO® sensors with long measuring arms). Interferometric optical fibre sensors are characterized by the modulation of the light signal phase propagated in the measuring system (Glisic & Inaudi, 2007). The measuring heads of SOFO® sensors, in the form of a single-mode optical fibre, may attain from a few centimetres to a dozen or so meters and are either integrated with the surface of the monitored object (e.g. by means of special tape - SMARTape®) or located inside the monitored structure (e.g. in the case of tested vessels they are buried in the composite). The sensors are designed to measure displacements (deformations). The measurement consists in analyzing the difference in the phases of the optical signals propagating in the two arms (the measuring arm and the reference arm) of the Michelson interferometer (Fig. 2).

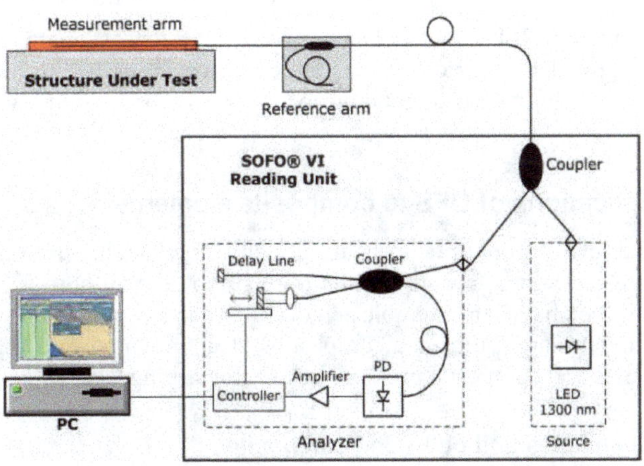

Fig. 2. Scheme of Smartec SOFO® interferometric measuring system (Glisic & Inaudi, 2007)

The measuring arm is in direct contact with the examined structure. The reference arm is separated mechanically from the monitored object, but it is still close enough for the temperature of both arms to be the same. Thanks to this the influence of temperature fluctuations on the measurement results is eliminated. The change in the phase of the light wave is a result of the change in the length of the optical fibre constituting the measuring head. Because of the interference of the two beams it is possible - through an analysis of interference fringes - to quantitatively determine the deformation of the sensors and consequently, the deformation of the monitored structure (Glisic & Inaudi, 2007). Since interferometric sensors belong to a group of sensors with long measuring arms (long-gage sensors), the measured deformations are averages for the whole length of the sensor.

3.3 EFPI sensors

Extrinsic Fabry-Perot interferometric (EFPI) sensors are built from two optical fibres facing each other, but with a small air gap (in the order of micrometers) between them, inside a capillary. The working principle of EFPI sensors is based on multi-reflection Fabry-Perot interference between optical signals reflected from air-glass interface mirrors (Leng & Asundi, 2003). The scheme of the EFPI principle of operation is shown in Fig. 3.

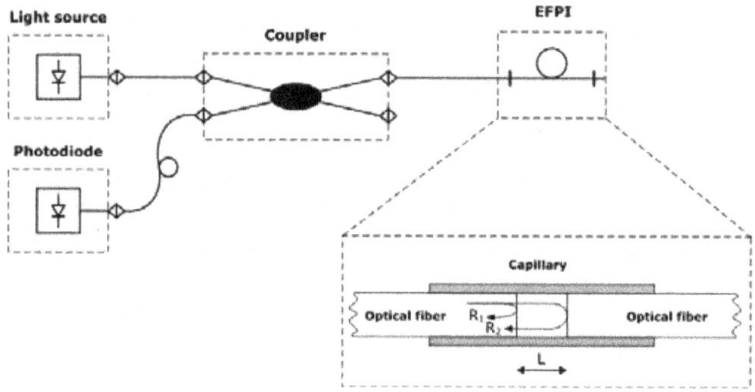

Fig. 3. Scheme of system based on extrinsic Fabry-Perot interferometric sensors

Changes in air gap length (L), which in fact are changes in sensor length, correspond to the variation in the applied mechanical strain. They can be measured by demodulating the interference signal, i.e. by means of coherent or low-coherent techniques (Glisic & Inaudi, 2007).

4. Selected applications of OFS to composite elements

There is a great variety of fibre-optic measuring methods which are used in the monitoring of composite structures. Therefore in the literature survey the authors decided to focus on the areas in which they have considerable achievements and experience. Among others, the monitoring of laminates and pultruded composites, structural aircraft components, SMART magnetic composites and composite pressure vessels for storing CH2 and CNG is discussed.

4.1 Reinforcing laminates and pultruded composites

Optical fibre sensors have found the widest application in the measurement of deformations of all kinds of laminates and pultruded composites. This is owing to several advantages which OFS have, such as:

- similar material properties of the sensor and the reinforcing material in the composite,
- OFS, even the ones complicated in shape, can be located directly in the material (sensor small size and weight and measuring head flexibility),
- resistance to electromagnetic interference (the sensor is a dielectric),
- a wide range of measured physical quantities and simultaneous measurement in many points,
- resistance to environmental conditions.

OFS, usually in the form of bare-fibre Bragg gratings or interferometric sensors, are located directly between the successive layers of the laminate or are interwoven with reinforcement bundles in the case of woven/drawn elements. Depending on the method of manufacturing the composite and its end use, the sensor may be impregnated with resin during the production of the material ("wet" methods such as pultrusion and winding) or later using the vacuum method.

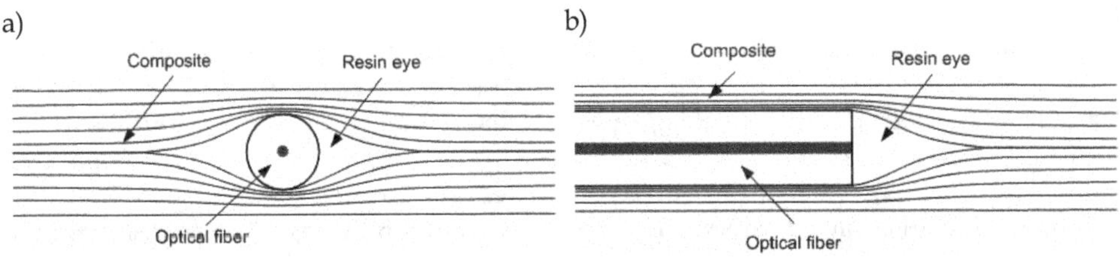

Fig. 4. Scheme showing how local defects arise in composite laminates with embedded optical fibre sensors (Balageas et al., 2006)

Nevertheless, one should also note potentially negative aspects which need to be taken into account or eliminated (or their effect minimized). One of the drawbacks is the sensor size. Even though the optical fibre diameter is relatively small (typically 125 μm), it is quite considerable in comparison with that of the fibres used to reinforce the composite (amounting to ~16 μm for glass fibres and ~8 μm for carbon fibres) and with the thickness of a single laminate layer. Thus by locating the optical fibre inside the composite one may introduce areas of reinforcement discontinuity (Lee et al., 1995). These areas are excessively filled with resin and are referred to as resin pockets or resin eyes (Fig. 4). The disturbances n stress field distribution in the material may initiate delamination of the latter (Balageas et al., 2006; Udd et al., 2005).

The size and shape of a local defect in the form a resin pocket depend on the orientation of the optical fibre relative to the reinforcement fibres. The laminate's mechanical parameters are least affected when the sensor is laid along the direction of the reinforcement and most affected when it is laid perpendicularly to the latter. If the sensor is laid parallel to the reinforcing fibres and the two ends of the optical fibre are led outside the area of the sample, the size of the additional defect in the laminate is close to zero (Balageas et al., 2000).

Another aspect which needs to be considered is the effect of the process of manufacture of composite laminate on the measuring properties of the FBG sensors located inside it. In (Kuang et al., 2001) the authors analyzed the spectrum reflected from the fibre Bragg grating in the reference environment (outside the composite before installation), comparing it with the response of a sensor embedded directly between the layers of the laminate. For some test samples they found the spectrum reflected from the sensor to be distorted and widened. This was due to the locally strong nonuniformity of the load acting on the sensor along its whole length (7 mm). Such local nonuniformities in the distribution of the strain field resulted mainly in an increase or decrease of the Bragg grating constant (Λ), which manifested itself in the splitting of the single Bragg wave peak into several peaks. This behaviour of the sensor was observed mainly for laminates consisting of several composite layers laid at different angles relative to each other. In the case of unidirectional composites, only a shift of the reflection (Bragg wave) spectrum towards lower values, due to the local compression or tension of the sensor as a result of the manufacturing process (e.g. resin curing in elevated pressure and temperature conditions), was observed. The possibility of fibre-optic monitoring of resin curing during the manufacture of high-strength laminates was described by, among others, Leng in (Leng & Asundi, 2003). Thanks to the use of FBG and EFPI sensors residual strains during curing can be determined and local material defects in the form of delaminations can be detected. The quoted authors prepared samples made up of 16 layers of carbon prepregs, between which optical fibre sensors were installed along the direction in which the reinforcement had been laid. In addition, a Teflon disk 300 mm in diameter was placed in one of the samples to simulate a local delamination. In order to harden them the samples were annealed at a temperature of up to 120°C. By analyzing the signals received from the sensors installed in the samples with and without the defect one could detect the local damage already during the curing process. It was also found that residual strains after the curing of the samples are greater along the direction of compression for the flawed sample. The difference between the signals received from the EFPI and FBG sensors for the samples during curing amounted to 5-10% (Leng & Asundi, 2003). Then in order to determine their bending strength the samples were subjected to the three-point bending test. The test results confirmed that the strength of the samples with the designed flaw was lower than that of the samples without the flaw and that the artificially introduced delamination reduced the stiffness of the structure (Leng & Asundi, 2003).

One of the main uses of laminates and pultruded composites is the reinforcement of the existing structures or structures under construction. The strength and load-bearing capacity of such structures as bridges, viaducts and buildings are commonly increased by introducing (through bonding) additional reinforcement made of materials characterized by high tensile strength and stiffness (e.g. laminates or pultruded composites in the form of strip or mat) glued to the surface of the structure being reinforced. Thanks to its relatively little weight the reinforcement does not additionally weight down the structure.

But composite reinforcements have drawbacks - they tend to creep and are brittle. For this reason research is underway aimed at the detection and control of internal material defects at the early stage in their development (Majumder et al., 2008). For example, a method of manufacturing glass fibre reinforced composites (GFRC) and carbon fibre reinforced composites (CFRC) through pultrusion with FBG and EFPI sensors located inside is presented in (Kalamkarov et al., 1999). The quasi-static and cyclic test measurement results received from the optical fibre sensors for the tested samples were in agreement with the results obtained from extensometric measurements. Moreover, it was shown that optical fibres with polymer cladding are characterized by better parameters of the sensor/material bond. Sensors in acrylic cladding tended to separate from the laminate and could not be used in composites which required manufacturing processes conducted at a temperature of above 85°C. In the case of the sensors in the polymer cladding the allowable temperatures were as high as 385°C (Kalamkarov et al., 1999).

The use FBG sensors located in pultruded epoxy-glass (GFRP-OFBG) and epoxy-carbon CFRP-OFBG) composites designed for reinforcing concrete beams is described in (Zhou et al., 2003). The composite elements prepared in this way were cast in concrete, cured and subjected to three-point bending. The FBG sensors registered the strains of the composite rods in a range of up to 200 microstrains (with an accuracy of 1÷2 με), registered the cracking of the concrete layer and informed about the slipping out of the reinforcements from the concrete. The authors showed that pultruded composite rods (with incorporated sensors) can be used both as elements increasing the load-bearing capacity of a structure (reinforcing the latter) and as measuring heads integrated directly with the monitored element.

a)

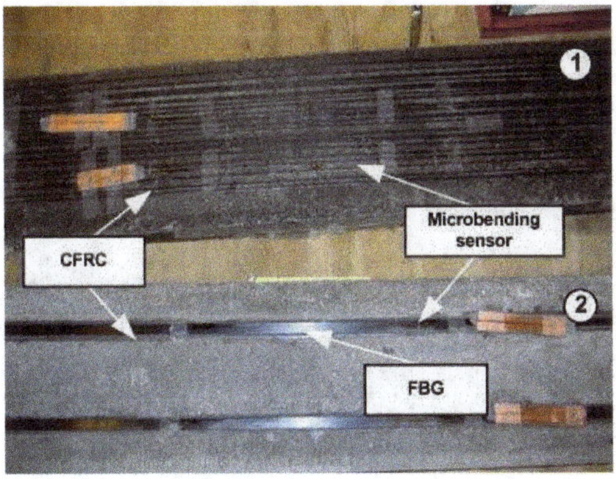

b)

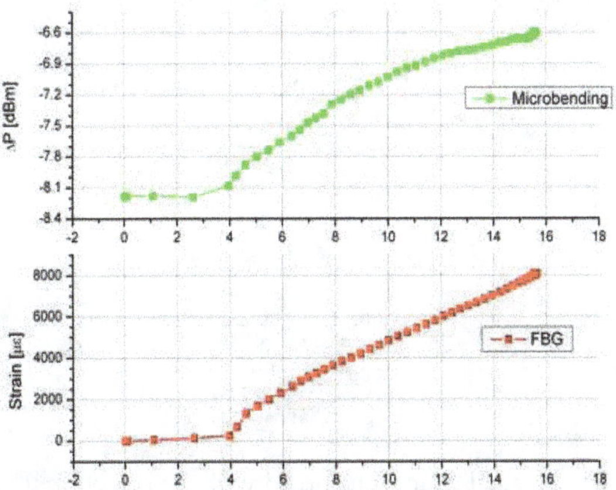

Fig. 5. Tested beams with composite reinforcements: (a) laboratory made in IMSAM (1) and by Polish Bridges company (2), and (b) characteristics registered by respectively microbending sensor and FBG sensor for beam with commercial composite reinforcement (Błażejewski et al., 2007)

Above one can find exemplary results of investigations into the monitoring of composite reinforcements of engineering structures, carried out by the present authors. Composite reinforcements, in the form of pultruded CFRC and GFRC strips, were glued to the tested concrete beams. Besides the reinforcements, two types of optical fibre sensors, i.e. FBG sensors and amplitude microbending measuring heads, were employed. Figure 5a shows two tested concrete beams with composite reinforcements (made in the laboratory in the Institute of Materials Science and Applied Mechanics (IMSAM) at Wrocław University of Technology and by the Polish Bridges (PB) company) and embedded optical fibre sensors.

During three-point bending in an MTS810 strength tester the following were registered as a function of time: the force, the strain (measured by the FBG sensors) and the change in optical power (registered by the microbending head).

a)

b)

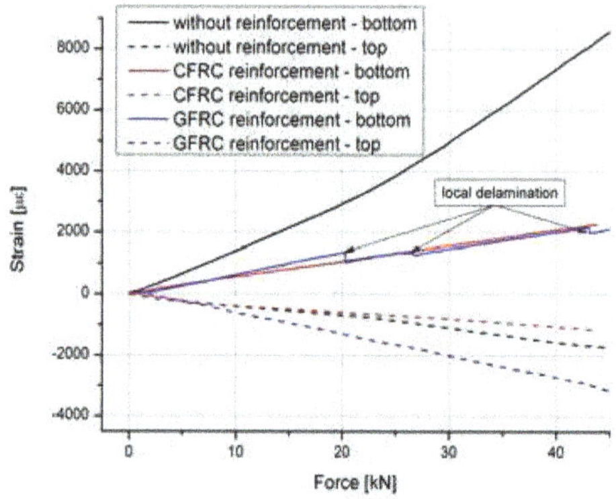

Fig. 6. Four-point bending of steel beam reinforced with unidirectional carbon-epoxy composite with integrated FBG sensors (a) and comparison of measurement results for different reinforcements (b)

Figure 5b shows exemplary measurement results for the microbending sensor and the FBG sensor, as a function of the force applied to the tested concrete beam with the Polish Bridges reinforcements. Similar measurements were carried out for concrete beams without reinforcement and with the laboratory made glass composite and carbon composite reinforcements. On this basis the effect of the type of reinforcement on the strength of the concrete beams was determined. The tests confirmed the suitability of optical fibre sensors for the testing of engineering structures reinforced with elements made of composite materials (Błażejewski et al., 2007).

Then tests were carried out for two types of reinforcement (pultruded carbon/glass reinforcement) with integrated FBG sensors, designed to increase the load-bearing capacity of steel frame structures. The reinforcement together with the sensor was glued to the beam's (tensioned) bottom. An additional reference (FBG) sensor was installed on the top (compressed) side. The reinforced beams were subjected to the static four-point bending test during which the force and the strain in two areas were registered.

The measurement results are presented in Fig. 6b. An additional measurement was performed on a steel beam without any reinforcements to determine the effect of the composite of the structure. It was found that the reinforcements applied made the beam stiffer whereby its load-bearing capacity increased as manifested by lower strains in comparison the beam without reinforcements under the same static load. Thanks to the reinforcement the elastic range of the whole beam increased (up to about 20 kN). In the strain-force diagram for the beam with glass reinforcement one can see local jumps in the registered strain values (at ~20.2 kN, 26.4 kN and 43 kN), indicating that the reinforcement locally gets unstuck from the structure. In the sample, this manifested itself in local whitening of the composite.

4.2 Aerospace structural components

Owing to their excellent mechanical properties, particularly high strength and stiffness, composite components are increasingly used in aerospace structures. Modern planes (e.g. Boeing B787) include many subassemblies made from composite materials, especially carbon fibre reinforced composites (CFRC). Their (volumetric) percentage in the total aircraft parts is as high as 80%, which amounts to 50% of the ready aircraft weight. However, the development of defects in composite materials is much more complicated than in metallic materials and it is still not fully explored (Diamanti et al., 2002). For this reason, in recent years much research has been devoted to structural health monitoring (SHM), also of aerospace structural components, by means of fibre-optic measuring techniques. The research projects are aimed at developing systems for monitoring aerospace structures not only during their everyday operation, but also during the processes of production (e.g. pressure and vacuum forming, resin impregnation and curing, machining, assembly, etc.) of the particular composite components.

In (Takahashi et al., 2010) the authors describe a method of developing a system for the continuous monitoring of the composite parts of the plane's fuselage (a partition with reinforcements) through a system controlling the state (distribution) of strain on its surface during the entire production stage and service life. The developed system was based on Bragg gratings integrated with the structure of a CFRC reinforced composite. Through the measurement of local strains the system can detect and locate defects and other damage arising in the structure's material. The authors also showed that it was possible to measure the strain of composite samples in the course of the whole process of pressure forming and described its effect on the tested object. In order to determine the optimum location of point FBG sensors on the reinforced composite partition a numerical FE model was created and loaded on its outer surface with a pressure corresponding to the maximum design pressure.

Two types of defects: reinforcement separation and local impact into the partition resulting in the delamination of the composite were implemented in the virtual model. Thanks to this it became possible to determine the optimum distance between FBG sensors, ensuring detection of defects in any area of the investigated object. It was estimated that the intersensor distance for sensors with a sensitivity of 10 $\mu\varepsilon$, installed on the reinforcements should not exceed 100 mm. In the case of the sensors controlling the surface of the partition, one FBG grating is capable of detecting a defect within an area 300 mm in diameter.

A major problem which designers of systems for continuous structural health monitoring encounter is the location of measuring transducers and their optimization with regard to the number of measuring points. The constraints put on the number of sensors are aimed at reducing the measuring system costs (e.g. costs of sensors, instrumentation, measuring units, system installation). They are also dictated by difficulties with the on-line processing of large quantities of measurement information. A large number of sensors does not always guarantee system reliability. Proper software with implemented data processing algorithms, capable of keeping the user informed about potential risks, is also needed.

Besides research on optimizing the arrangement and number of sensors (e.g. through numerical modelling and the hybrid (model + experiment) approach), purely experimental research on implementing the very wide networks of sensors (comprising a few thousand and more sensors) is being conducted. The aim is to create SHM systems which, similarly as the human nervous system (nerves/sensors, the brain/the information processing unit), enable the monitoring of entire engineering structures (Balageas et al., 2001).

Exemplary research on implementing an SHM system comprising a large number of sensors is presented in the paper (Childers et al., 2001). The authors describe a measuring system consisting of 3000 measuring points based on fibre Bragg gratings, for monitoring the strain of a transport plane wing. Static tests on 12 m long composite wing made of carbon-epoxy composite with Kevlar inserts were carried out. A system consisting of 466 foil strain gauges was used as the reference. A novel system for handling many FBG sensors in a single fibre had been specially designed for the tests. The system is based on the optical frequency domain technique (OFDR) and is capable of simultaneously reading signals from 4 measuring heads. Each of the heads is 8 m long and contains 800 sensors spaced at every 10 mm.

The results obtained from the two different measuring systems were very similar. Moreover, it was found that thanks to the use of the fibre-optic system the total weight of the measuring system could be considerably reduced and at the same the number of measuring points could be increased. As a result, the behaviour of the structure under real loads could be more accurately represented. In addition, owing to the small number of cables the system based on Bragg gratings turned out to be much easier to install than the strain gauge system (Childers et al., 2001).

4.3 SMART magnetic composite elements

SMART magnetic materials (SMM) are active materials which in recent years have been increasingly often and wider used in the aerospace industry, the navy, the car industry, civil engineering and in many other fields (Schwartz, 2002). The so-called giant magnetostrictive materials (GMMs) form a major subgroup in SMM. They enable the conversion of magnetic energy into mechanical energy and vice versa with high efficiency. A typical representative of GMMs is Terfenol-D which offers many advantages but its low tensile strength and the occurrence of high eddy currents detract from them. Attempts are made to eliminate the drawbacks by creating composites. For several years now research projects, e.g. (Kaleta et al., 2009; Gąsior et al., 2009; Yin Lo et al., 2006), targeted at creating magnetostrictive materials capable of assuming the form of, e.g., rods or laminates and effectively replacing solid Terfenol-D, have been conducted. Such composites offer the possibility of new applications, in e.g. SHM (tagging).

a)

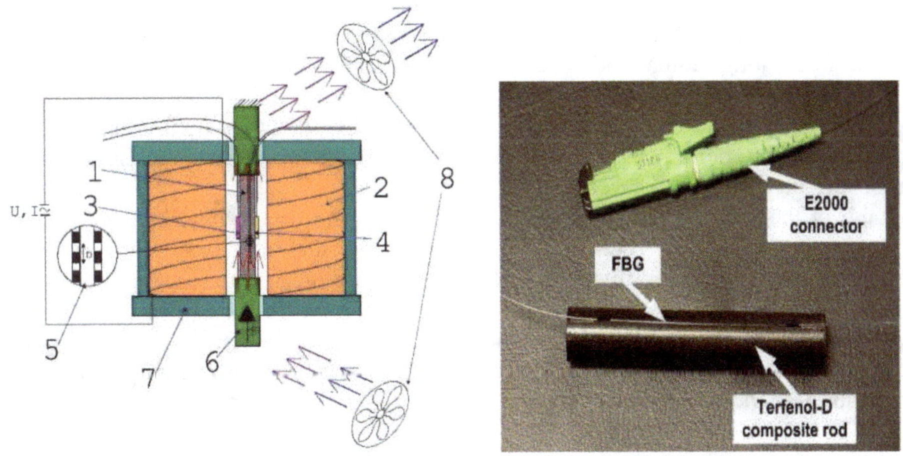

b)

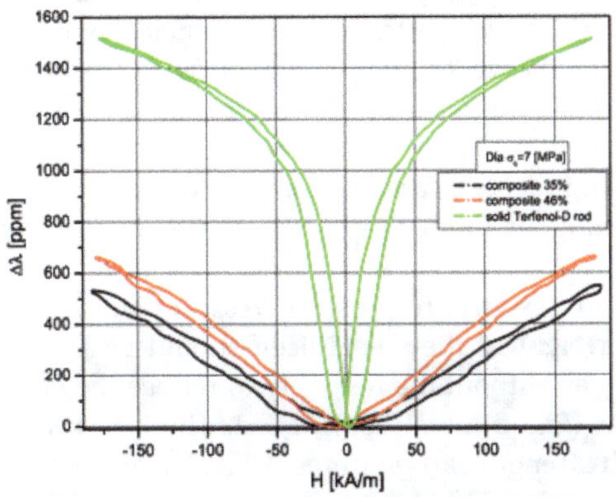

Fig. 7. Composite sample containing powdered Terfenol-D, with installed fibre-optic sensor and test stand for magnetostrictive composites (a) and exemplary change in longitudinal magnetostriction under influence of applied magnetic field for different composite and solid samples, registered by FBG sensor (b) (Kaleta et al., 2009)

Monolithic (solid) GMMs and composites incorporating them are tested mainly with regard to their magnetic properties. The parameters describing these properties are magnetostriction (the deformation of a ferromagnetic under the influence of an external magnetic field) or the opposite effect (the Villary effect). In both cases, the conventional methods of measuring strains (e.g. by means of extensometers or strain gauges) are practically inapplicable since GMMs are inside coils (or even constitute their core) and the sharp fluctuations in E-M fields there disturb the measurement.

For this reason, optical fibre sensors are an ideal solution for measuring the mechanical parameters of magnetic materials (including GMMs). Owing to their dielectric nature they are resistant to all electromagnetic interferences and in addition, enable strain registration with an accuracy as high as 1 $\mu\varepsilon$.

In (Kaleta et al., 2009; Gąsior et al., 2009) the authors used FBG sensors to measure the deformation of solid Terfenol-D rods and composite rods based on epoxy resin and powdered Terfenol-D. The fibre-optic sensors were installed on the surface of the samples (Fig. 7a) in the longitudinal and circumferential directions whereby magnetostriction could be measured in two directions. During the bonding of the sensors to the samples the latter were prestressed. As a result, strains could be measured under both compression and tension. This is important since in order to obtain optimum operating conditions for these composites they need to be prestressed (Kaleta et al., 2009).
The aim of the static tests described in the above papers was to determine the magneto-mechanical properties of rods made of GMM composites and solid Terfenol-D. Their magnetostriction was determined depending on the percentage of the magnetic phase in the material (Fig. 7b). The test results showed that although the magnetostriction is three times lower in comparison with that of the solid material, it is about 30 times higher than that of nickel (the reference material). But the influence of eddy currents was eliminated and the tensile strength increased (Kaleta et al., 2009).

4.4 Composite pressure vessels

Today composite pressure vessels are used in many fields, such as: the car industry, the aerospace industry, rescue services, etc. They owe their popularity to the considerable reduction in the weight of the vessel in comparison with steel vessels, and to their high mechanical strength. Also their excellent resistance to corrosion is not without significance. In order to illustrate the subject, a case of a composite hydrogen vessels is presented below. The use of hydrogen to feed fuel cells and for combustion in conventional engines requires a safe and expensive fuelling method. Storing hydrogen in pressure vessels of the CH2 (compressed hydrogen) type has become the predominant technology, especially in cars. The latest (4th generation) designs are pressure vessels with a nominal working pressure (NWP) of 70 MPa. The weight of the vessel itself should be low enough in order for the latter to be installed in a car without the necessity to reduce the boot space. For this reason, wholly composite vessels (Fig. 8) must be installed. The load-bearing wrapping of such a vessel is made from high-strength carbon fibres (CFRC) while the liner (ensuring tightness) is made of high-quality plastics. The burst test pressure (BTP) at a safety factor of 2.35 (for CFRC) has to be above 164.5 MPa. In the case of glass composites (GFRC) the safety factor increases to 3.65 whereby BTP is higher.

a) b)

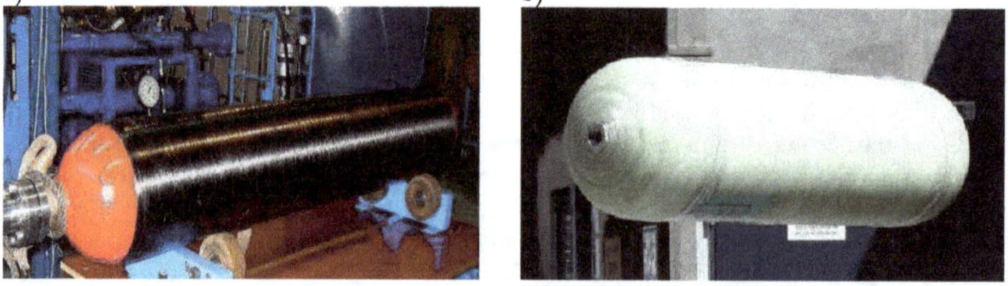

Fig. 8. Examples of pressure vessels for storing CNG and CH2: of type 2 (a) and type 4 (b), (Błażejewski et al., 2007)

Because of the stringent requirements the CH2 vessel is an integrated high-tech product which can withstand extreme strains, and all its building components, i.e. the materials, the calculations, the design, the lining and wrapping technologies and the test methods, represent the cutting edge of research and engineering. The problem of SHM for such a vessel is discussed below. The discussion is based on the present authors' experience gained in carrying out research projects, among others, within the 6th EU Framework Program (Storage of Hydrogen - StorHy) and the 7th UE FP (Integrated GAS Powertrain - InGas).

The combination of the highly-strained composite load-bearing structure of the vessel and the hazardous substance (which compressed hydrogen fuel is) necessitates the monitoring of the structural health of such objects. It is estimated that the danger area in the case of explosion of a full car methane tank is 250 m large in diameter and in the case of hydrogen it may be as large as 1000 m in diameter (Mair, 2007). Therefore it is of critical importance to ensure operational safety over the entire ten-year service life.

At the moment the standards require that pressure vessels be subjected to routine tests certifying them fit for further use, but it is not necessary to permanently install SHM systems. Because of such regulations pressure vessel manufacturers prefer to increase the amount of the material reinforcing the structure to increase its safety than to optimize the the vessel design. This leads to higher production costs and a higher final product price. However, currently efforts are made, e.g. (Sulatisky et al., 2010), to exert pressure on the legislative bodies to introduce regulations reducing the current high safety factors for the manufacture of pressure vessels when a continuous structural health monitoring is employed. As already mentioned, standard NDE methods are not suitable for the on-line monitoring of such objects and can be used only for periodic inspections (Foedinger et al., 1999; Degrieck et al., 2001). A system of sensors should ensure the assessment of the technical condition of the high-pressure cylinder during both its production and (above all) its long use.

The research in this field, e.g. (Foedinger et al., 1999; Degrieck et al., 2001; Kang et al., 2006; Hernández-Moreno et al., 2009; Sulatisky et al., 2010; Błażejewski et al., 2010; Glisic & Inaudi, 2004; Frias et al., 2010), shows that the best of the offered solutions is to use optical fibre sensors for this purpose, since they are characterized by resistance to electromagnetic interference, non-sparking safety, ease of integration with the structure of the composite material and high sensitivity in a wide measuring range. Such sensors, permanently installed on the surface of the vessel or incorporated into the structure of the composite material, enable the continuous monitoring of the structure's technical condition over its whole service life.

In one of the first works (Foedinger et al., 1999) dealing with the application of optical fibre sensors to the monitoring of the structural health of standard testing and evaluation bottles (STEBs), the authors used FBG sensors which were integrated with the composite load-bearing layer as it was being wound during its manufacturing. The sensors were to measure the strain and temperature of the vessel during the curing of the epoxy resin and quasi-static pressure tests. A micrographic analysis of the cross sections of the composite load-bearing shell containing the optical fibre was carried out and showed good interaction between the two as well as no defects in the areas incorporating the sensors. The latter were located in several areas of the pressure vessel, e.g. in its cylindrical part and in its bottoms. Each time they were positioned in the direction consistent with the direction of the reinforcement, which means that they were wound at different angles to the principal axis of the vessel. In addition, resistance strain gauges and thermocouples were employed as the reference sensors. The strain gauges were installed on the vessel's outer surface (after resin curing) in the neighbourhood of the fibre Bragg gratings. The investigations showed good agreement between the sensor measurement results and the strain gauge measurement results (depending on the winding direction and the location, the difference between the FBG sensor results and the strain gauge results ranged from 1.2% to 24%) and the numerical FE model results.

In (Degrieck et al., 2001), the authors investigated the possibility of using fibre Bragg gratings to monitor the strains of components made of composite materials, including laminates and pressure vessels made by winding. The research was carried out in several stages. First tensile tests and temperature tests were carried out on a bare optical fibre with an FBG sensor. Then carbon-epoxy laminate with a fibre Bragg grating placed between the layers was made. The laminate was subjected to the static three-point bending test. Finally, a pressure vessel was made by winding a glass fibre, in which an optical fibre together with a Bragg grating was wound (in the circumferential direction) with the final reinforcement layer, parallel to the latter. The vessel was then subjected to tests. In the first test it was statically loaded with an internal pressure of 0-16 bar and then cyclically with 0-3 bar. In both tests a change in pressure resulted in a shift of the Bragg wave reflected from the sensor incorporated into the structure of the composite load-bearing layer. It was found that being located directly in the material structure, FBG sensors are ideal for measuring strains in composite materials and in addition, they represent a promising technique for nondestructive testing and evaluating the structural health of composite elements (Degrieck et al., 2001).

The suitability of optical fibre (FBG) sensors for monitoring composite pressure vessels, especially taking into account the possibility of integrating such sensors with the load-bearing layer during the manufacture of vessels by winding, is the subject of the paper (Kang et al., 2006). The authors describe the difficulties they encountered while embedding measuring heads (made by welding FBG sensor together) in a composite material. Only one of the four measuring heads in a pressure vessel with the sensors prepared in this way survived the manufacturing process (winding and high-temperature curing). It turned out that because of their low resistance to lateral stresses in the weld, the welded fibres are poorly resistant to the conditions prevailing during vessel manufacturing whereby the risk of optical fibre damage during reinforcement winding increases considerably. Measuring heads consisting of several sensors produced directly in one optical fibre segment additionally coated with a thin protective layer proved to be much better. In the latter case, the percentage of sensors which after the vessel manufacturing process were still fit for measurements increased to ~79% (11 out of the 14 sensors were available). By comparing the reflection spectrum for the pressure vessel before annealing with the spectrum for the ready vessel one could determine the local residual strain. In all the measuring points local compression amounting to several hundred microstrains (με) was registered, but the strains at the vessel bottoms were relatively larger (Kang et al., 2006).

Then the vessel was subjected to the pressure test until failure. For reference purposes, strain gauges were installed on the vessel's surface. The gauges were damaged before the vessel burst. The maximum registered strains amounted to 0.95% for the pressure of 2900 psi (~200 bar). Also the splitting of the Bragg wave into several peaks was observed – probably due to the displacement of a sensor relative to its optimal position fixed during winding (the sensor was not situated parallel to the reinforcement fibres), (Kang et al., 2006).

The reliability and safety of composite vessels with an integrated SHM system is also the subject of the research conducted by Hao and others (Hao et al., 2007). They proposed to create a system which would monitor the circumferential and longitudinal strain of a tested vessel made of glass-epoxy composite, by means of three FBG sensors installed on its surface. In addition the authors used two strain gauges to locally measure strains in the circumferential and longitudinal directions. The tested vessel (with a 3.2 mm thick load-bearing wall) was statically loaded with a pressure of 0-200 bar. The obtained measurement results showed good convergence for the strains registered in the longitudinal direction, particularly in a low pressure range (up to 8 bars), and large scatters for the circumferential strains. The differences between the FBG sensors and the strain gauges amounted to as much as 60%, even though they were separated by a distance of maximum 12 mm. According to the quoted authors this is due to the different distances of the sensors from the defect around which the vessel burst (Hao et al., 2007).

The use of fibre Bragg gratings for the monitoring of the state of stress of a composite pressure vessel was also the subject of research carried out by Frias and others (Frias et al., 2010). In their paper the authors describe a measuring system based on six FBG sensors and two piezoelectric polyvinlyidene fluoride (PVTF) sensors, integrated with the composite vessel structure. The sensors were located between the steel liner and the load-bearing layer made of glass fibre in thermoplastic resin. The liner was made by welding the two bottoms and the circumferential FBG sensors were located in the direct vicinity of the weld. The other sensors (two Bragg gratings and two piezoelectric sensors) were located in the cylindrical part, and registered longitudinal strains. The vessel was subjected to cyclic testing within which 9 test series with 30 cycles in each were carried out. The upper cycling pressure amounted to 5-40 bar with a 5 bar step in the successive series. The test results showed good convergence for the different measuring systems. Moreover, the liner in the weld region was found to plasticize quickly under the cyclic loading, as indicated by the circumferential FBG sensors. The latter showed that the vessel behaved nonlinearly in the circumferential direction in the weld area and linearly in the longitudinal direction outside this area (Frias et al., 2010).

Glisic and his co-workers in (Glisic & Inaudi, 2004) presented the application possibilities of a system for monitoring the structural health of 4th-type composite (the so-called full composite, a polymer liner and carbon-epoxy reinforcement) pressure vessels for storing compressed methane in cars, with a nominal working pressure of 350 bar. The proposed measuring system is based on interferometric SOFO® sensors. The latter are characterized by relatively long measuring arms which average the measured strain over their length. The aim of the research carried out within the EU ZEM Project (the 5th EU Framework Program) was to develop a topology of sensors distribution on a vessel, and a proper algorithm analyzing the measurement data, enabling the detection of local defects which could endanger the safety of the users (Glisic & Inaudi, 2004).

SMARTape® sensors were integrated with the vessel structure and located between the last carbon fibre layer and the first glass fibre layer. The interferometric SOFO® sensors measured strains in the longitudinal direction (four sensors spaced at every 90° on the circumference) and in the circumferential (helical) direction (2 sensors with opposite lead angles).

Glisic and Inaudi proposed an algorithm (a monitoring strategy) for measurement data analysis which made it possible to detect different anomalous behaviours of the vessel and defects arising in its structure. It was assumed that in the case of damage the vessel would become more deformable (under a constant pressure), which meant that the slope of the strain-pressure line would change. It was also assumed that a local defect could result in local deformations (e.g. ovalization, bending) of the vessel, which would be registered by pairs of SOFO® sensors distributed symmetrically on its surface. Thanks to the latter assumption the correlation of measurements between selected sensors could be analyzed. In the case of a defect, the linear dependence between the strains registered by one sensor and the ones registered by another sensor will be disturbed and will change (Glisic & Inaudi, 2004).

In order to prove the above assumptions a series of static and quasi-static tests were carried out on new pressure vessels and vessels with designed flaws (cuts differing in their geometrical dimensions and delaminations), subjected to a pressure of 0-350 bar. The authors of (Glisic & Inaudi, 2004) show that an increase in such parameters as deformability or the relative compliance coefficient and a change in the reciprocal linear dependence between the particular sensors are indications of vessel damage.

5. Structural health monitoring of composite pressure vessels

5.1 Concept of composite pressure vessel monitoring system

The on-line monitoring of a pressure vessel is not limited to only the measurement of the mechanical parameters (e.g. strain fields) of its composite load-bearing layer. In order to determine the vessel's structural health and to detect critical defects which may pose a risk to the safety of the users it is necessary to employ proper algorithms to analyze the acquired measurement data. An ideal solution to this problem is a hybrid method in which the acquired measurement data are compared with the numerical model of the monitored object. Moreover, a good numerical model should indicate the primary sensor locations to be taken into account when installing sensors. However, this approach is not always possible, usually due to the lack of sufficient information needed to create a reliable numerical model of the investigated object. In addition, the use of the hybrid method necessitates the use of separate numerical models for each type of vessel. Any change in the design must be properly taken into account in the modelling process.

In the case of pressure vessels for which there is no precise information on the geometry of the load-bearing wrapping or the materials used, the present authors proposed to use a method consisting in analyzing local changes in the strain field (measured in many places) during the service life of the vessel and comparing them for particular pressure values. Any deviations from the assumed uniform strain field distribution in the vessel's cylindrical part indicate its potential damage (Błażejewski et al., 2008).

In order to create a fully functional system of monitoring the structural health of composite pressure vessels a special coefficient called ABS (absolute value) was calculated and compared with the threshold value. The ABS coefficient value was calculated as the absolute value of the difference between the coefficients of the directional strain-pressure lines for the successive vessel stress cycles. In addition, in order to locate possible defects the ABS coefficient values were compared between the particular measuring points (sensors). For a pressure vessel with no defects the dependencies between the particular sensors should be the same or very similar (considering the uniform state of stress in the load-bearing layer). When damage (e.g. a delamination or a crack) arises, the strain field is no longer uniform and the dependencies between the sensors are disturbed (Błażejewski et al., 2010).

5.2 Cyclic tests of pressure vessels

A pressure vessel with a steel liner, for storing hydrogen was subjected to cyclic tests at room temperature in accordance with the Draft ECE Compressed Gaseous Hydrogen Regulation, 2003. The vessel was loaded with pressures in the maximum range of 20-875 bar on a test rig in the Institute of Materials Science and Applied Mechanics at Wrocław University of Technology.

a)

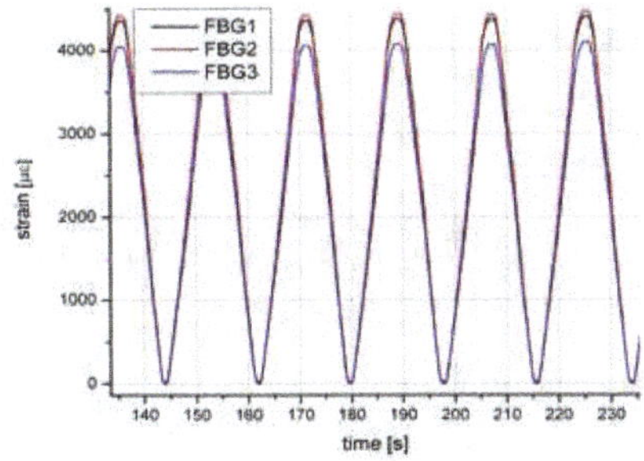

b)

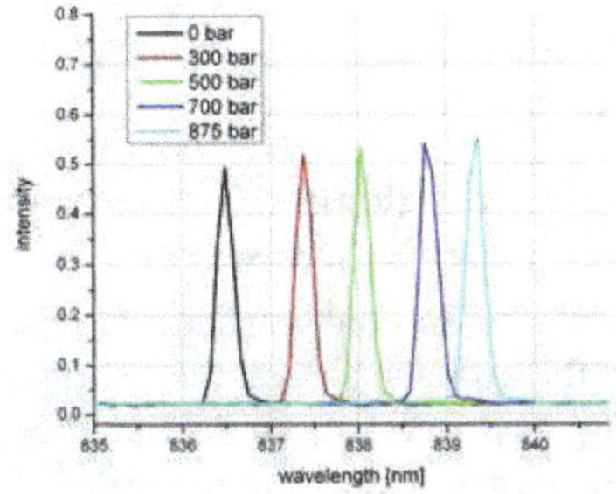

Fig. 9. Hoop strain characteristic registered by fibre Bragg gratings in different places of vessel (a) and Bragg wave shift registered by sensor FBG1 for different internal pressures (b) (Błażejewski et al., 2007)

Optical fibre sensors (FBG) for local strain measurement were used for the monitoring of the structural health of the vessel. In addition, acoustic emission sensors were employed for reference measurement purposes (Błażejewski et al., 2007). In the course of the basic tests (20-875 bar) about 700 cycles simulating vessel operation (filling-emptying) were carried out. The diagram in Fig. 9 shows exemplary vessel hoop strains registered by sensors: FBG1, FBG2 and FBG3 during cycle loading and the variation in the length of the Bragg wave registered by the FBG1 sensor in a pressure range of 0-875 bar. It should be noted that the vessel's circumferential strains for the pressure of 875 bar are relatively high (ranging from ~4000 to 4500 με depending on the type of sensor), (Gąsior et al., 2007).

Figure 10 shows the maximum strains (at internal pressure P_{max}=875 bar) registered by the FBG sensors installed in the circumferential direction, versus the number of vessel loading cycles. It should be noted that an increase in the number of cycles results in a local increase in registered strain values (enlarged fragment A in Fig. 10) leading to the failure of the vessel. This is due to the fatigue (cyclic softening) of the steel from which the liner was made. The bursting of the vessel was preceded by rapid increase in locally registered circumferential strains (enlarged fragment B in Fig. 10). By analyzing the slope of the strain-number of cycles curve in its linear region one can determine how long the vessel can safely remain in service, i.e. predict its life. This is particularly important in the case when pressure vessels are to stay in service for many years.

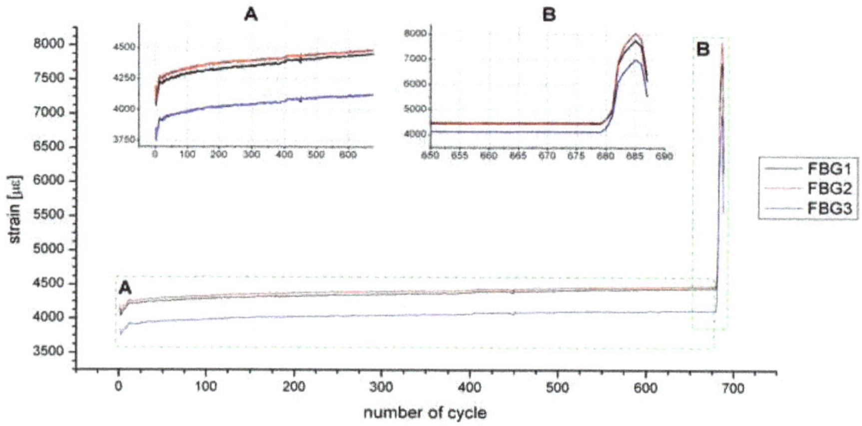

Fig. 10. Maximum circumferential strain (for P = 875 bar) versus number of vessel loading cycles and close-ups of marked areas, (Błażejewski et al., 2007)

Figure 11 shows signals registered by the particular sensors (the circumferential FBG sensors, the strain gauges and the acoustic emission sensors) at the moment of failure of the vessel (the broken line in Fig. 11). It is apparent that the failure was preceded by the previously mentioned rapid increase in strains (in the order of 1500 ÷ 3000 με depending on the sensor and its distance from the defect) registered in the circumferential direction. Moreover, there is good correlation between the measured changes in the strain field on the vessel's outer surface and the acoustic signals registered by the AE sensors. At the moment of failure a considerable increase in the root mean square (RMS) acoustic emission signal occurs, which indicates a high degree of damage to the composite layer (Gąsior et al., 2007). Figure 12 shows an analysis of local circumferential strains for the last 8 cycles of the tests to which the pressure vessel was subjected. The analysis was carried out in accordance with the procedure described in sect. 5.2, consisting in the determination of the ABS coefficient. Its value was calculated as the difference between the directional coefficients for the strain-pressure lines obtained from the reference measurement and the next measurements. The first of the measurements was adopted as the reference point. Figure 12 shows a marked jump in ABS coefficient values between cycles 3 and 4 for all the analyzed measuring points. Moreover, the differences between the particular ABS coefficient values increase, which is a symptom of nonuniformity of the strain field distribution on the external surface of the vessel. This situation is indicative of the strong development of defects in the load-bearing structure of the composite pressure vessel (Błażejewski et al., 2008).

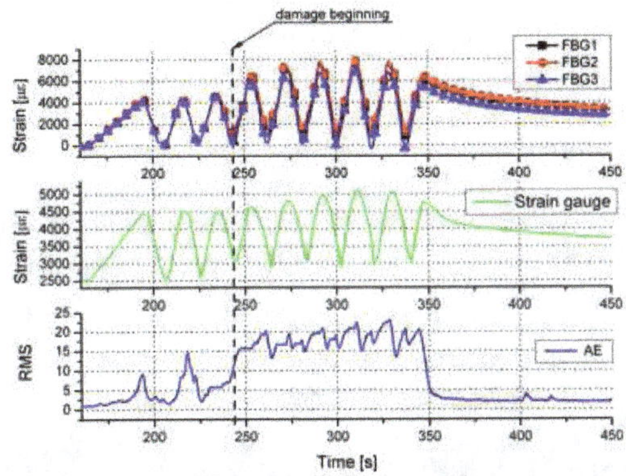

Fig. 11. Local circumferential vessel strains registered by FBG sensors (upper diagram) and strain gauges (middle diagram), and acoustic emission signal (bottom diagram) at instant of vessel failure (Gąsior et al., 2007)

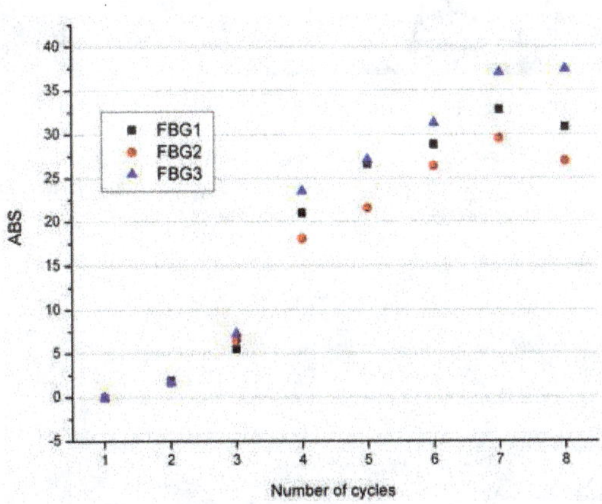

Fig. 12. Analysis of measurement results for final vessel loading cycles (detection of defects in structure of pressure vessel), (Gąsior et al., 2007)

5.3 Cyclic tests of pressure vessels with designed flaws

A major problem in the monitoring of extremely strained composite structures is the early detection and identification of defects arising in the structure of the material. The defects can be the result of material fatigue under cyclic loads (e.g. filling and emptying of the vessel) or other events, e.g. external impacts or the exceedance of the allowable stress (pressure) and working temperature.

The standards relating to the testing of pressure vessels for storing high-pressure fuels (ECE Regulation No. 110, 2008; ISO/DIS15869.3, 2008) distinguish two main types of defects: flaws and delaminations. According to the cited regulations, a vessel with a flaw should meet the requirements of, among others, the cyclic tests at room temperature for a vessel without visible flaws.

Tests of vessels with flaws are referred to in the standards as composite flaw tolerance tests. Depending on the standard, the size of a flaw can be defined as "larger than the limit for the visual inspection, specified by the vessel manufacturer" and so no smaller than:

- length 25mm x depth 1.25mm,
- length 200mm x 0.75mm depth.

The flaws are made (by cutting) in the vessel's composite shell in the direction parallel to its principal axis (Fig. 13). As already mentioned, after the flaws are made the vessel is subjected to the standard cyclic test whose result is deemed positive if the vessel does not fail (loss of airtightness or burst).

a) b)

Fig. 13. Composite pressure vessels with designed flaws (cuts) (Błażejewski et al., 2008)

A pressure vessel for storing hydrogen at NWP = 207 bar (3000 psi) with 12 fibre Bragg gratings installed and two designed flaws in the form of cuts (Fig. 14) was subjected to a cyclic test at room temperature. The test consisted of over 4050 cycles in a pressure range of 20-258.5 bar (1,25 x NWP).

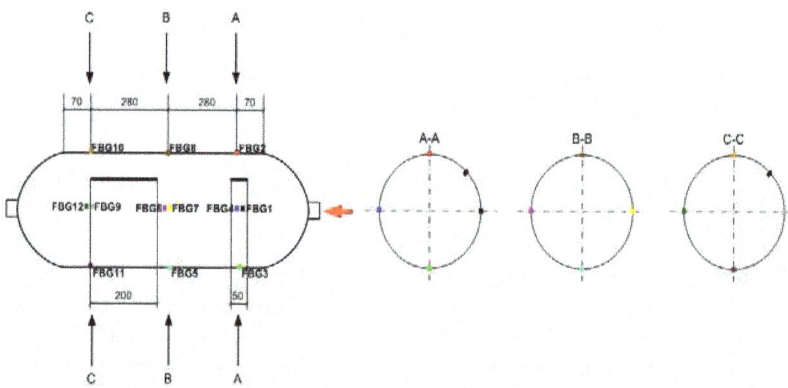

Fig. 14. Arrangement of designed flaws and sensors on tested pressure vessel (Błażejewski et al., 2008)

The measured local strains in a cross section with the shorter flaws (cross section A-A) are shown in Fig. 15. Figure 15a shows the minimum (ε_{min}) and maximum (ε_{max}) strains registered by sensors FBG1-FBG4 for pressures 20 and 285.bar. Fig. 15b shows the variation in strain amplitude (ε_{max} - ε_{min}) versus vessel load cycles. It should be noted that in the analyzed cross section as the number of load cycles decreases so does the strain values registered by sensors FBG1 and FBG2 located closest to the flaw. At the same time the amplitude of the strains registered by the other sensors increases. This is evidence of local defects accumulation in the load-bearing shell of the composite vessel and of the growth of the designed flaw.

a)

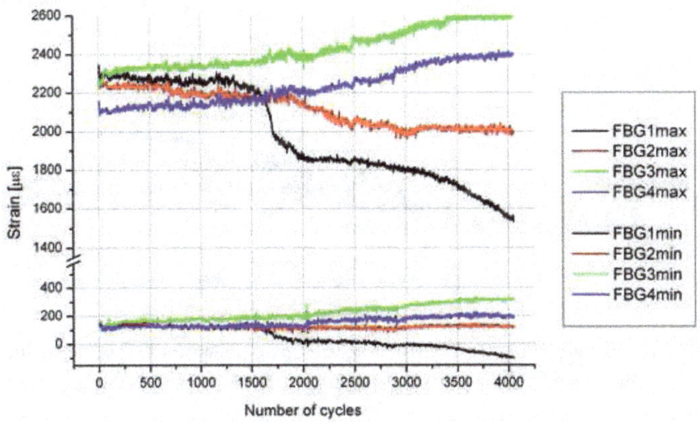

b)

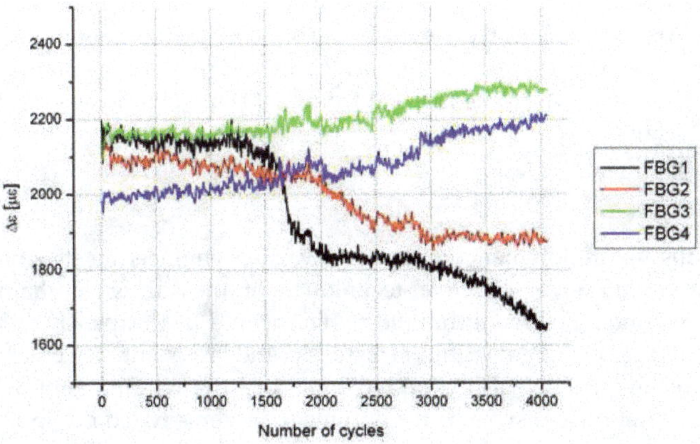

Fig. 15. Local strains of pressure vessel with designed flaws, measured by fibre Bragg gratings in cross section A-A (a) for pressure of 20 bar (min) and 258.5 bar (max), and amplitude of locally measured strains (b) versus number of vessel load (20-285.5 bar) cycles (Błażejewski et al., 2008)

The fluctuations of the maximum and minimum strain values are due to the specific character of the cyclic tests, i.e. due to the actual fluctuations of the maximum and minimum internal pressure around the fixed value (the switching pressure).

Figure 16 shows an analysis of the ABS coefficients for each of the FBG sensors installed on the tested pressure vessel, carried out in accordance with the description in sect. 5.2. It should be noted that as the number of pressure cycles increases so does the value of each of the analyzed ABS coefficients. This indicates the growth (accumulation) of defects in the tested vessel.

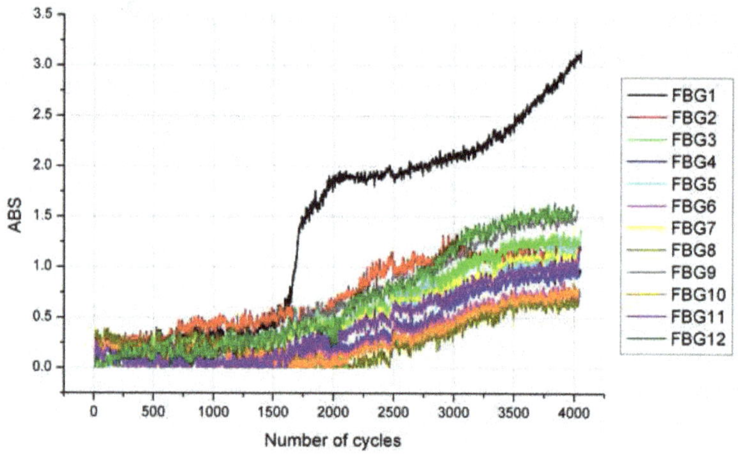

Fig. 16. Analysis of ABS coefficients versus number of pressure cycles for all FBG sensors installed on tested vessel with designed flaws (Błażejewski et al., 2008)

The largest changes were registered by sensors FBG1, FBG2, FBG11 and FBG 12, i.e. the sensors located closest to the introduced designed flaws. Moreover, one can see that the variation of the ABS coefficient, registered by sensor FBG1 after about 1600 cycles sharply increased during the next cycles. This indicates intensive local accumulation of defects, whose increase slowed down in the interval from ~2000 to ~3000 cycles, and then began to sharply increase again.

6. Conclusions

- The increasing use of polymer composites in various engineering structures necessitates the application of novel effective techniques of measuring mechanical quantities. Preferred are nondestructive techniques which would be inexpensive, reliable over the long service life of engineering structures and resistant to external influences. Measuring techniques are also expected to satisfy the requirements which structural health monitoring (SHM) systems have to meet. Systems based on optical fibre sensors (OFS) satisfy most of the requirements.
- The OFS technology offers the possibility of implementing "nervous systems" for infrastructure elements that allow health and damage assessment. Provided proper procedures are followed, optical fibre sensors can be located between the reinforcement fibres inside the composite material without affecting the mechanical properties of the composite element. Optical fibre based sensors can be used for measurements in adverse conditions (e.g. strong electromagnetic interference, high temperature and humidity) in which conventional methods cannot be used. Measuring heads based on fibre Bragg gratings have proved to be particularly suitable for composite structures. Also interferometric sensors (e.g. EFPI, SOFO®) can be used to some extent for this purpose.

- The increased interest in the novel measuring methods is also due to the many advantages of OFS over the conventional measuring methods. In practical applications measuring heads based on fibre Bragg gratings have been used most often.
- A major advantage of optical fibre sensors is that they can be used to build SHM systems comprising several thousand measuring points. This is particularly valuable when the object to be monitored is large or when it is not possible to determine the weak (critical) points in its structure. Thanks to the installation of many sensors the calculation methods can be simplified and the possibility that important information concerning the operational safety will be overlooked can be reduced. In addition, such systems are lightweight (in comparison with, e.g., strain gauge systems) and require little cabling.
- OFS can be used for on-line structural health monitoring. A major field of their application are extremely strained pressure vessels for fuel gas (methane, hydrogen). It was demonstrated that the OFS system is capable of detecting and controlling the growth of defects arising in the structure of composite materials whereby the safe service life of pressure vessels can be determined. Thanks to the way in which optical fibre sensors are installed it is possible to diagnose the health of structures during both their manufacture and long service life.
- Exemplary applications to building elements (reinforced concrete beams), composite aerospace structures and new generation of composites stimulated with a magnetic field (SGMM composites) were described.

7. References

Balageas, D., Fritzen, C. i Guemes, A. (2006). *Structural Health Monitoring.* London/Newport Beach, Great Britain/USA: ISTE.

Błażejewski, W., Czulak, A., Gasior, P., Kaleta, J. i Mech, R. (2010). SMART composite high pressure vessels with integrated optical fiber sensors. *Sensors and Smart Structures Technologies for Civil, Mechanical, and Aerospace Systems 2010, Proc. of SPIE Vol. 7647 764712-1.* San Diego.

Błażejewski, W., Gąsior, P., Kaleta, J. i Sankowska, A. (2007). Optical fiber sensors integrated with composite material based constructions. *Lightguides and Their Applications III, Proc. of SPIE 66081L-1-66081L-10.*

Błażejewski, W., Gąsior, P., Kaleta, J. i Sankowska, A. (2008). Optical Fiber Sensors as NDT methods for strain state monitoring of high pressure composites vessels. Comparison of different types of OFS. *8th International Conference on Durability of Composite Systems - Duracosys.* Porto.

Childers, B., M.E., F., Allison, S., Moore, T., Hare, D., Batten, C. i inni. (2001). Use of 3000 Bragg grating strain sensors distributed on four eight-meter optical fibers during static load tests of a composite structure. *Smart Structures and Materials 2001, Proceedings of the SPIE Conference.* San Diego.

Degrieck, J., De Waele, W. i Verleysen, P. (2001). Monitoring of fibre reinforced composites with embedded optical fibre Bragg sensors, with application to filament wound pressure vessels. *NDT&E International, 34,* pp. 289-296.

Diamanti, K., Hodgkinson, J. i Soutis, C. (2002). Damage detection of composite laminates using PZT generated Lamb waves. *Proc. of the First European Workshop on Structural Health Monitoring,* (pp. 398-405). Paris.

Draft ECE Compressed Gaseous Hydrogen Regulation, R. 1. (2003).

Foedinger, R., Rea, D., Sirkis, J., Baldwin, C., Troll, J., Grande, R. i inni. (1999). Embedded fiber optic sensor arrays for structural health monitoring of filament wound composite pressure vessels. *Smart Structures and Materials 1999: Sensory Phenomena and Measurement Instrumentation for Smart Structures and Materials, Proc.of SPIE Vol. 3670*, (pp. 289-301). Newport Beach.

Frias, C., Faria, H., Frazão, O., Vieira, P. i Marques, A. (2010). Manufacturing and testing composite overwrapped pressure vessels with embedded sensors. *Materials and Design* (31), pp. 4016–4022.

Gąsior, P., Kaleta, J. i Sankowska, A. (2007). Optical fiber sensors in health monitoring of composite high pressure vessels for hydrogen. *Optical Measurement Systems for Industrial Inspection V, Proc. of SPIE 66163G-1-66163G-10, 6616*. Munich.

Gąsior, P., Kaleta, J., Lewandowski, D. i Mech, R. (2009). Fiber Bragg gratings application for strain measurements of Terfenol-D based composites. *SMART'09: Proceedings of the IV ECCOMAS thematic conference on Smart Structures and Materials*. Porto.

Glisic, B. i Inaudi, D. (2004). Health monitoring of a full composite CNG tanks using long-gage fiber optic sensors. *11th SPIE's Annual International Symposium on Smart Structures and Materials, Proc. of Spie Vol. 5384-7*. San Diego.

Glisic, B. i Inaudi, D. (2007). *Fibre Optic Methods for Structural Health Monitoring*. John Wiley and Sons Ltd.

Hao, J., Leng, J. i Wei, Z. (2007). Non-destructive Evaluation of Composite Pressure Vessel by Using FBG Sensors. *Chinese Journal of Aeronautics, 20*(2), pp. 120-123.

Hernández-Moreno, H., Collombet, F., Douchin, B., Choqueuse, D. i Davies, P. (2009). Entire Life Time Monitoring of Filament Wound Composite Cylinders Using Bragg Grating Sensors III In-Service External Pressure Loading. *Appl Compos Mater*(16), pp. 135–147.

ISO/DIS15869.3. (2008). *Gaseous hydrogen and hydrogen blends – Land vehicle fuel tanks*.

Kalamkarov, A. L., Fitzgerald, S., MacDonald, D. i Georgiade, A. (1999). On the processing and evaluation of pultruded smart composites. *Composites*, B(30), pp. 753-763.

Kaleta, J., Lewandowski, D., Mech, R. i Gąsior, P. (2009). Magnetomechanical properties of Terfenol-D powder composites. *Solid State Phenomena*, 154, pp. 35-40.

Kang, D., Kim, C. i Kim, C. (2006). The embedment of fiber Bragg grating sensors into filament wound pressure tanks considering multiplexing. *NDT&E International, 39*, pp. 109–116.

Kuang, K., Kenny, R., Whelan, M., Cantwell, W. i Chalker, P. (2001). Embedded fibre Bragg grating sensors in advanced composite materials. *Composites Science and Technology*(61), pp. 1379–1387.

Lee, D., Lee, J. i Yun, S. (1995). The mechnaical characterristics of smart composite structures with embedded optical fiber sensors. *Composite Structure*(32), pp. 39-50.

Leng, J. i Asundi, A. (2003, November). Structural health monitoring of smart composite materials by using EFPI and FBG sensors. *Sensors and Actuators, A*(103), pp. 330-340

Mair, G. (2007). *Potentials and Procedures of the Probabilistic Approach*. Available at: http://www.storhy.net/pdf/PotentialProbabilisticApproach_Mair.pdf

Majumder, M., Gangopadhyay, T., Chakraborty, A., Dasgupta, K. i Bhattacharya, D. (2008). Fibre Bragg gratings in structural health monitoring - Present status and applications. *Sensors and Actuators, A*(147), pp. 150–164.

ECE Regulation No. 110 (2008).

Schwartz, M. (2002). *Encyclopedia of Smart Materials.* John Wiley & Sons Ltd.

Sloan, J. (2010, March). *Carbon Fiber: UP!* Available at: http://www.compositesworld.com/articles/carbon-fiber-up

Sulatisky, M., Hay, B. i Mourre, D. (2010). Damage Detection in High-Pressure Cylinders. *18th World Hydrogen Energy Conference (WHEC 2010).* Essen.

Takahashi, I., Sekine, K., Takeya, H., Iwahori, Y., Takeda, N. i Koshioka, Y. (2010). Life cycle structural health monitoring of airframe structures by strain mapping using FBG sensors. *Sensors and Smart Structures Technologies for Civil, Mechanical, and Aerospace Systems 2010, Proc. of SPIE Vol. 7647 764723-1.* San Diego.

Udd, E., Winz, M., Kreger, S. i Heider, D. (2005). Failure Mechanisms of Fiber Optic Sensors Placed in Composite Materials. *Smart Structures and Materials 2005: Smart Sensor Technology and Measurement Systems, Proc. of Spie Proceedings Vol. 5758,* (pp. 409-416). San Diego.

Yin Lo, C., Wing Or, S. i Lai Wa Chan, H. (2006). Large Magnetostriction in Epoxy-Bonded Terfenol-D Continuous-Fiber Composite With [112] Crystallographic Orientation. *IEEE Transactions on Magnetics, 42*(10), pp. 3111-3113.

Yu, F. i Yin, S. (2002). *Fiber Optic Sensors.* New York: Marcel Dekker Inc.

Zhou, Z., Ou, J. P. i Wang, B. (2003). Smart FRP-OFGB bars and their application in reinforced concrete beams. *Proceedings of 1st International Conference on Structural Health Monitoring and Intelligent Infrastructure,* (pp. 861-866). Tokyo.

4

Design, Manufacture and Performance Evaluation of Non-Asbestos Sealing Composites

Boqin Gu, Ye Chen and Jianfeng Zhou
Nanjing University of Technology,
China

1. Introduction

Asbestos was once considered to be a "miracle mineral". This naturally occurring silicate has many desirable characteristics, including resistance to fire, heat, and corrosion. It is strong, durable and flexible. Asbestos is inexpensive because it is available in abundant quantities. Its versatility has led to its use as a component of a variety of products in numerous industries (American Academy of Actuaries, 2007). The development of fiber-reinforced elastomer gaskets began in the 1880s and led to the patent application for Klingerit in 1904 (Piringer & Rustemeyer, 2004). Since more than one hundred years ago, these kinds of non-metallic gaskets, which were made of compressed asbestos fibers (CAF) materials, have been the most widely used sealing elements with a maximal yield.

Up till the 1970s, the health hazards of asbestos were recognized. Several diseases have been linked to asbestos exposure, including mesothelioma, lung cancer, other cancers, asbestosis, and pleural changes (American Academy of Actuaries, 2007). Due to the restriction on the use of asbestos, the pressure was on scientists and engineers to develop non-asbestos gasket material replacements, and asbestos was replaced by alternative fibers and fillers. These alternatives were employed in an attempt to replicate the product properties of the former CAF materials. These substitutes were developed using reinforcing fibers like aramid, glass and carbon fibers to achieve high strength, and additives like inorganic materials (e.g., clays, precipitated silica, graphite etc.). Since the 1990s, a great number of worldwide famous sealing materials manufacturers have put significant efforts onto the development of a variety of novel non-asbestos sealing composites (NASC), such as Garlock in USA, Klinger in Austria, Kempchen in Germany, and Valqua and Pilar in Japan. Collaborating with some international organizations and research institutions, they conducted a series of experimental and application researches, and obtained many valuable results (Payne & Bazergui, 1990). These investigations have laid a foundation for further studies on performance evaluations and industrial applications.

Aramid was used as reinforcing fibers in the earliest non-asbestos gasket materials, because it provided processing advantages, especially during calendaring. However, it is very expensive and has poor thermal stability. Most commonly used fibers and fillers in non-asbestos gasket materials do not achieve the outstanding stability as CAF materials do when the binding elastomers become aged. More recently, novel formulations and manufacturing processes have been developed based on cheap reinforcing fibers, the hybrid effect of different reinforcing fibers or an adapted material structure. These non-asbestos gasket materials not only are significantly cheaper but also have a better performance at high temperatures. Gu & Chen developed two kinds of sealing composite materials. One was reinforced with aramid and pre-oxidized hybrid fibers and prepared by molding preparation method and the other was reinforced with carbon and glass hybrid fibers and manufactured by using calendar preparation method. The effect of different surface treatment methods of the fibers on the heat resistance of the composite materials was studied. The optimum prescriptions of the composites were obtained by regression design method (Gu & Chen, 2007; Chen & Gu, 2008). Gao & Chen carried out explorative investigations on the preparation method of nanometer calcium carbonate filled modifications of rubber-based sealing composite materials and on the influence of nanometer filled modifications on the mechanical properties of the prepared gasket materials (Gao & Chen, 2009). A novel material concept for compressed fiber materials was proposed to significantly decelerate the ageing of elastomer bound gasket materials by the use of special elastomers and an adapted material structure, i.e., the multi-layer structure. The steam testing, as opposed to the standard gasket testing, has been used to demonstrate this improvement (Piringer & Rustemeyer, 2004).

It is essential that gasket materials possess good mechanical performances and sealability. Mechanical performances include compressibility, resilience and stress relaxation property. They are usually used to evaluate abilities to cause gasket material deformation into flange face irregularities under assembly condition, to hold sealing surfaces of joints contacted tightly under internal pressures, and of stress retention at high temperatures, respectively. Sealability is a comprehensive performance of gaskets and indicates the ability to prevent the sealed fluid from leakage through the joints. Sealability can also be called tightness and can be measured quantitatively by the leakage rate.

In the mid 1980s, the American Pressure Vessel Research Committee (PVRC) set up "Room-temperature Mechanical Test Procedure" and "Hot Mechanical Test Procedure" for estimating the probable long-term performance and potential fire survivability of non-asbestos gaskets to guide the qualification and selection of non-asbestos spiral wound, jacketed and sheet gaskets for petroleum and petrochemical plant services in the range of 423-866K. Procedures, test fixtures, and typical test results for several process plant gaskets were discussed. An aged exposure parameter was introduced that correlates cumulative damage with exposure time and temperature for materials that degrade over time (Payne et al., 1989a; Payne et al., 1989b; Payne & Bazergui, 1990). The change in properties of some compressed sheet gasket materials subjected to temperature exposure for periods of up to one year was investigated by Marchand & Derenne. This resulted in a better understanding of the long term effect of thermal degradation on the properties of elastomeric sheet gasket materials. An improved qualification protocol based on the obtainment of the thermal endurance graph of a sheet gasket material was proposed for the extrapolation of the long term service temperature (Marchand & Derenne, 1996). Tsuji et al. researched experimentally the effect of aging time on sealing performance of non-asbestos spiral wound gaskets at elevated temperatures under either the stress controlled condition or the strain controlled condition. The tightness parameter T_p at different elevated temperatures was obtained. The results indicated that the non-asbestos spiral wound gasket had the same sealing performance as the substitute for the asbestos spiral wound gasket between 483 and

693 K (Tsuji et al., 2004). Xie et al. investigated the compressive and resilient performances and relaxation property of the developed compressed non-asbestos fiber reinforced rubber sheet gasket materials, and discussed in detail the effect of the non-asbestos fiber on the properties of the sheet gasket materials (Xie & Cai, 2002; Xie & Xie, 2004).

Systematic researches on gasket performances and their characterizations have been carried out in the Fluid Engineering and Sealing Technology Laboratory at Nanjing University of Technology since the 1980s. The compressive-resilient performance, creep and stress relaxation properties, and leakage behaviour of some types of gaskets were investigated. The representation of gasket performances was put forward, and the formulae for expressing gasket performances were obtained by means of the proposed models and regression analysis of experimental data (Gu et al., 1999; Gu et al., 2000; Gu et al., 2001; Gu, 2002; Zhu et al., 2007; Zhu et al., 2008; Gu et al., 2010). A novel tightness concept was presented, and the tightness analysis and design methods of gasket sealing joints based on the criterion of the maximum allowable leakage rate were developed. Relatively accurate predictions of leakage rates of some gasket sealing connections were also obtained (Gu & Zhu, 1988; Gu & Huang, 1997; Gu et al., 2004; Gu & Chen, 2006; Gu et al., 2007a; Gu et al., 2007b).

Up to now, much work has been fulfilled on the development, performance evaluation and engineering applications of the sealing composites reinforced with non-asbestos fibers, and many results have been achieved. However, relatively comprehensive and systematic reports on design, manufacture and performance evaluation of NASC are still very scarce, especially on the characterization of micro structural parameters, and the mechanical analysis and macro performance prediction of these materials according to the theories of micromechanics and viscoelastic mechanics.

In this chapter, manufacturing technology, surface treatment methods for reinforcing fibers, and formulation design methods of NASC are introduced. Measurements and characterizations of some micro structural parameters of fibers including their aspect ratio, orientation and distribution are investigated. A micromechanical model of single fiber cylindrical cell and a model of compressive type single-fiber cell are established, on the basis of which the methods are proposed for evaluating macro-mechanical performances of NASC, such as tension, compression, and stress relaxation. A leakage model for predicting non-asbestos gasket leakage rates is presented and verified experimentally. Furthermore, the performances of some developed NASC are also evaluated.

2. Manufacturing technology of NASC

2.1 Molding process

The molding process of NASC is similar to that of the traditional rubber based composites, as shown in Fig. 1. The plastication should be carried out using a mill mixer or an internal mixer, to improve plasticity of raw rubber. Before this step, raw rubber may be roasted in a hot chamber within 333-343 K to decrease its hardness and improve manufacturability. The roll temperature of the mill mixer should be controlled below 343 K, and the roller space is about 0.5-1 mm.

The gross rubber is obtained in the mixing step where all accessory ingredients and filling materials are evenly dispersed in broken-down rubber by extrusion and shearing actions of rollers repeatedly.

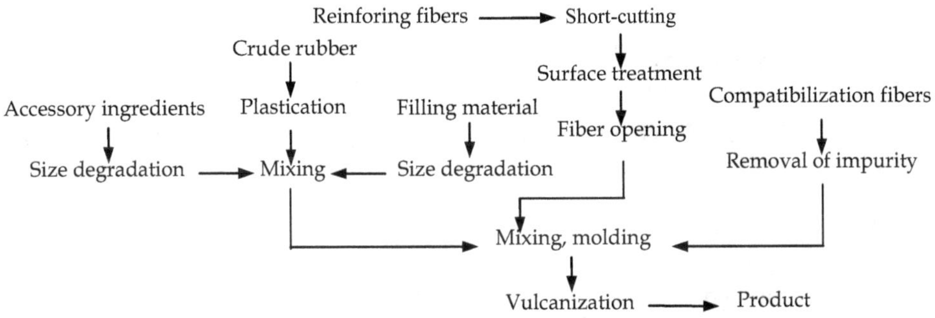

Fig. 1. A molding process of NASC

The mixing step can also be carried out in a mill mixer or an internal mixer. During mixing, the reinforcing fibers can be evenly dispersed in gross rubber, and the aspect ratio of the fibers is reduced to an appropriate value. In the vulcanization step, the rubber and a vulcanizing agent will chemically crosslink at sulfurizing pressures and temperatures. The effect of sulfurization parameters on the transverse tensile strength of aramid fiber reinforced NASC is illustrated in Fig. 2.

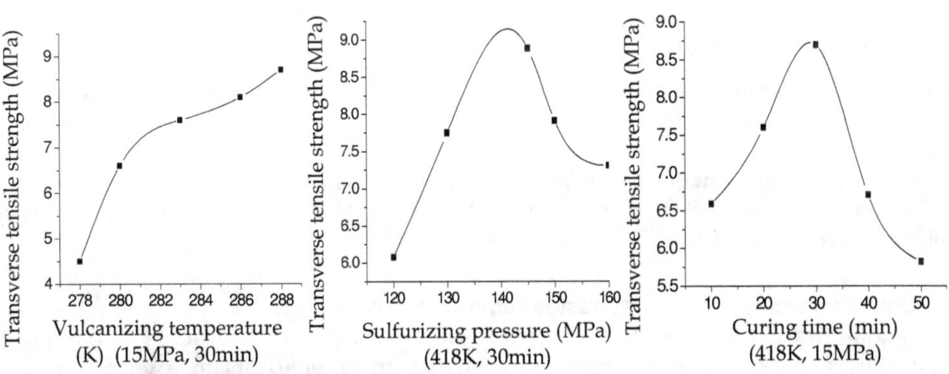

Fig. 2. The effect of sulfurization parameters on transverse tensile strength of aramid fiber reinforced NASC

2.2 Calender process

A calender process flow diagram is illustrated in Fig. 3. In this process, rubber is dissolved by an organic solvent and then mixed with other constituents. The NASC sheet is molded in a roller-type calender.

Preparation of rubber paste is fulfilled in three steps including plasticizing, mixing and dissolving. The mixing and plasticizing steps are the same as those in the molding process. The gross rubber should be separated into small pieces, and then dissolved into the rubber paste by organic solvents, such as gasoline, benzene, methylbenzene and ethyl acetate.

In the mixing step, rubber paste, filling material and reinforcing fibers are mixed, and good quality gross rubber particles with suitable diameter, humidity and hardness are prepared.

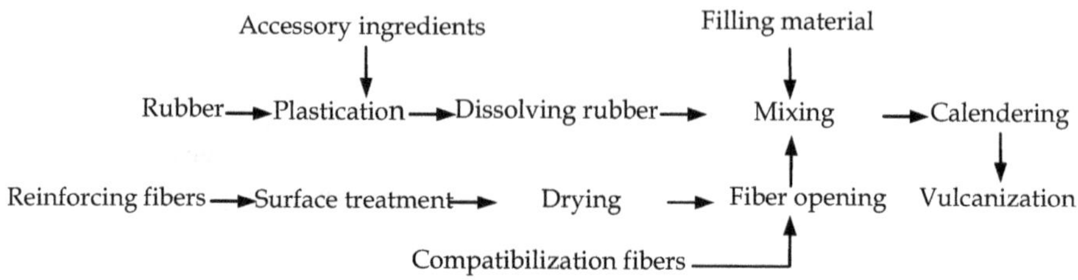

Fig. 3. A calender process of NASC

The calender process is carried out with two rollers rotating in the opposite directions. Gross rubber particles are fed into the clearance between rollers, and manufactured into NASC sheets. The operating principle of a roller-type calender is shown in Fig. 4.

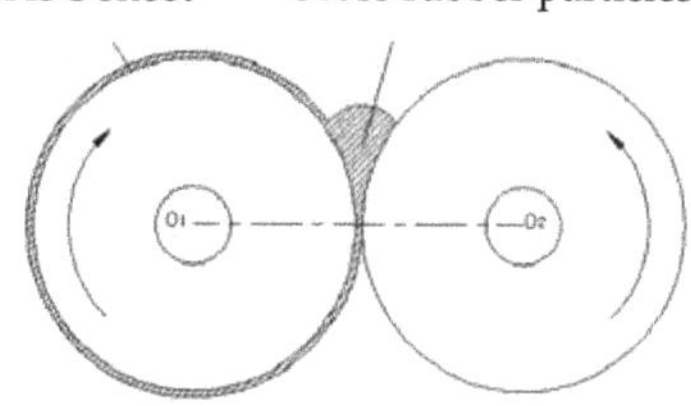

Fig. 4. The operating principle of a roller-type calender

The operating parameters of the calender process mainly include speed ratio of rollers, calender speed (roller surface speed), roller temperature, feed speed, roller clearance and extrusion pressure.

The two rollers are usually thermally stabilized by steam and cool water, respectively. Accordingly, they are called hot roller and cold roller. The appropriate speed ratio of two rollers is in the range of 1.04-1.06.

2.3 Surface treatment of non-asbestos reinforcing short fibers

The surface activity of most non-asbestos fibers is very poor, which leads to the lower wettability between fibers and rubbers. Therefore, in order to obtain a good interface bonding strength, the reinforcing fibers must be pretreated by suitable surface treatment methods. The commonly used surface treatment methods include surface activating treatment and dipping treatment.

Some NASC were developed by the molding process with aramid short fiber as reinforcing fiber, and NBR and NR as elastic binding material. Three surface treatment methods were adopted to pretreat aramid fiber; they are RFL latex dipping, HRH binding agent treatment and silane coupling agent dipping. The results of SEM observation of the developed NASC are shown in Fig. 5.

It can be seen that the surfaces of the aramid fibers without treatment are very smooth, and there are only a few rubber particles adhering on the fiber surfaces. After surface treatments, the adhesion effect between fiber and rubber matrix was obviously improved.

Carbon short fiber can be treated by coupling agent dipping process or epoxide resin coating process after pretreating by air oxidation process or low temperature plasma process. The transverse tensile strength of NASC developed by the calender process was tested after aging treatment at 623 K, and the results are shown in Fig. 6. It can be seen that surface treatments have improved the reinforcing effects of carbon fiber.

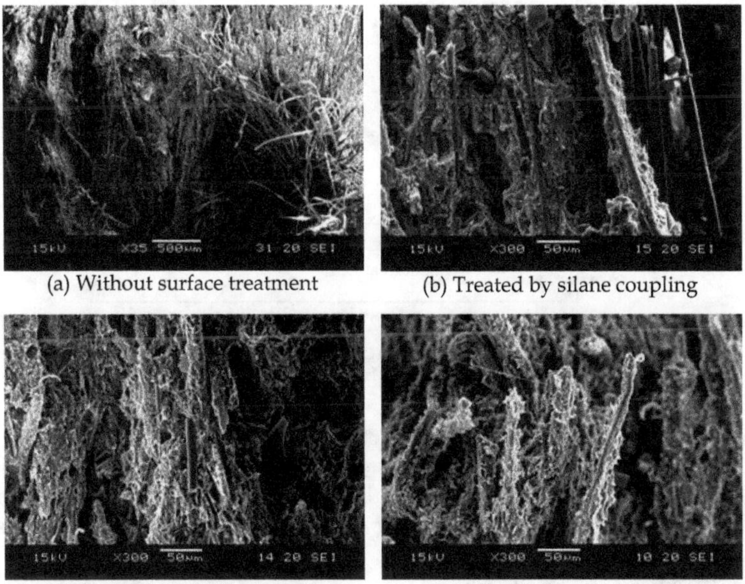

(a) Without surface treatment (b) Treated by silane coupling

(c) Treated by HRH binding (d) Treated by RFL latex

Fig. 5. SEM photos of transverse tensile fracture sections of aramid fiber reinforced NASC

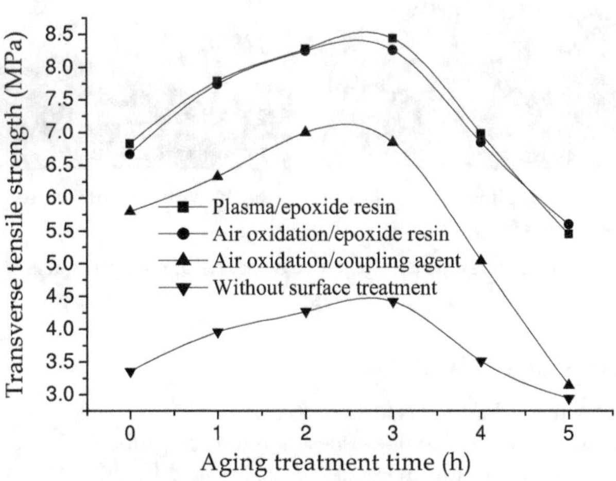

Fig. 6. Transverse tensile strength of carbon fiber reinforced NASC after aging at 623 K

The treatment effect of glass fiber by coupling agent is also distinctive, as illustrated in Fig. 7. The glass fiber should be firstly dipped in a silicon coupling agent water solution for pre-treatment, and then immersed into an epoxide resin methylbenzene solution (2wt%) for further coating treatment. Finally, they must be dried at 423 K to solidify impregnated layer.

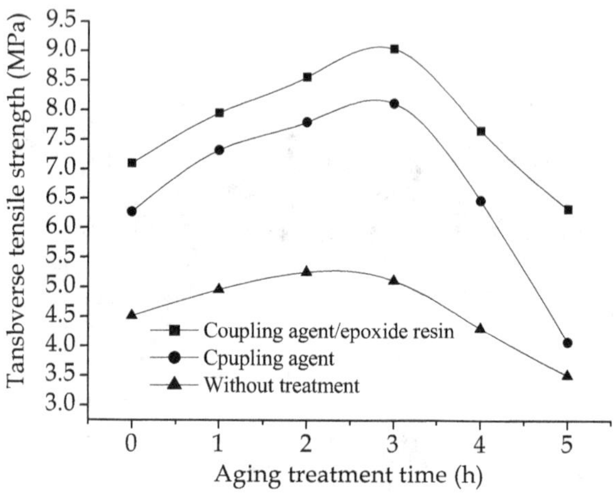

Fig. 7. Transverse tensile strength of glass fiber reinforced NASC after aging at 623 K

The SEM photos of transverse tensile fracture sections of carbon/glass hybrid fiber reinforced NASC aged at 623 K for 5 hours are shown in Fig. 8. It can be seen that the intertwined and the rubber coating conditions of fibres are obviously improved after the surface treatment of the fibers by the coupling agent/epoxide resin coating process.

(a) Treated by coupling agent/epoxide resin coating process

(b) Without surface treatment

Fig. 8. SEM photos of transverse tensile fracture sections of carbon/glass hybrid fiber reinforced NASC

3. Formulation design of NASC

3.1 Selection of raw materials for NASC

The main constituents of NASC are non-asbestos reinforcing fibers, elastic binding materials, filling materials and chemical additives. There are great differences of the physicochemical properties between non-asbestos fibers and asbestos fibers. Therefore, in the design of NASC, the effects of the properties and the content of the non-asbestos reinforcing fibers on the performances and manufacturability of the NASC should be paid more attention.

The suitable non-asbestos reinforcing fibers are usually selected according to the performance requirements of NASC. Heat resistance, aspect ratio, medium-resistance, compatibility with rubber, performance-price ratio and source of the reinforcing fibers should be considered, too. The commonly used non-asbestos reinforcing fibers include carbon fiber, aramid fiber, glass fiber, mineral wool, plant fiber, and so on.

Elastic binding material binds reinforcing fibers and other filling materials together, makes up elastic network structure, and provides compression-resilience and medium-resistance performances of NASC. Natural rubber, nitrile rubber, styrene-butadiene rubber, neoprene rubber, ethylene propylene rubber, fluorine rubber and silicone rubber can all be used as the elastic binding materials of NASC. In consideration of the product performances and price, natural rubber and nitrile rubber are preferably selected.

The filling materials of rubber products include compatibilization filler, reinforcing filler and functional filler. Commonly used filling materials include brucite fiber or sepiolite fiber, carbon black, carbonate, sulphate, metallic oxides, and silicon oxides. In NASC, large grained filler must pair up with short grained filler to maintain proper inter space among rubber bulk material, fibers and filler particles, and ensure sufficient combination between rubber and filler particles.

The selectable accessory ingredients include vulcanizing agents, vulcanization accelerators, vulcanizing activators, anti-aging agents, plasticizing agents and colouring agents etc.. The mixture of dibenzothiazyl disulfide (40 wt%) and tetramethylthiuram disulfide (60 wt%) can be used as a vulcanization accelerator. Zinc oxide (3-5 wt%) and stearic acid (0.5-2 wt%) can be selected as a vulcanizing activator. N-phenyl-n-isopropyl ursol (4010NA) is usually used as the anti-aging agent of rubbers, and the amount is about 1-4wt% (related to rubber weight). Plasticizing agents, colouring agents, solvent, blowing agents, dusting agents and reinforcing resins etc. can also be adopted as accessory ingredients.

3.2 Formulation design of NASC

The uniform design method can be used to further decrease test number in formulation optimization. The approximate formulation of composites can be obtained by uniform design, but the principal and subordinate factors which affect material performances cannot be analyzed by the design table. Furthermore, the regression design can also be adopted to optimize the formulation of NASC. In the mixing regression design, a small quantity of test points will be selected to obtain enough experimental data. According to these data, the regression equation can be derived from the relationship between the test index and the different constituent contents. Finally, the optimization point can be obtained.

3.2.1 Formulation design of aramid fiber reinforced NASC

The formulation design of the aramid fiber reinforced NASC can be defined as the design problem with three components which include reinforcing fiber (aramid fiber), elastic binding material (NBR/NR) and filling material (compatibilization fiber). In consideration of the attribute for processing of the material, the content of components in NASC is limited by Eq. (1).

$$7\% \le x_1 \le 22\%, \quad 40\% \le x_2 \le 55\%, \quad 10\% \le x_3 \le 25\% \tag{1}$$

where x_1, x_2 and x_3 are the weight contents of aramid fiber, compatibilization fiber (brucite fiber or sepiolite fiber) and elastic biding material, respectively.

The contents of other ingredients, such as accessory ingredient, can be determined according to the contents of three components mentioned above. By selecting 5 factor levels for each component, the uniform design scheme for test points is listed in Table 1.

Test points	x_1 (wt%)	x_2 (wt%)	x_3 (wt%)
①	7	43	19
②	10	49	25
③	13	55	17
④	16	40	23
⑤	19	46	15
⑥	22	52	21

Table 1. Experiment scheme according to the uniform design method

According to the test points in Table 1, aramid fiber reinforced materials were developed by the molding process. It can be seen in Table 2 that the transverse tensile strength of the materials developed according to , and test points meets the strength requirement (≥7.0 MPa). The strength of the materials prepared according to and test points are obviously higher than that of scheme . Too much reinforcing fiber in the material of formulation and led to the poor processing manufacturability and low production efficiency, and therefore, formulation was chosen to be the preliminary scheme.

Test points	Transverse tensile strength (MPa)
①	3.55
②	5.26
③	8.81
④	6.75
⑤	10.69
⑥	10.87

Table 2. Transverse tensile strength of aramid fiber reinforced NASC

Results of the single factor experiment about the usage of elastic binding material and aramid fiber are given in Tables 3 and 4, respectively. The aramid fiber and the rubber contents are suggested to be 13 wt% and 20 wt%, respectively.

Performances	Rubber content (wt%)				
	15%	17%	20%	23%	26%
Transverse tensile strength (MPa)	9.22	8.89	8.7	6.87	6.54
Compression rate (%)	4.75	10.06	11.2	13.66	11.4
Resilience rate (%)	43.20	52.15	52.23	51.04	45.7
Stress relaxation rate (%)	28.9	29.4	32.35	37.06	36.15

Table 3. The effect of rubber content on performances of NASC (The amount of aramid fiber is 13 wt%)

Performances	Aramid fiber content (wt%)			
	7%	10%	13%	15%
Transverse tensile strength (MPa)	6.57	7.77	8.70	8.86
Compression rate (%)	15.40	11.36	11.20	9.08
Resilience rate (%)	37.78	48.48	52.23	50.36
Stress relaxation rate (%)	41.42	37.80	32.35	29.42

Table 4. The effect of aramid fiber content on performances of NASC (The amount of elastic binding material is 20 wt%)

The optimization formulation of aramid fiber reinforced NASC is listed in Table 5.

Gradient	Weight content (wt%)	Gradient		Weight content (wt%)
NBR-26	13.0	Filling materials	Calcium carbonate	7.9
NR	7.0		Kaolin clay	3.4
Aramid fiber	13.0		White carbon black	3.4
Compatibilization fiber	45.3		Soap stone powder	2.3
Vulcanizing agent	2.5	Other accessory ingredients		1.8
Curing catalyst	0.4			

Table 5. Formulation of aramid fiber reinforced NASC

3.2.2 Formulation design of carbon/glass hybrid fiber reinforced NASC

The regression design of carbon/glass hybrid fiber reinforced NASC can be defined as the design problem with three components which include reinforcing hybrid fiber (carbon/glass hybrid fiber), elastic binding material (NBR) and filling material (compatibilization fiber and other stuffings). In consideration of the attribute for processing of the material, the component contents in NASC are in the following ranges:

$$0.12 \le X_1 \le 0.25,\ 0.4 \le X_2 \le 0.8,\ 0.1 \le X_3 \le 0.5 \tag{2}$$

where X_1, X_2 and X_3 are the contents of the elastic binding material, the reinforcing fiber and the filling material, respectively.

The stress relaxation rate y of NASC after ageing treatment at 573 K is used as an evaluating indicator, and the regression design of NASC is to obtain the regression equation by the experiments under the following conditions:

$$a_i \le X_i \le b_i, (i = 1,2,3) \quad \text{and} \quad X_1 + X_2 + X_3 = 1 \tag{3}$$

where a_i and b_i are the constraint conditions described in Eq. (2).

A symmetrical complex Z is established. The coordinate system $x_1x_2x_3$ represents the actual design space of the components of NASC, and the coordinate system $z_1z_2z_3$ represents the encoding space. Accordingly, the relationship between Z and X can be expressed by:

$$X_1 = a_1 + (b_1 - a_1)z_1/B,\quad X_2 = a_1 + (b_2 - a_2)z_2/B,\quad X_3 = 1 - (X_1 + X_2) \tag{4}$$

where B=max(z_{ij}), i is the test number, and j the variable number. The coordinates of the actual test points are listed in Table 6, where the stress relaxation rates of the materials after ageing treatment at 573 K are also listed.

The stress relaxation rate of NASC after ageing at 573 K can be expressed by the coordinates of the encoding point, and the regression equation holds:

$$\hat{y} = 63z_1 + 77z_2 + 120.8z_3 - 137.6z_1z_2 + 0.49z_1z_3 - 261.6z_2z_3 \tag{5}$$

where $\hat{y}$ is the predicted value of the stress relaxation rate of the material, and z_1, z_2 and z_3 are the coordinates of the encoding point.

NO.	Encoding content of components			Actual content of components			Relaxation rate y (%)
	z_1	z_2	z_3	X_1	X_2	X_3	
1	0	1/2	1/2	0.12	0.667	0.213	51.6
2	1/2	0	1/2	0.207	0.4	0.393	32.9
3	1/2	1/2	0	0.207	0.667	0.127	43.9
4	1/4	1/4	1/2	0.163	0.533	0.303	34.6
5	1/4	1/2	1/4	0.163	0.667	0.17	42.3
6	1/2	1/4	1/4	0.207	0.533	0.26	31.3
7	3/4	1/8	1/8	0.25	0.467	0.283	58.4
8	1/8	3/4	1/8	0.142	0.8	0.058	52.4
9	1/8	1/8	3/4	0.142	0.467	0.392	43.1

Table 6. Experimental scheme and test results

The purpose of prescription optimization is to improve the heat resistance and the attribute for processing of NASC, to reduce the self-cost of the raw materials of the product, and to ensure the stress relaxation rate of the material satisfying the index prescribed in the product standard. The prescription optimization of NASC is to obtain the optimal values of z_1, z_2 and z_3 under the following conditions:

$$z_i \geq 0,\ \sum_{i=1}^{3} z_i = 1,\ z_1 = z_{1\min},\ z_2 = z_{2\min}, \hat{y} \leq 35\ (i = 1,2,3) \tag{6}$$

According to Eq. (5), the coordinates of the best theoretic design point of the contents of the components in NASC can be obtained, as z_1=0.38, z_2=0 and z_3=0.63. From Eq. (4), the formulation is x_1=0.186, x_2=0.4 and x_3=0.414.

Some NASC were prepared when the weight ratio (carbon fiber to glass fiber) equalled to 4:0, 3:1, 2:2, 1:2, 1:3 and 0:4 respectively. The experimental results of the transverse tensile strength and the stress relaxation rate of these materials after ageing treatment at 573 K for 5 hours are shown in Fig. 9. It can be found that the strength of the material is the highest and the relaxation rate is relatively low when the weight ratio is 2:3 (weight content of carbon fiber in hybrid fibers is 40%).

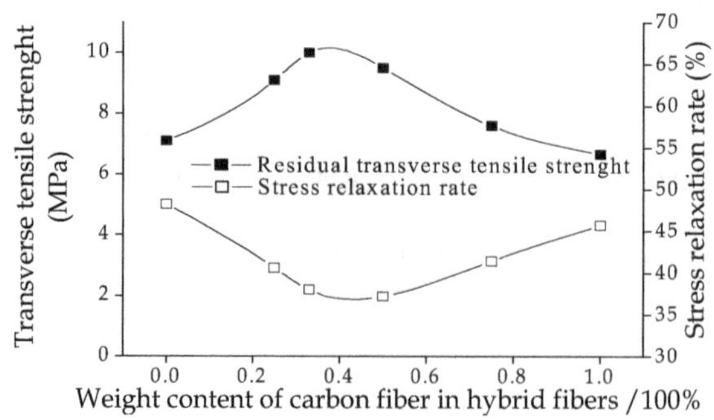

Fig. 9. The effect of carbon fiber content in hybrid fibers on transverse tensile strength and stress relaxation rate of NAFC

Therefore, the optimization formulation of the NASC reinforced with carbon/glass hybrid fibers was obtained, as listed in Table 7.

<table>
<tr><th>Component</th><th>Content (wt%)</th><th colspan="2">Component</th><th>Content (wt%)</th></tr>
<tr><td>NBR</td><td>18.6</td><td colspan="2">Accelerating agent</td><td>0.22</td></tr>
<tr><td>Carbon fiber</td><td>16.0</td><td colspan="2">Zinc white</td><td>0.48</td></tr>
<tr><td>Glass fiber</td><td>24.0</td><td colspan="2">Stearinic acid</td><td>0.18</td></tr>
<tr><td>Sepiolite fiber</td><td>10.3</td><td colspan="2">Antiaging protective</td><td>0.6</td></tr>
<tr><td>Phenolic resin</td><td>8.38</td><td rowspan="4">Filling material</td><td>Lime carbonate</td><td>6.31</td></tr>
<tr><td>Hyperfine graphite powder</td><td>7.44</td><td>Kaolin clay</td><td>3.2</td></tr>
<tr><td>Vulcanized agent</td><td>0.35</td><td>Soap stone powder</td><td>1.6</td></tr>
<tr><td></td><td></td><td>Carbon black</td><td>2.34</td></tr>
</table>

Table 7. Formulation of carbon/glass hybrid fibers reinforced NASC

4. Measurement and characterization of micro structures of NASC

4.1 Measurement and characterization of aspect ratio

The aspect ratio of short fiber determines the transfer and distribution of stress in material, and its distribution regularity can be obtained by measuring the radius and the length of fibers in the NASC products. Some samples of NASC were prepared. Using organic solvents, such as gasoline or toluene, to dissolve the matrix rubber, the short fibers different in size are left and dried. Considering the fiber radius is about micron, a digital camera with high-resolution is used to take pictures of fibers, and the length and radius of each short fiber are obtained by numerically analyzing the pictures. Then the distribution of aspect ratios can be obtained by counting the aspect ratios of 500 to 1000 fibers. The typical histograms of aspect ratios of aramid fibers can be seen in Fig. 10.

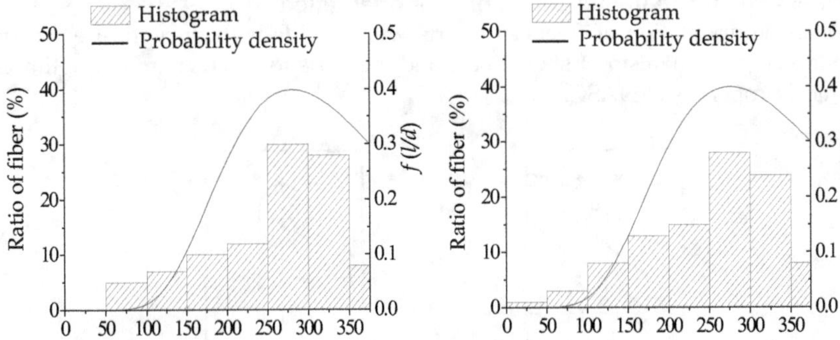

(a) Fiber content 5%

(b) Fiber content 10%

Fig. 10. Aspect ratio distribution of aramid fibers with the initial length of 4-6 mm

The distribution of aspect ratios in NASC is influenced by the initial aspect ratio, the rapture characteristic and the content of fibers, the matrix characteristic and the preparation process.

If the aspect ratio varies in a relatively small range, it can be characterized by the mean aspect ratio. If it distributes dispersedly, however, the characterization with a distribution function is appropriate. Furthermore, a histogram can also be used to express the distribution of aspect ratios without considering the dispersity of aspect ratios.
The log-normal density function is usually used to characterize the aspect ratios, as Eq. (7) expresses.

$$f(l/d) = \frac{1}{\sqrt{2\pi}} \exp\left[-\frac{\left[\ln(l/d) - a\right]^2}{2b^2} \right] \tag{7}$$

where $a = \sqrt{2\left[\ln(\bar{l}/d) - \ln(l_{\text{mod}}/d)\right]/3}$, $b = \left[2\ln(\bar{l}/d) + \ln(l_{\text{mod}}/d)\right]/3$, $\bar{l}/d$ is the mean aspect ratio and l_{mod}/d the maximum probability density of the aspect ratio.
The arithmetical mean of aspect ratios can be calculated by Eq. (8).

$$\bar{l}_s/d = \frac{1}{m}\sum_{n=1}^{m}(l_n/d) \tag{8}$$

where $\bar{l}_s/d$ is the average value of aspect ratios, m is the number of short fibers and l_n/d is the aspect ratio of the number n short fiber. The arithmetical mean aspect ratio can also be calculated by Eq. (9) if the histogram of aspect ratios is known.

$$\bar{l}_s/d = \sum p_i(l_i/d)/\sum p_i \tag{9}$$

where l_i/d is the aspect ratio of fibers in the interval i, and p_i is the ratio of the number of fibers whose aspect ratios are in the interval i to the total number of fibers.

4.2 Measurement and characterization of fiber orientation

Investigating the orientation of fibers in matrix will provide both theoretical explanations for the reinforcing mechanism of short fibers and the basis for prediction and optimization of mechanical properties of NASC.

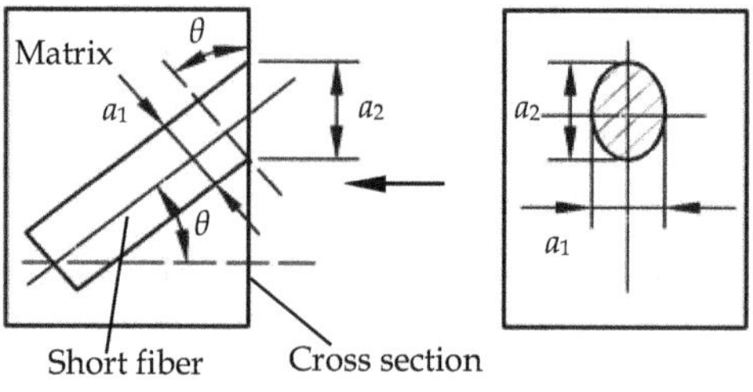

Fig. 11. A schematic diagram of the section-analysis method

A proprietary technology was developed to obtain the fiber orientation in matrix. Four steps are involved. Firstly, cut a rectangular block of material and hold it in position by a clamp in the slicer. Secondly, slice the cross section of the material block at -423 K using the slicer knife to machine a smooth end face, of which the normal direction should be along the rolling direction. Thirdly, visually observe the end face using a microscope associated with a high-resolution digital camera. Finally, analyze the photos and find out the configuration of the elliptical cross section of fibers. The orientation angle can be determined from the long axis and the short axis of the ellipse, as illustrated in Fig. 11.

The normal direction of the cross section is in accordance with the reference direction, and the orientation angle θ is the separation angle between the axial direction of a fiber and the reference direction. The relationship among θ and the long axis length a_1 and the short axis length a_2 is in the form

$$\theta = \arccos\left(a_1/a_2\right) \tag{10}$$

The orientation of a fiber can be characterized by a histogram, a distribution function or modified coefficients. The orientation expressed by an exponential density function $f(\theta)$ is

$$f(\theta) = \xi \exp(-\lambda\theta)/1 - \exp(-\pi\lambda/2) \tag{11}$$

where ξ is the fiber orientation parameter. In order to utilize the results of the fiber orientation characterization to predict the macro mechanical properties of NASC, the modified coefficients are introduced. The longitudinal modified coefficient $f_{\theta 1}$ can be calculated by Eq. (12) based on the distribution function.

$$f_{\theta 1} = \int_0^{\frac{\pi}{2}} f(\theta)\cos^2\theta \mathrm{d}\theta \tag{12}$$

$f_{\theta 1}$ is employed to evaluate the influence of fiber orientations on longitudinal mechanical property. The transversal mechanical property is evaluated by the transverse orientation coefficient $f_{\theta 2}$ which can also be obtained based on a distribution function or a histogram given by Eqs. (13) and (14), respectively.

$$f_{\theta 2} = \int_0^{\frac{\pi}{2}} f(\theta)\sin\theta\cos\theta \mathrm{d}\theta \tag{13}$$

$$f_{\theta 2} = \sum_{n=1}^{m_\theta} f_\theta(\theta_n)\sin\theta_n\cos\theta_n \tag{14}$$

4.3 Fiber distribution in matrix

Theoretical work on NASC is usually based on the assumption that fibers are stiff and straight in matrix. The real fibers, however, bend and wind together. With the reduce of fiber contents or the increase of fiber lengths, fiber bending and winding are enhanced. Due to the forming process, the bending, winding and uneven distribution of fibers always exist randomly and are very difficult to be characterized.

4.4 Thickness of interface

For most fiber reinforced composites, the thickness of interface between a fiber and the matrix is very small and hence is usually not considered. For NASC, however, to enhance the reinforcing effect of fibers and to optimize the bonding capability of fiber surface, short fibers should be pretreated using surface conditioning agents before mixing, and the interface has great influence on mechanical properties of NASC. The interface thickness can be easily determined by calculating the change in the fiber radius before and after pretreatment.

4.5 Porosity

Some pores will occur during the preparation process of NASC, especially in the sulfuration process. Most of these pores exist in the interface and they will change the stress transfer mode. The porosity p, which is the ratio of the total volume of all pores to the volume of the composite, can be used to evaluate the volume fraction of pores in the composite.

5. A micromechanical model and prediction of macro performances

Loads exerting on NASC will be transferred to fibers through matrix, and the mechanism relates to constitutive equations and mechanical behaviors of fibers, matrix and interface phase. According to the thermo-viscoelastic theory and the shear-lag theory of fibrous composites, a micromechanical model of a single fiber cylindrical cell, which includes a fiber, the matrix and their interphase, is established.

5.1 A micromechanical model of NASC

Fig. 12 illustrates a micromechanical model of NASC, where l is the fiber length, L the matrix length and $L \approx l$. The volume content of fiber is V_f.

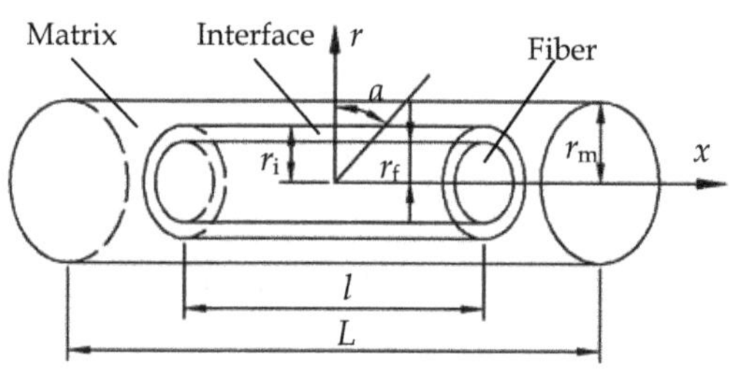

Fig. 12. A micromechanical model of NASC

The stress transfer procedure can be seen in Fig. 13.

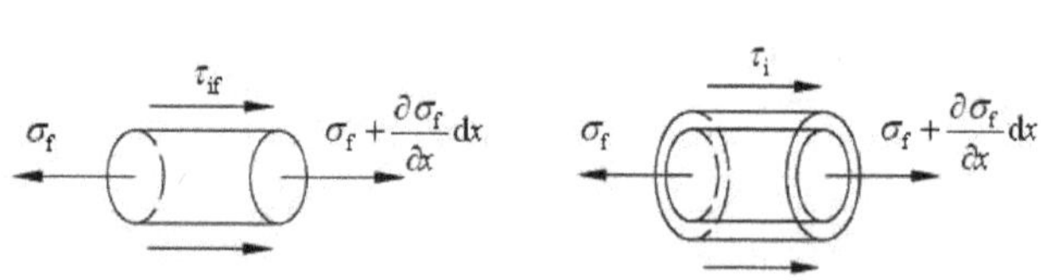

Fig. 13. Micromechanical stress transfer in NASC

When the single fiber cylindrical cell is only subjected to the tension load along the x axis, the circumferential and the radial stresses in the model are much less than the axial stress, and they can be neglected.

The equilibrium relationship of the stresses shown in Fig. 13 can be expressed by

$$\frac{\partial \sigma_{\mathrm{f}}(x)}{\partial x} = -\frac{2\tau_{\mathrm{if}}(x)}{r_{\mathrm{f}}} \tag{15}$$

where $\sigma(x)$ is the axial stress in fiber and $\tau_{if}(x)$ is the shear stress on fiber surface. The axial shear stress in matrix $\tau_m(r, x)$ can be expressed by

$$\tau_{\mathrm{m}}(r,x) = \frac{\left(r_{\mathrm{m}}^2 - r^2\right) r_{\mathrm{i}}}{\left(r_{\mathrm{m}}^2 - r_{\mathrm{i}}^2\right) r} \tau_{\mathrm{im}}(x) \tag{16}$$

where $\tau_{im}(x)$ is the shear stress at the interface boundary near the matrix.

The shear stress in the matrix is a time dependent physical quantity, and it holds the following relationship with the shear strain.

$$\frac{\partial u_{\mathrm{m}}(r,x,t)}{\partial r} = \int_{0^-}^{t} J_{\tau\mathrm{m}}(t-s) d\tau_{\mathrm{m}}(r,x,t) \tag{17}$$

where $u_m(r, x, t)$ is the axial displacements of the matrix, and $J_{\tau m}(t\text{-}s)$ is the creep function. If the bond of both the matrix and the fiber with interface ideally, the axial displacement of the matrix can be expressed by

$$u_{\mathrm{m}}(r,x,t) = u_{\mathrm{i}}(r_{\mathrm{f}},x,t) + \frac{r_{\mathrm{f}}}{\mu_{\mathrm{i}}} \ln\left(\frac{r_{\mathrm{i}}}{r_{\mathrm{f}}}\right) \tau_{\mathrm{if}}(x,t) + \frac{\left[2r_{\mathrm{m}}^2 \ln(r/r_{\mathrm{i}}) - (r^2 - r_{\mathrm{i}}^2)\right] r_{\mathrm{i}}}{2\left(r_{\mathrm{m}}^2 - r_{\mathrm{i}}^2\right)} \int_{0^-}^{t} J_{\tau\mathrm{m}}(t-s) d\tau_{\mathrm{im}}(x,t) \tag{18}$$

The strain of the matrix $\varepsilon_m(r, x, t)$ holds

$$\begin{aligned} \varepsilon_{\mathrm{m}}(r,x,t) &= \frac{\partial u_{\mathrm{m}}(r,x,t)}{\partial x} = \frac{\sigma_{\mathrm{f}}(x)}{E_{\mathrm{f}}} + \alpha_{\mathrm{f}} \Delta T - \frac{r_{\mathrm{f}}^2}{2\mu_{\mathrm{i}}} \ln\left(\frac{r_{\mathrm{i}}}{r_{\mathrm{f}}}\right) \frac{\partial^2 \sigma_{\mathrm{f}}(x,0)}{\partial x^2} \\ &- \frac{\left[2r_{\mathrm{m}}^2 \ln(r/r_{\mathrm{i}}) - (r^2 - r_{\mathrm{i}}^2)\right] r_{\mathrm{f}}^2}{4\left(r_{\mathrm{m}}^2 - r_{\mathrm{i}}^2\right)} \left[J_{\tau\mathrm{m}}(t) \frac{\partial^2 \sigma_{\mathrm{f}}(x,0)}{\partial x^2} - \int_{0^+}^{t} J_{\tau\mathrm{m}}(t-s) \frac{\partial^2 \dot{\sigma}_f(x,s)}{\partial x^2} ds \right] \end{aligned} \tag{19}$$

According to the viscoelastic theory, the shear stress of the matrix $\tau_{if}(x, t)$ holds

$$\tau_{\mathrm{if}}(x,t) = \frac{u_{\mathrm{m}}(r_{\mathrm{m}},x,t) - u_{\mathrm{i}}(r_{\mathrm{f}},x,t) - \frac{r_{\mathrm{f}}}{\mu_{\mathrm{i}}} ln\left(\frac{r_{\mathrm{i}}}{r_{\mathrm{f}}}\right) \tau_{\mathrm{if}}(x,t)}{\left[2r_{\mathrm{m}}^2 \ln(r_{\mathrm{m}}/r_{\mathrm{i}}) - (r_{\mathrm{m}}^2 - r_{\mathrm{i}}^2)\right] r_{\mathrm{f}} / \left[2\left(r_{\mathrm{m}}^2 - r_{\mathrm{i}}^2\right)\right]} dG_{\tau\mathrm{m}} \tag{20}$$

where $G_{\tau m}$ is the relaxation function of $\tau_{if}(x, t)$. Differentiating $\sigma_f(x, t)$ by x gives

$$\frac{\partial^2 \sigma_{\mathrm{f}}(x,t)}{\partial x^2} = -\frac{2\left[\frac{\partial u_{\mathrm{m}}(r_{\mathrm{m}},x,t)}{\partial x} - \frac{\partial u_{\mathrm{i}}(r_{\mathrm{f}},x,t)}{\partial x} + \frac{r_{\mathrm{f}}^2}{2\mu_{\mathrm{i}}} \ln\left(\frac{r_{\mathrm{i}}}{r_{\mathrm{f}}}\right) \frac{\partial^2 \sigma_{\mathrm{f}}(x,t)}{\partial x^2}\right]}{\left[2r_{\mathrm{m}}^2 \ln(r_{\mathrm{m}}/r_{\mathrm{i}}) - \left(r_{\mathrm{m}}^2 - r_{\mathrm{i}}^2\right)\right] r_{\mathrm{f}}^2 / \left[2\left(r_{\mathrm{m}}^2 - r_{\mathrm{i}}^2\right)\right]} dG_{\tau\mathrm{m}} \tag{21}$$

Considering $u_i(r_f,x,t) = u_f(r_f,x,t)$, $\frac{\partial u_i(r_f,x,t)}{\partial x} = \frac{\partial u_f(r_f,x,t)}{\partial x} = \varepsilon_f = \frac{\sigma_f(x,t)}{E_f} + a_f \Delta T$ and $\frac{\partial u_m(r_m,x,t)}{\partial x} = \varepsilon_m(r_m,t)$, Eq. (22) is obtained.

$$\begin{aligned}
&\frac{\partial^2 \sigma_f(x,t)}{\partial x^2} + \frac{2\frac{r_f^2}{2\mu_i} ln\left(\frac{r_i}{r_f}\right)}{\left[2r_m^2 \ln(r_m/r_i) - (r_m^2 - r_i^2)\right] r_f^2 / \left[2(r_m^2 - r_i^2)\right]} \left[\frac{\partial^2 \sigma_f(x,t)}{\partial x^2} G_{\tau m}(0) + \int_{0^+}^{t} \frac{\partial^2 \sigma_f(x,s)}{\partial x^2} \dot{G}_{\tau m}(t-s) ds\right] \\
&- \frac{2}{\left[2r_m^2 \ln(r_m/r_i) - (r_m^2 - r_i^2)\right] r_f^2 E_f / \left[2(r_m^2 - r_i^2)\right]} \left[\sigma_f(x,t) G_{\tau m}(0) + \int_{0^+}^{t} \sigma_f(x,s) \dot{G}_{\tau m}(t-s) ds\right] \\
&+ \frac{2\left[\varepsilon_m(r_m,t) - a_f \Delta T\right]}{\left[2r_m^2 \ln(r_m/r_i) - (r_m^2 - r_i^2)\right] r_f^2 / \left[2(r_m^2 - r_i^2)\right]} dG_{\tau m} = 0
\end{aligned} \quad (22)$$

Eq. (22) is the micromechanical model of NASC which expresses the stress transfer regularity among the matrix, the fiber and their interface.

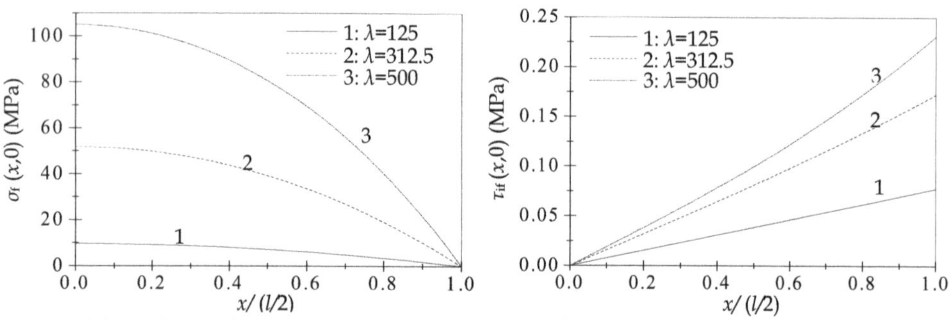

(a) Axial stress distribution in fibers (b) Axial stress distribution in interface

Fig. 14. Influence of the fiber aspect ratio on the stress distribution

The stresses in fibers and interface are illustrated in Fig. 14, where the coordinate x represents the axial distance from the midst (x=0) to the end (x=$l/2$) of the fibers. Fig. 14(a) reveals that the axial stress in the fibers σ_f reduces from the midst to the end, and with the increasing aspect ratio λ, σ_f at the fiber end increases. Fig. 14(b) illustrates the shear stress distribution in interface at the fiber side, and the shear stress increases from the midst to the end.

5.2 Prediction of mechanical performances of NASC

5.2.1 Prediction of the stress-strain relationship

Since the module of interface is much smaller than that of fibers and the interface thickness is much smaller than the matrix, the load is borne mainly by the fibers and the matrix. When the aspect ratio of the fiber is no larger than a critical value, the stress-strain relationship can be expressed by

$$\sigma_c(\varepsilon_c) = f_{\theta 1} f_i f_f V_f \overline{\sigma}_f(\varepsilon_c) + V_m \overline{\sigma}_m(\varepsilon_c) \quad (23)$$

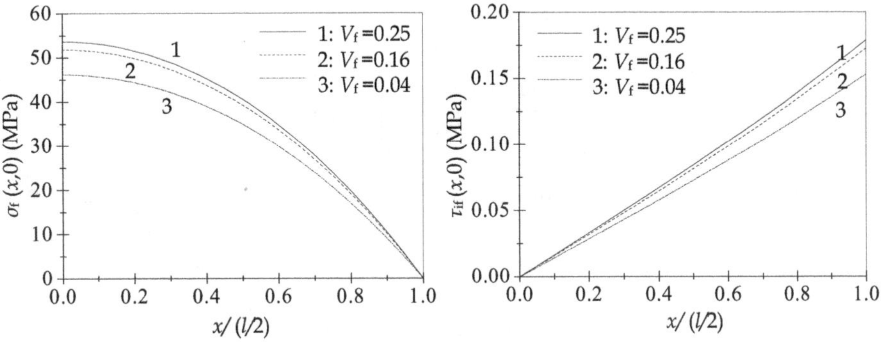

(a) Axial stress distribution in fiber (b) Axial stress distribution in interfce

Fig. 15. Influence of the fiber volume fraction on the stress distribution

where $f_{\theta 1}$=1 when all fibers are parallel to the load direction, otherwise $f_{\theta 1}$=0.5; f_i is the correction coefficient for interface performance obtained by experiments; $f_f = (1 - V_f^2)$ and

$$\bar{\sigma}_f(\varepsilon_C) = \left[1 - \frac{\tanh(\beta_0\lambda/a_0)}{\beta_0\lambda/a_0}\right] E_f(\varepsilon_c - a_f\Delta T),\ \ \bar{\sigma}_m(\varepsilon_c) = E_m\varepsilon_c \tag{24}$$

When the aspect ratio of the fibers is larger than the critical value, the relationship between stress and strain can be expressed by

$$\sigma_c(\varepsilon_c) = \begin{cases} f_{\theta 1} f_i f_f V_f \bar{\sigma}_f(\varepsilon_c, \bar{\lambda}) + V_m \bar{\sigma}_m(\varepsilon_c, \bar{\lambda}) & \varepsilon_c \le \varepsilon_{cf} \\ f_{\theta 1} f_i f_f V_f \bar{\sigma}_f(\varepsilon_c, \bar{\lambda}/2) + V_m \bar{\sigma}_m(\varepsilon_c, \bar{\lambda}/2) & \varepsilon_c > \varepsilon_{cf} \end{cases} \tag{25}$$

where ε_{cf} is the tensile strain of the fibers at the time it breaks, which holds

$$\varepsilon_{cf} = \sigma_{fu}^T \Big/ \left\{ E_f\left[1 - \frac{1}{\cosh(\beta_0\lambda/a_0)}\right]\right\} + a_f\Delta T \tag{26}$$

5.2.2 Prediction of the tensile module

Due to the effect of fibers, the maximum permissible deformation of NASC is only about 10% of that of rubber and the bearing capacity of NASC is greatly improved. The tensile module can be used to characterize the bearing capacity of NASC. The longitudinal tensile module E_{cL} can be calculated by Eq. (27) if the aspect ratios of fibers are relatively concentrated, otherwise it can be calculated by Eq. (28).

$$E_{cL} = f_p\left(f_{\theta 1} f_i f_f V_f \frac{\bar{\sigma}_f}{\varepsilon_c} + V_m E_m\right) \tag{27}$$

$$E_{cL} = f_p\left(\sum_{n=1}^{m} f_{\theta 1} f_i f_f V_{fn} \frac{\bar{\sigma}_{fn}}{\varepsilon_c} + f_m V_m E_m\right) \tag{28}$$

where $\overline{\sigma}_{\text{fn}}$ can be calculated according to Eq. (29).

$$\overline{\sigma}_{\text{fn}} = \left[1 - \frac{\tanh\left(\beta_0 \overline{\lambda}_{\text{n}} / \alpha_0\right)}{\beta_0 \lambda / \alpha_0}\right] E_{\text{f}} \left(\varepsilon_{\text{c}} - \alpha_{\text{f}} \Delta T\right) \tag{29}$$

where $\overline{\lambda}_{\text{n}}$ is the mean aspect ratio of fibers with the length l_n, V_{fn} is the volume content of fibers with the length l_n, and f_p=1-1.9p+0.9p^2. The Halpin-Tsai equation is employed to predict the transverse tensile module of NASC E_{cH} with single orientation.

$$E_{\text{cH}} = \frac{1 + 2\eta V_{\text{f}}}{1 - \eta V_{\text{f}}} E_{\text{m}} \tag{30}$$

where

$$\eta = \frac{E_{\text{f}} / E_{\text{m}} - 1}{E_{\text{f}} / E_{\text{m}} + 2} \tag{31}$$

In consideration of the effect of fiber length and orientation and porosity of material on the transverse tensile module, Eq. (32) is established.

$$E_{\text{cH}} = f_{\text{p}} \left(\frac{1+2V_{\text{f}}}{1-V_{\text{f}}} E_{\text{m}} + f_{\theta 2} f_{\text{i}} f_{\text{f}} V_{\text{f}} \frac{\overline{\sigma}_{\text{f}}}{\varepsilon_{\text{c}}} \right) \tag{32}$$

5.2.3 Prediction of the longitudinal tensile strength

The temperature-related longitudinal tensile strength σ_{cL}^T of NASC can be calculated by Eq. (33).

$$\sigma_{\text{cL}}^T = f_{\theta 1} f_{\text{i}} f_{\text{f}} V_{\text{f}} \overline{\sigma_{\text{f}}^T} + f_{\text{d}} V_{\text{m}} \overline{\sigma_{\text{m}}^T} \tag{33}$$

where $f_{\text{d}} = (1 - V_{\text{f}})$, $\overline{\sigma_{\text{f}}^T}$ is the mean stress in fibers when tensile failure occurs and it can be obtained from the micromechanical model of NASC, $\overline{\sigma_{\text{m}}^T}$ is the mean stress in the matrix when tensile failure occurs and $\overline{\sigma_{\text{m}}^T} = E_{\text{m}}^T \varepsilon_{\text{cu}}^T$, and f_d is the weakening coefficient of rubber induced by fibers.

When the aspect ratio λ is smaller than the critical value λ_c, the debonding between fibers and the matrix will occur. However, when $\lambda > \lambda_c$, the fiber breaks. Therefore, the temperature-related longitudinal tensile strength can be calculated according to Eq. (34).

$$\sigma_{\text{cL}}^T = f_{\theta 1} f_{\text{i}} f_{\text{f}} V_{\text{f}} \left[\int_{\lambda_{\min}}^{\lambda_{\text{c}}} f(\lambda) \lambda \left(\tau_{\text{s}}^T - \Delta \tau_{\text{i}}^T\right) \mathrm{d}\lambda + \int_{\lambda_{\text{c}}}^{\lambda_{\max}} f(\lambda) (1 - 0.5\lambda_{\text{c}} / \lambda) \left(\sigma_{\text{fu}}^T - \Delta \sigma_{\text{f}}^T\right) \mathrm{d}l \right] + f_{\text{d}} V_{\text{m}} \overline{\sigma_{\text{mm}}^T} \tag{34}$$

where $\Delta\sigma_{\text{f}}^T$ is the stress in fibers induced by temperature change and $\Delta\tau_{\text{i}}^T$ is the shear stress in fibers, $\Delta\sigma_{\text{f}}^T = \left[1 - \frac{\tanh\left(\beta_0 \lambda / \alpha_0\right)}{\beta_0 \lambda / \alpha_0}\right] E_{\text{f}} \alpha_{\text{f}} \Delta T$ and $\Delta\tau_{\text{i}}^T = \frac{E_{\text{f}} \alpha_{\text{f}} \Delta T}{\lambda}$.

5.2.4 Prediction of the transverse tensile strength

The transverse tensile strength is a main evaluating indicator for NASC in engineering, and its prediction may be performed according to 3 cases, as follows.

1. Debonding of interface from fiber

In this case, the transverse tensile strength σ_{cH}^{T} can be calculated by

$$\sigma_{cH}^{T} = \sqrt{\pi V_f}\tau_{if}^{T} + 2\frac{t_i}{r_f}\sqrt{\frac{V_f}{\pi}}\sigma_{iH}^{T} + \left(1 - 2\frac{t_i + r_f}{r_f}\sqrt{\frac{V_f}{\pi}}\right)\sigma_{mH}^{T} \tag{35}$$

Eq. (35) can be used to predict the transverse tensile strength assuming that all fibers orientate in the same direction at the moment the matrix debonds from the fibers. If the fibers orientate in different directions, some will bear the longitudinal load. In this case, the transverse tensile strength is larger than that calculated by Eq. (35), and it can be calculated by Eq. (36).

$$\sigma_{cH}^{T} = f_{\theta 3}\left(\sqrt{V_f \pi}\tau_{if}^{T} + 2\frac{t_i}{r_f}\sqrt{\frac{V_f}{\pi}}\sigma_{iH}^{T}\right) + f_{\theta 2} f_i f_f V_f \overline{\sigma_f^{T}} + \left(1 - 2\frac{t_i + r_f}{r_f}\sqrt{\frac{V_f}{\pi}}\right)\sigma_{mH}^{T} \tag{36}$$

2. Debonding of interface from matrix

If all fibers orientate in the same direction at the moment the interface debonds from the matrix, the transverse tensile strength σ_{cH}^{T} can be calculated by Eq. (37).

$$\sigma_{cH}^{T} = \frac{t_i + r_f}{r_f}\sqrt{V_f \pi}\tau_{im}^{T} + \left(1 - 2\frac{t_i + r_f}{r_f}\sqrt{\frac{V_f}{\pi}}\right)\sigma_{mH}^{T} \tag{37}$$

If fibers orientate in different directions, σ_{cH}^{T} can be calculated by Eq. (38).

$$\sigma_{cH}^{T} = f_{\theta 3}\frac{t_i + r_f}{r_f}\sqrt{\pi V_f}\tau_{im}^{T} + f_{\theta 2} f_i f_f V_f \overline{\sigma_f^{T}} + \left(1 - 2\frac{t_i + r_f}{r_f}\sqrt{\frac{V_f}{\pi}}\right)\sigma_{mH}^{T} \tag{38}$$

3. Damage of interface

The transverse tensile strength can be calculated by

$$\sigma_{cH}^{T} = \frac{F_H}{WL} = \sqrt{V_f \pi}\tau_{is}^{T} + 2\frac{t_i}{r_f}\sqrt{\frac{V_f}{\pi}}\sigma_{iH}^{T} + \left(1 - 2\frac{t_i + r_f}{r_f}\sqrt{\frac{V_f}{\pi}}\right)\sigma_{mH}^{T} \tag{39}$$

Using Eq. (39), the transverse tensile strength σ_{cH}^{T} can be predicted assuming that all fibers orientate in the same direction at the moment the interface is damaged. If fibers orientate in different directions, σ_{cH}^{T} can be calculated by Eq. (40).

$$\sigma_{cH}^{T} = f_{\theta 3}\left(\sqrt{V_f \pi}\tau_{is}^{T} + 2\frac{t_i}{r_f}\sqrt{\frac{V_f}{\pi}}\sigma_{iH}^{T}\right) + f_{\theta 2} f_i f_f V_f \overline{\sigma_f^{T}} + \left(1 - 2\frac{t_i + r_f}{r_f}\sqrt{\frac{V_f}{\pi}}\right)\sigma_{mH}^{T} \tag{40}$$

5.2.5 Prediction of compression-resilience performances

NASC is used to make gaskets, and the direction of compressive loads acting on it is normal to the orientation direction of fibers. The model of a compressive type single-fiber cell (Fig.

16) is established to derive the compression-creep constitutive equation of NASC with the assumption $L \approx l$. The cell model is divided into three parts along compression-resilience direction; the first part consists of matrix, interface and fiber (0-1), the second part consists of matrix and interface (1-2), and the third part includes only matrix (2-3).

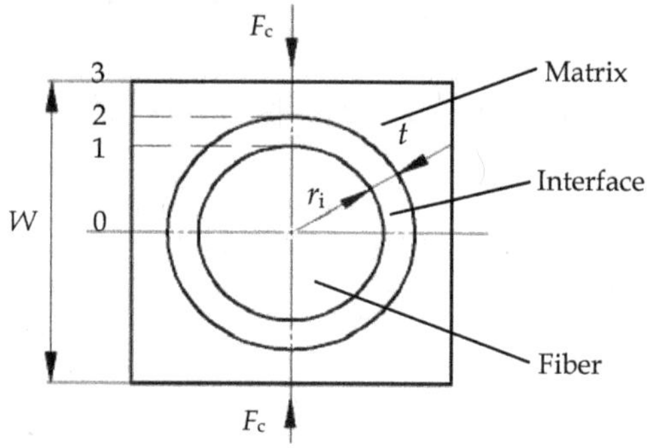

Fig. 16. A model of a compressive type single-fiber cell

The total strain of the material ε_{cc} under the load F_c holds

$$\varepsilon_{cc} = \frac{2}{W}\left[\varepsilon_{c0\text{-}1} r_f + \varepsilon_{c1\text{-}2} t_i + \varepsilon_{c2\text{-}3}\left(\frac{W}{2} - t_i - r_f\right)\right] \tag{41}$$

The compressive stress and strain of material relate to time. The compression-creep constitutive equation of NASC is in the form

$$\begin{aligned}\varepsilon_{cc}(t) = &\left(\frac{2}{\sqrt{2}E_f} + \frac{2t_i}{\sqrt{t_i^2 + 2t_i r_f}\,E_i}\right)\sigma_{cc}(t) + 2\left[\sigma_{cc}(0) J_{cm}^T(t) + \int_{0^+}^{t} J_{cm}^T(t-\varsigma)\frac{d\sigma_{cc}(\varsigma)}{d\varsigma}d\varsigma\right] \\ &\times\left(\frac{1}{2} - \frac{t_i\sqrt{V_f/\Pi}}{r_f} - \sqrt{V_f/\Pi}\right) - \left[2\alpha_f\sqrt{V_f/\Pi} + \frac{2\alpha_i t_i}{r_f}\sqrt{V_f/\Pi} + 2\alpha_m\left(\frac{1}{2} - \frac{t_i\sqrt{V_f/\Pi}}{r_f} - \sqrt{V_f/\Pi}\right)\right]\Delta T\end{aligned} \tag{42}$$

The compression rate of NASC can be calculated by

$$\begin{aligned}R_c = &\left(\frac{2}{\sqrt{2}E_f} + \frac{2t_i}{\sqrt{t_i^2 + 2t_i r_f}\,E_i}\right)\sigma_{cc}(t) + 2\int_{0^+}^{t} J_{cm}^T(t-\varsigma)d\varsigma\left(\frac{1}{2} - \frac{t_i\sqrt{V_f/\Pi}}{r_f} - \sqrt{V_f/\Pi}\right)\frac{\Delta\sigma_{cc}(t)}{\Delta t} \\ &- \left[2\alpha_f\sqrt{V_f/\pi} + \frac{2\alpha_i t_i}{r_f}\sqrt{V_f/\Pi} + 2\alpha_m\left(\frac{1}{2} - \frac{t_i\sqrt{V_f/\Pi}}{r_f} - \sqrt{V_f/\Pi}\right)\right]\Delta T\end{aligned} \tag{43}$$

Substituting ΔT=0 into Eq. (43) gives the compressive rate of NASC at room temperature. The compressive strain can be calculated by

$$\varepsilon_{cc} = \left(\frac{2}{\sqrt{2}E_f} + \frac{2t_i}{\sqrt{t_i^2 + 2t_i r_f}\,E_i}\right)\sigma_{cc}(t) + \left[X + Y(t)\right]\left(1 - \frac{2t_i\sqrt{V_f/\Pi}}{r_f} - 2\sqrt{V_f/\Pi}\right)\frac{\Delta\sigma_{cc}(t)}{\Delta t} \tag{44a}$$

$$-\left[2a_{\mathrm{f}}\sqrt{V_{\mathrm{f}}/\pi}+\frac{2a_{\mathrm{i}}t_{\mathrm{i}}}{r_{\mathrm{f}}}\sqrt{V_{\mathrm{f}}/\pi}+2\alpha_{\mathrm{m}}\left(\frac{1}{2}-\frac{t_{\mathrm{i}}\sqrt{V_{\mathrm{f}}/\pi}}{r_{\mathrm{f}}}-\sqrt{V_{\mathrm{f}}/\pi}\right)\right]\Delta T \tag{44b}$$

where $X=\int_{0^+}^{t}J_{\mathrm{cm}}^{T}(t-\varsigma)\mathrm{d}\varsigma$ and $Y(t)=\int_{0^+}^{t}J_{\mathrm{cm}}^{T}(t-\varsigma)\mathrm{d}\varsigma$.

After unloading the compressive load, the residual compressive strain is

$$\varepsilon_{\mathrm{cb}}=Y(t)\left(1-\frac{2t_{\mathrm{i}}\sqrt{V_{\mathrm{f}}/\pi}}{r_{\mathrm{f}}}-2\sqrt{V_{\mathrm{f}}/\pi}\right)\frac{\sigma_{\mathrm{cc}}(t)}{t} \tag{45}$$

The resilient rate of NASC can be obtained by

$$R_{\mathrm{c}}=\frac{\varepsilon_{\mathrm{bb}}}{\varepsilon_{\mathrm{cc}}} \tag{46}$$

where

$$\begin{aligned}\varepsilon_{\mathrm{bb}}=&\left(\frac{2}{\sqrt{2}E_{\mathrm{f}}}+\frac{2t_{\mathrm{i}}}{\sqrt{t_{\mathrm{i}}^{2}+2t_{\mathrm{i}}r_{\mathrm{f}}E_{\mathrm{i}}}}\right)\sigma_{\mathrm{cc}}(t)+X\left(1-\frac{2t_{\mathrm{i}}\sqrt{V_{\mathrm{f}}/\pi}}{r_{\mathrm{f}}}-2\sqrt{V_{\mathrm{f}}/\pi}\right)\frac{\Delta\sigma_{\mathrm{cc}}(t)}{\Delta t}\\&-\left[2\alpha_{\mathrm{f}}\sqrt{V_{\mathrm{f}}/\pi}+\frac{2a_{\mathrm{i}}t_{\mathrm{i}}}{r_{\mathrm{f}}}\sqrt{V_{\mathrm{f}}/\pi}+2\alpha_{\mathrm{m}}\left(\frac{1}{2}-\frac{t_{\mathrm{i}}\sqrt{V_{\mathrm{f}}/\pi}}{r_{\mathrm{f}}}-\sqrt{V_{\mathrm{f}}/\pi}\right)\right]\Delta T\end{aligned} \tag{47}$$

Substituting ΔT=0 into Eq. (46) gives the resilient rate of NASC at room temperature.

5.2.6 Prediction of creep behavior

The creep behavior equation of NASC can be expressed by Eq. (48).

$$\begin{aligned}\varepsilon_{\mathrm{cc}}(t)=&\left(\frac{2}{\sqrt{2}E_{f}}+\frac{2t_{\mathrm{i}}}{\sqrt{t_{\mathrm{i}}^{2}+2t_{\mathrm{i}}r_{\mathrm{f}}E_{\mathrm{i}}}}\right)\sigma_{\mathrm{cc}}(t)+2\sigma_{\mathrm{cc}}(0)J_{\mathrm{cm}}^{T}(t)\left(\frac{1}{2}-\frac{t_{\mathrm{i}}\sqrt{V_{\mathrm{f}}/\pi}}{r_{\mathrm{f}}}-\sqrt{V_{\mathrm{f}}/\pi}\right)\\&-\left[2\alpha_{\mathrm{f}}\sqrt{V_{\mathrm{f}}/\pi}+\frac{2\alpha_{\mathrm{i}}t_{\mathrm{i}}}{r_{\mathrm{f}}}\sqrt{V_{\mathrm{f}}/\pi}+2\alpha_{\mathrm{m}}\left(\frac{1}{2}-\frac{t_{\mathrm{i}}\sqrt{V_{\mathrm{f}}/\pi}}{r_{\mathrm{f}}}-\sqrt{V_{\mathrm{f}}/\pi}\right)\right]\Delta T\end{aligned} \tag{48}$$

Because the strain in both fiber and interface is much smaller than that in matrix, only the strain in matrix is considered under compressive load. Then the stress-strain relationship of the third part in Fig.6 resulted from the resilience constitutive equation is in the form

$$\sigma_{\mathrm{cm}}(t)=\varepsilon_{\mathrm{c2\text{-}3}}(0)G_{\mathrm{cm}}^{T}(t)+\int_{0^+}^{t}G_{\mathrm{cm}}^{T}(t-\varsigma)\left[\frac{\mathrm{d}\varepsilon_{\mathrm{c2\text{-}3}}(\varsigma)}{\mathrm{d}\varsigma}+\alpha_{\mathrm{m}}\frac{\mathrm{d}T(\varsigma)}{\mathrm{d}\varsigma}\right]\mathrm{d}\varsigma \tag{49}$$

If the compressive strain ε_{cc} is constant during the course of stress-resilience, Eq. (50) is obtained.

$$\sigma_{\mathrm{cm}}(t)=\varepsilon_{\mathrm{c2\text{-}3}}(0)G_{\mathrm{cm}}^{T}(t)+\int_{0^+}^{t}G_{\mathrm{cm}}^{T}(t-V)\alpha_{\mathrm{m}}\frac{\mathrm{d}T(V)}{\mathrm{d}V}\mathrm{d}V \tag{50}$$

The stress-resilience equation of NASC is in the form

$$\sigma_{cc}(t)=\frac{\varepsilon_{cc}(0)G_{cm}^{T}(t)}{1-\frac{2t_i\sqrt{V_f/\pi}}{r_f}-2\sqrt{V_f/\pi}}+\int_{0^+}^{t}G_{cm}^{T}(t-\varsigma)\alpha_m\frac{dT(\varsigma)}{d\varsigma}d\varsigma \tag{51}$$

The detailed derivation of Eqs. (7)-(51) can be found elsewhere (Zhu, 2008).

6. Performance evaluation of NASC

6.1 Mechanical performances of non-asbestos sealing composites

The properties of NASC specified in current standards include mainly density, transverse tensile strength, plasticity, oil resistance, and so on. The performances of non-asbestos sealing gaskets can be divided into two types. One is related to the performance indexes of gasket products, which are the main items of product inspection prescribed in gasket product standards. These performance indexes are the important basis for evaluating the quality of gasket products. They involve compressibility, resilient rate, stress relaxation rate and leakage rate (AQSIQ, 2003; AQSIQ, 2008). The other is correlated with the gasket constants in current code design guidelines, such as gasket yield factor y which is defined as the minimum gasket stress required to cause the gasket material to deform into the flange face irregularities and gasket factor m which is defined as the ratio of the minimum gasket stress needed to hold the joint sealed under the operating conditions to the internal pressure p (ASME, 2007). In this section, the theoretic prediction and experimental evaluation of some performances of NASC and gasket products are presented.

6.1.1 Test apparatus and procedure

A series of tests for tensile property, compressibility, resilient rate, and stress relaxation property were conducted on an INSTRON 3367 universal testing machine. Samples were made of the developed NASC reinforced with aramid fiber, as mentioned in section 3.2.
For tensile tests, the dumbbell-shaped samples were adopted, and the tests were performed according to the Chinese standard GB/T 1447-2005 "Fiber-reinforced plastics composites/determination of tensile properties". The test conditions are listed in Table 8.

Sample Quantity	Sample Dimension (mm)	Fiber Length (mm)	Fiber Content (wt%)	Tensile speed (mm/min)	Temperature (K)
36	20×10×2.5	1-3, 4-6,7-9	5, 10, 15, 20, 30, 40	200	293, 373, 423

Table 8. The conditions for tensile tests

The compressibility and resilient rate tests were carried out according to the Chinese standard GB/T 12622-2008 "Standard test method for compressibility and recovery of gaskets for pipe flanges". The test conditions are listed in Table 9.

Sample Quantity	Sample Dimension (mm)	Fiber Content (wt%)	Loading Speed (mm/min)	The Maximum Stress (MPa)	Temperature (K)
8	10×10×2.5	20, 40	200	10, 20	293

Table 9. The conditions for compressibility and resilient rate tests

For the tests of the stress relaxation property, the samples were placed in the chamber of the testing machine and heated to the desired test temperatures. After the temperature remained constant for about 5 minutes, the compressive loads were applied to the samples. The change in compressive stress of the samples was measured over time under the constant sample deformation. The test conditions are listed in Table 10.

Sample Quantity	Sample Dimension (mm)	Fiber Content (wt%)	Loading Speed (mm/min)	The Maximum Stress (MPa)	Temperature (K)
8	10×10×2.5	20, 40	200	10, 20	293, 423

Table 10. The conditions for stress relaxation tests

6.1.2 Results and discussions

Figs. 17 and 18 show the stress-strain curves of the samples at room temperature with different fiber mean aspect ratios $\bar{\lambda}$ and fiber volume contents V_f, respectively.

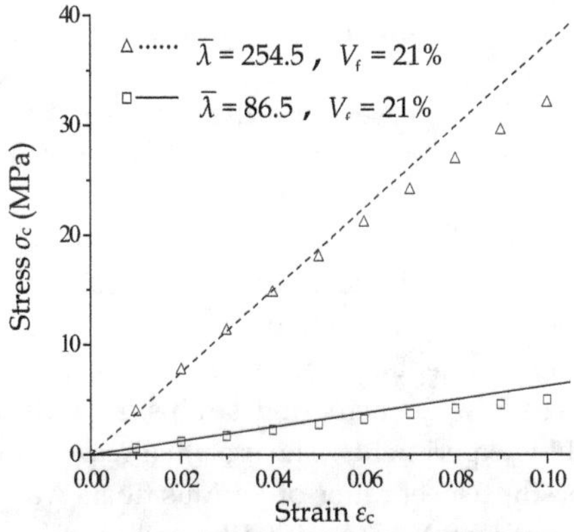

Fig. 17. Stress-strain curves of NASC with different fiber aspect ratios

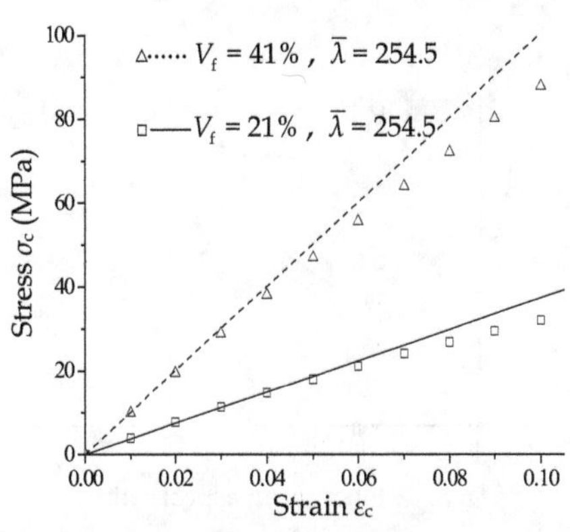

Fig. 18. Stress-strain curves of NASC with different fiber volume contents

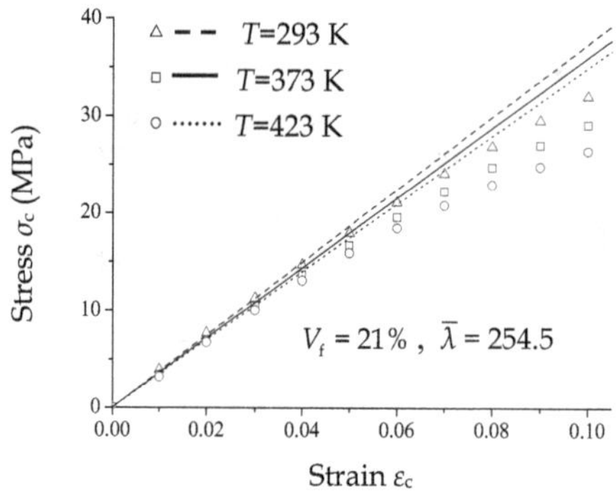

Fig. 19. Stress-strain curves of NASC at different test temperatures

Fig. 19 illustrates the stress-strain relationships of the samples at different test temperatures with a known fiber mean aspect ratio and a given fiber volume content. In these figures, symbols "Δ, □, o" represent experimental data, and the curves are plotted according to the prediction formulae proposed in section 5.2. It can be seen from the Figs. 17-19 that the slope of stress-strain curves of NASC increases with the increase of fiber mean aspect ratio and fiber volume content. Aramid fiber has a good reinforcing effect on elastomer matrix composites. The larger the fiber mean aspect ratio, the more obvious is the reinforcing effect. Because the modulus of aramid fiber is much larger than those of elastomer matrix material and compatibilization fiber like sepiolite fiber, the increase in the aramid fiber content results in the significant augment of the strength of NASC.

The error between the experimental data and values predicted using the derived constitutive equation increases with increasing strains, and all the predicted stresses are larger than those obtained by experiments. The essential reason for the error of prediction is that the difference between the aspect ratios of various fibers was neglected. The changes in the transverse tensile strength of NASC with fiber mean aspect ratio and fiber volume content are shown in Figs. 20-21.

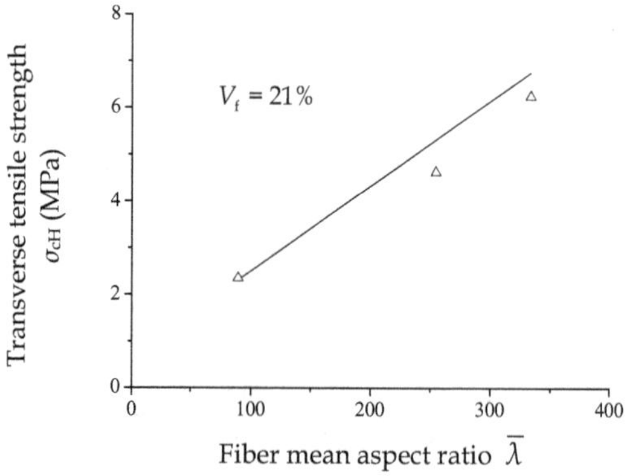

Fig. 20. Relationship between transverse tensile strength and fiber mean aspect ratio

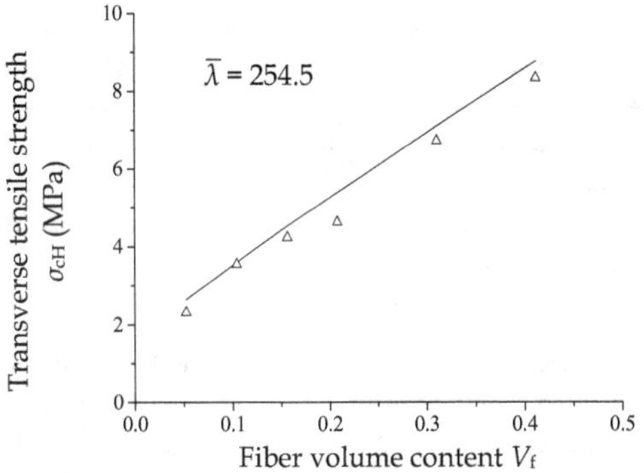

Fig. 21. Relationship between transverse tensile strength and fiber volume content

Fig. 22 illustrates the effect of test temperature on the transverse tensile strength. In these Figures, symbol "Δ" represents experimental data, and the curves represent the prediction results. The transverse tensile strength increases with the increase in the fiber mean aspect ratio and the fiber volume content. The transverse tensile strength of NASC is interrelated with temperature, and it deceases with increasing temperature. The temperature induces thermal stresses in the fiber, matrix and fiber-matrix interface, which affect the temperature-dependent tensile strength of the composite finally. Because the coefficient of thermal expansion of the elastomer matrix is much lager than that of the fiber, big thermal stresses both in the fiber and in the matrix will be generated with variation of temperature, and a small shear stress in the fiber-matrix interface is produced as well (Zhu et al., 2008). These factors result in the decrease of transverse tensile strength with increasing temperature.

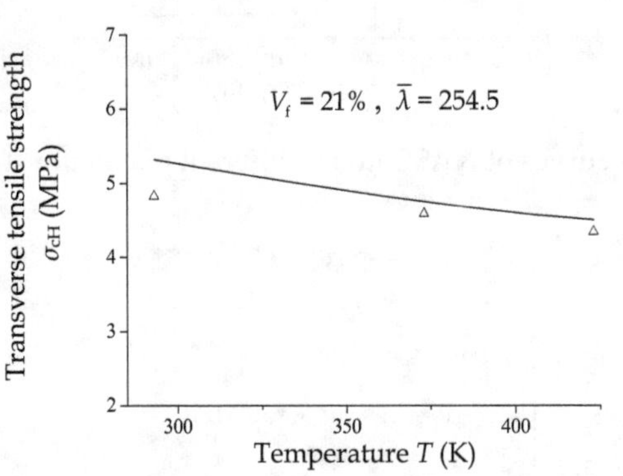

Fig. 22. Effect of temperature on transverse tensile strength

The compressibility and resilient rate of NASC obtained by tests and those predicted using Eqs. (43) and (46) are listed in Table 11. It can be found that the compressibility of NASC increases with increasing the maximum compressive stress and decreasing the fiber volume content, while the resilient rate decreases with the increase in the maximum compressive stress and the fiber volume content.

Fiber Content (wt%)	The Maximum Compressive Stress (MPa)	Compressibility (%)			Resilient Rate (%)		
		Test Data	Prediction Result	Relative Error (%)	Test Data	Prediction Result	Relative Error (%)
21	10	31.1	30.1	3.3	69.3	62.5	10.3
21	20	36.2	33.8	6.9	62.7	55.3	12.5
41	10	23.1	21.6	6.7	54.8	48.0	13.2
41	20	28.3	25.9	8.9	51.2	46.6	9.4

Table 11. Compressibility and resilient rate of NASC

Figs. 23-25 present the test and prediction results of stress relaxation property of NASC, in which symbols "Δ, □" represent experimental results, and the curves are plotted according to the prediction values of stress relaxation property using Eq. (51). The stress relaxation rate increases with increasing initial stress and temperature and decreasing fiber volume content. The fiber mean aspect ratio and orientation have less evident influence on the stress relaxation property of NASC. The experimental and prediction results are basically identical, and the maximum error is about 8.5%.

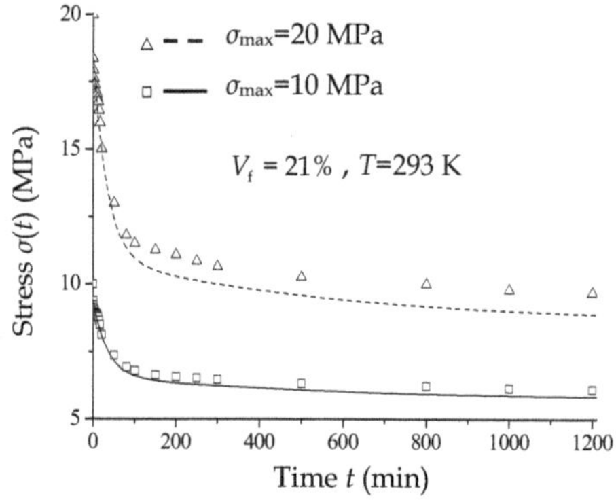

Fig. 23. Stress relaxation curves of NASC under different maximum stresses

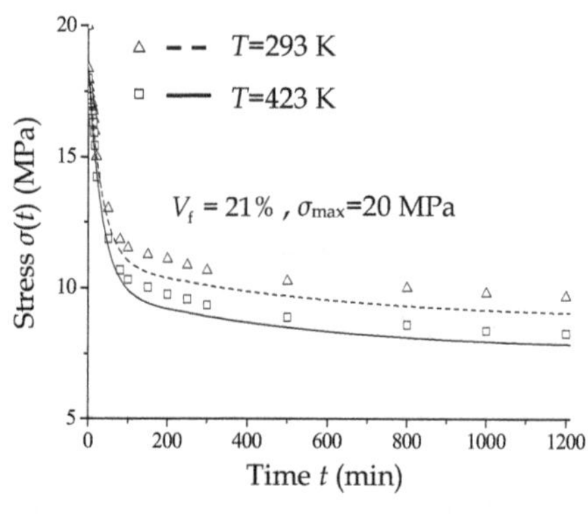

Fig. 24. Stress relaxation curves of NASC at different temperatures

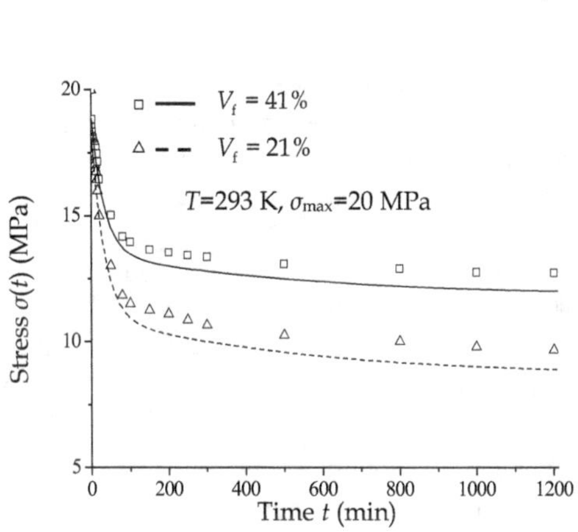

Fig. 25. Stress relaxation curves of NASC with different fiber contents

6.2 Sealing behaviour of non-asbestos sealing composites

6.2.1 Leakage model of non-asbestos gasket sealing

NASC can be assumed to be a kind of porous media. The flow of gas through porous media includes molecular transfer and convection transfer. The leakage rate L of gases flowing through porous media can be calculated using the capillary model (Gu et al., 2007b).

$$L = \sum_{i=1}^{k}\left(\frac{\pi r_i^4}{8\eta c l_m} p_m + \frac{4 r_i^3}{3 c l_m}\sqrt{\frac{2\pi RT}{M}}\right)(p_1 - p_2) \tag{52}$$

where r_i and l_m are the radius and the average length of capillaries, respectively, c is the bending coefficient of capillaries, η the dynamic viscosity of gases, T the Kelvin temperature of gases, M the gas molecular weight, R the gas constant, p_m the average pressure, $p_m=(p_1+p_2)/2$, and p_1 and p_2 are the pressures at the entrance and exit of capillaries, respectively.

There exists the following relationship between the number k and the radius r_i of leakage paths and the gasket compressive stress σ:

$$k = f_1\left(\sigma^{-m}\right) \tag{53}$$

$$r_i = f_2\left(\sigma^{-n}\right) \tag{54}$$

According to Eqs.(52)-(54), the leakage rate of gas through gasket sealing joints L can be calculated by the following formula:

$$L = \left[A_L \frac{1}{\eta}\left(\frac{\sigma}{\sigma_0}\right)^{-n_L} p_m + A_M\sqrt{\frac{T}{M}}\left(\frac{\sigma}{\sigma_0}\right)^{-n_M}\right]\frac{(p_1 - p_2)}{l} \tag{55}$$

where l is the gasket effective width, A_L, A_M, n_L and n_M are regression coefficients which can be obtained from experimental data, and σ_0 is the gasket seating stress prescribed in the corresponding gasket standards (AQSIQ,2003, AQSIQ,2008).

6.2.2 Test verification and leakage rate prediction method

Tests were performed on a fully automatic testing machine for gasket performances, and the specific details on the test machine and test procedure can be found in the reference (Gu et al., 2007b). The test conditions are listed in Table 12.

Sample Quantity	Sample Dimension (mm)	Test Medium	Gasket Seating Stress (MPa)	Gas Pressure Difference (106 Pa)	Test Temperature (K)
3	ϕ109×ϕ61×2	99.9% nitrogen	20, 30, 40, 50, 60	1, 2, 3, 4.5, 6	293

Table 12. Conditions for the leakage test

The relationship between medium pressure difference p_1–p_2 and leakage rate L under different gasket seating stresses σ_S is shown in Fig. 26, in which symbols represent experimental data, and curves are plotted according to the leakage model. The experimental results show that the leakage rate increases obviously with the increase of the test medium pressure when the gasket seating stresses are smaller. On the contrary, when the gasket seating stresses are larger, this tendency is not obvious. The variation of leakage rates with pressure difference across gasket width follows precisely the curves represented by the leakage model, as indicated by the high value of the correlation coefficient R which is nearly equal to unity.

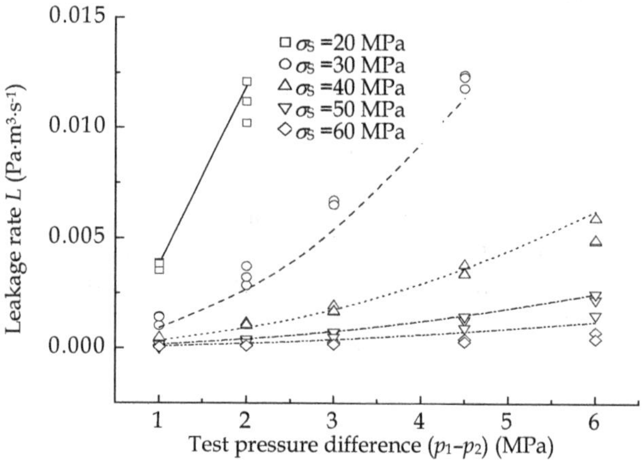

Fig. 26. Relationship between leakage rate and test pressure difference under different seating stresses

The coefficients A_L, A_M, n_L and n_M in Eq. (55) can be obtained by regression analysis of the experimental data, and they are listed in Table 13. According to Eq. (55) and the regression coefficients, the relationships between the pressure difference and the leakage rate under different gasket compressive stresses are obtained, and they are also illustrated in Fig. 26.

A_L	A_M	n_L	n_M	R
8.52×10^{-23}	2.79×10^{-12}	4.08	3.06	0.984

Table 13. Regression coefficients in Eq. (55)

According to Eq. (55) and the regression coefficients in Table 13, the leakage rates of other gases leaking through non-asbestos gaskets with different sizes and used under various working conditions can also be predicted. It can be assumed that the leakage paths in gaskets are distributed uniformly in the circumferential direction. The larger the gasket perimeter, the more the leakage paths are. Therefore, when the leakage rate prediction is performed for the gaskets with different sizes, the correction of Eq. (56) is necessary, and hence, the leakage prediction formula holds:

$$L=\left[A_L\frac{1}{\eta}\left(\frac{\sigma}{\sigma_0}\right)^{-n_L}p_m+A_M\sqrt{\frac{T}{M}}\left(\frac{\sigma}{\sigma_0}\right)^{-n_M}\right]\frac{(p_1-p_2)}{l}\frac{D_2}{D_1} \tag{56}$$

where D_1 and D_2 are diameters of the tested gasket and the gasket on which the leakage rate will be predicted, respectively.

It can be seen from Eq. (55) and Eq. (56) as well that the total leakage rate of gases through non-asbestos gaskets is the sum of the laminar flow rate and the molecular flow rate. The leakage is mainly the molecular flow when the gasket seating stress is large and the gas pressure is low, and the leakage rate is directly proportional to the pressure difference (p_1-p_2). On the contrary, when the gasket seating stress is small and the gas pressure high, the leakage is predominantly the laminar flow. In this case, the leakage rate is in direct proportion to the square difference of pressures (p_1^2-p_2^2), and the influence of the molecular flow can be ignored. The latter situation represents the leakage state of gasket sealing joints in most pressure vessels and piping.

7. Conclusions

Two kinds of non-asbestos sealing composites were developed by using aramid, glass, and pre-oxidized carbon fibers as reinforcing fibers, and nitrile rubber and natural rubber as binders. Design of these composites as well as preparation techniques were discussed. The surface treatment effects of fibers by using such methods as immersion coating, surface oxidation and plasma treatments were investigated. The results of the transverse tensile test and the SEM analysis reveal that the surface pretreatment can largely enhance the associative strength of the interface between fiber and matrix.

Both the molding and the calendaring preparation processes of NASC were presented. Contents of main compositions of two kinds of NASC were determined by uniform design and regression design methods in which the transverse tensile strength, compressibility, resilient rate and stress relaxation rate of the composites were used as evaluation indexes. According to the data resulted from mixing regression design experiments, the regression equation was obtained.

Methods for measuring and characterizing the micro structural parameters including the aspect ratio, orientation, distribution of short fibers, interphase thickness, and porosity of NASC were discussed. The short fiber aspect ratio can be obtained by measuring the dimensions of the fibers that were separated from the composite, and results can be evaluated by the mean aspect ratio method, the histogram method and the distribution function method. The short fiber orientation can be tested by the section-analysis method and the plane-observation method, and results can be evaluated by the distribution function method, the histogram method and the modified coefficient method.

The micromechanical model of a single fiber cylindrical cell, which includes fiber, matrix and their interphase, was established. The stress transfer mode among the fiber, the matrix and the interphase was investigated, and the stress distributions in them were obtained. A model of a compressive type single-fiber cell was established, and the prediction methods for the stress-strain relationship, the tensile module, the longitudinal tensile strength, the transverse tensile strength, the compression-resilience performance and the creep behavior were proposed.

A series of tests for tensile property, compressibility, resilient rate, and stress relaxation property of NASC were conducted. The experimental results were compared with the predicted values. The predicted stress-strain relationship of NASC is in good agreement with the experimental result when the strain of the composite is relatively small, while the error is obvious when the strain is relatively large. The transverse tensile strength increases with the increase of the fiber volume content and the mean aspect ratio and decreases with the increase in the test temperature. All the experimental results are smaller than those predicted, and the maximum error between them is about 13%. The compressibility increases with increasing the maximum compressive stress and decreasing the fiber volume content, while the resilient rate decreases with the increase in the maximum compressive stress and the fiber volume content. They are nearly irrelevant to the fiber mean aspect ratio and orientation. The errors both between the predicted compressive rates and experimental data and between the predicted resilient rates and the experimental results are less than 15%. At the beginning of the stress relaxation, the stress in the composite decreases quickly, the stress relaxation speed decreases essentially after 80 to 100 minutes, and the stress relaxation curve trends to be a straight line. The stress relaxation rate increases with increasing initial stress and temperature and decreasing fiber volume content. Such microstructure parameters as fiber mean aspect ratio and orientation have less evident influence on the stress relaxation property of NASC. The experimental and predicted results are basically identical, and the maximum error is about 8.5%, which indicates that the stress relaxation property of NASC can be well evaluated by the stress relaxation equation presented in this chapter.

A leakage model was developed to predict non-asbestos gasket leakage rates with different gasket sizes and various gases under different working conditions. The model is constructed on the base of the gasket leakage tests conducted over a nitrogen pressure range of 1 to 6 MPa and a gasket seating stress range of 20 to 60 MPa. The established model gives relatively accurate predictions, but it is necessary to conduct additional tests before predicting leakage rate of other types of gaskets. High temperature will cause the deterioration of sealing materials and result in the increase of leakage rates, which has not been taken into consideration in the present model, and this subject should be studied.

8. Acknowledgement

This work was supported by the National Natural Science Foundation of China (Grant No. 10872088) and the Doctoral Foundation of the Ministry of Education of China (Grant No. 20070291004 and No. 20093221120009). The authors of this chapter gratefully acknowledge the corresponding government organizations for the funds that made this project possible. We would also like to thank those postgraduate students who participated in our previous work throughout these years and made the contribution in the aspect of material preparation, performance test, data processing, and so on.

9. References

American Academy of Actuaries (2007). *Overview of Asbestos Claims Issues and Trends*, Mass Torts Subcommittee of the American Academy of Actuaries, Washington, USA

AQSIQ (2003). The General Administration of Quality Supervision, Inspection and Quarantine of the People's Republic of China, GB/T 9129-2003, *Specifications of Non-metallic Flat Gaskets for Pipe Flanges*. Standards Press of China, ISBN 155066.1-19884, Beijing, China (in Chinese)

AQSIQ (2008). The General Administration of Quality Supervision, Inspection and Quarantine of the People's Republic of China, GB/T 17727-2008, *Nonmetal material gaskets for marine flange*. Standards Press of China, ISBN 750665591, Beijing, China (in Chinese)

ASME (2007). The American Society of Mechanical Engineers, *ASME Boiler and Pressure Vessel Code 2007, Sec VIII, Division 1, Rules for construction of Pressure Vessels*. New York, USA

Chen, Y. & Gu B.Q. (2008). Development of a new kind of sealing composite material reinforced with carbon and glass hybrid fibers. *Advanced Materials Research*, Vols. 44-46, 651-658, ISSN 1022-6680

Gao, P. & Chen Y. (2009). The effect of nanometer modification on the performances of short fiber reinforced rubber-based sealing composite material. *Lubrication Engineering*, Vol.34, No.8, 66-69, ISSN 0254-0150 (in Chinese)

Gu B. Q. & Chen, Y. (2007). Development of a new kind of sealing composite material reinforced with aramid and pre-oxidized fibers. *Key Engineering Materials*. Vols. 353-358, 1243-1246, ISSN 1013-9826

Gu, B. Q. (1999). Application of model of gases flowing through porous media to gasket sealing. *Journal of Nanjing University of Chemical Technology*, Vol. 21, No. 1, 19-22, ISSN 1007-7537 (in Chinese)

Gu, B. Q., Shi, L. X. & Lu, X. F. (2000). Investigation of high temperature performances of spiral wound gaskets with flexible graphite filler. *China Petroleum Machinery*, Vol. 28, No. 2, 13-16, ISSN 1001-4578 (in Chinese)

Gu, B. Q., Lu, X. F. & Xu, G. P. (2001). Determination of performance indexes of non-asbestos fiber-reinforced elastomer gasket. *Proceedings of the 5th National Conference on Pressure Vessel Technology*, 116-120, ISSN 1007-7472.0.2000-05-023, Nanjing, China, Setember, 2001, Subcommittee on Pressure Vessels, Chinese Society of Mechanical Engineers, Hefei, China (in Chinese)

Gu, B. Q. (2002). Performances and application of non-asbestos gaskets. *Process Equipment & Piping*, Vol. 39, No. 1, 5-9, ISSN 1009-3281 (in Chinese)

Gu, B. Q., Li, Y. Y., Sun, Z. G., Huang, X. L., Zhou, J. F. & Shao, C. L. (2010). Study on time-correlated leakage model of nonmetallic gaskets. *Advanced Materials Research*, Vols. 97-101, 629-633, ISSN 10226680

Gu, B. Q. & Zhu, H. S. (1988). Tightness analysis of bolted flanged connections. *Petro-chemical Equipment*, Vol. 17, No. 1, 16-20, ISSN 1000-7466 (in Chinese)

Gu, B. Q. & Huang, X. L. (1997). Dichtheitsklassen und entsprechende Leckratenmeßverfahren. *3R international*, 1997, Vol.36, No.4/5,198-201, ISSN 0340-3386

Gu, B. Q., Li, X. H. &Tian, Z. (2004). *Static seal design Technology*. Standards Press of China, ISBN 7-5066-3430-9, Beijing, China (in Chinese)

Gu, B. Q. & Chen, Y. (2006). Tightness assessment method of bolted flanged connections at elevated temperature. *Lubrication Engineering*, Vol. 178, No. 6, 39-41, ISSN 0254-0150 (in Chinese)

Gu, B. Q., Chen, Y., Zhang, Y. & Zhu, R. S. (2007a). Tightness prediction of flanged joints with elastic washers. *Proceedings of the Ninth International Conference on Engineering Structural Integrity Assessment*, 1512-1515, ISBN 978-7-111-05002-5, Beijing, China, October, 2007, China Machine Press, Beijing, China

Gu, B. Q., Chen, Y. & Zhu, D. S. (2007b). Prediction of leakage rates through sealing connections with nonmetallic gaskets. *Chinese Journal of Chemical Engineering*, Vol. 15, No. 6, 837-841, ISSN 1004-9541

Marchand, L., Derenne, M. (1996). Long term performance of elastomeric sheet gasket materials subjected to temperature exposure. *Proceedings of the International Conference on Pressure Vessel Technology*, 107-123, ISBN 079181789X, Montreal, Can, July, 1996, ASME, New York, NY, USA

Payne, J. & Bazergui, A. (1990). *Summary Report on Elevated Temperature Tests for Asbestos-free Gasket Materials*, NACE for MTI., ISBN 1-877914-10-X

Payne. J. R., Mueller, R. T., Bazergui, A. (1989a). Gasket qualification test scheme for petrochemical plants. Part I. Test methods and application results. *Application of Modal Analysis Techniques to Seismic and Dynamic Loadings*, Vol. 158, 53-58, ISSN 0277027X, Honolulu, HI, USA, July, 1989, ASME, New York, NY, USA; JSME, Tokyo, Jpn

Payne. J. R., Mueller, R. T., Bazergui, A. (1989b). Gasket qualification test scheme for petrochemical plants. Part II. Quality criteria and evaluation schemes. *Application of Modal Analysis Techniques to Seismic and Dynamic Loadings*, Vol. 158, 59-69, ISSN 0277027X, Honolulu, HI, USA, July, 1989, ASME, New York, NY, USA; JSME, Tokyo, Jpn

Piringer, S. & Rustemeyer U. (2004). Improving performance of fiber-based gaskets at elevated temperatures-the sandwich approach. *Sealing Technology*, Vol. 2004, No. 3, 6-11, ISSN 1350-4789

Tsuji, H., Kitagawa, H., Kodaira, N. (2004). Effect of aging time on sealing performance of non-asbestos gasket at elevated temperature. *Analysis of Bolted Joints*, 209-214, ISBN 0791846814, San Diego, CA, United States, July, 2004, ASME, New York, NY, USA

Xie, S. J. & Cai R. L. (2002). Research on mechanical behaviour of the compressed nonasbestos fiber reinforced rubber sheet. *Fluid Machinery*, Vol. 30, No. 12, 4-6, ISSN 1005-0329 (in Chinese)

Xie, S. J. & Xie, J. M. (2004). Investigation into the effect of nonasbestos fibers on the performances of compressed nonasbestos sealing sheets. *Non-Metallic Mines*, Vol. 27, No. 1, 50-52, ISSN 1000-8098 (in Chinese)

Zhu, D. S., Gu, B. Q. & Chen, Y. (2007). Study on Tensile Strength of Short-fiber-reinforced elastomer matrix composites. *Proceedings of the Ninth International Conference on Engineering Structural Integrity Assessment*, 803-807, ISBN 978-7-111-05002-5, Beijing, China, October, 2007, China Machine Press, Beijing, China

Zhu, D. S., Gu, B. Q. & Chen, Y. (2008). Study on Temperature-Dependent Tensile Strength of Short-Fiber-Reinforced Elastomer Matrix Composites. *Advanced Materials Research,* Vols. 44-46, 97-104, ISSN 1022-6680

Zhu, D. S. (2008). Prediction of Mechanical Performances of Short Fiber Rubber Matrix Composite. A dissertation for the doctoral degree at Nanjing University of Technology (in Chinese)

5

Properties of Polymer Composites with Cellulose Microfibrils

Denis Mihaela Panaitescu, Adriana Nicoleta Frone, Marius Ghiurea, Catalin Ilie Spataru, Constantin Radovici and Michaela Doina Iorga
National Institute of Research and Development in Chemistry and Petrochemistry Romania

1. Introduction

Plastic materials are indispensable in our lives but they are an important source of environmental pollution. In order to reduce the environmental load generated from the disposal of used plastic materials, a growing interest has been focused on biodegradable polymers based materials. These polymers can be completely converted into water and carbon dioxide by the action of microorganisms after disposal.

Owing to their good mechanical properties, polymer composites with micro or nano cellulose fibers are promising substitutes for glass-fiber-containing composites in many industrial fields like automotive, construction, electronics, biomedicine, cosmetics, and last but not least, the packaging industry. Among the advantages of using cellulose fibers as reinforcements in polymer composites renewability, low cost, low density, low energy consumption, high specific strength and modulus, high sound attenuation, low abrasivity and relatively reactive surface are the most important. Cellulose fibers modified at nanometer size induce much higher mechanical properties to polymer matrices as regard to common cellulose fibers because of their higher crystallinity and mechanical properties combined with higher surface area and better interfaces.

The production of nano-scale cellulose fibers and their application in composite materials has gained increasing attention in the past two decades (Hubbe et al., 2008). Two main methods have been used to generate nano-scale cellulose fibers: the chemical way, mainly by strong acid hydrolysis, which removes the amorphous regions of cellulose fiber and produces nano-size fibrils and the mechanical method, by high pressure refiner treatment (Chakraborty et al., 2005), grinder treatment (Taniguchi and Okamura, 1998), microfluidizer (Zimmermann et al., 2004), and high-pressure homogenizer treatment (Istvan and Plackett, 2010).

In this chapter the discussion is focused on the physical and mechanical properties of polyvinyl alcohol (PVA) as a biodegradable matrix reinforced with three types of cellulose fibers prepared by different methods. This chapter will first discuss the most used methods for cellulose nanofibers preparation and the reinforcing effect of these cellulose fibers in different polymer matrices. This will be followed by the experimental results obtained in our laboratory on PVA composites prepared with cellulose fibers isolated by mechanical treatment, acid hydrolysis and ultrasound treatment. The isolated cellulose fibers were characterized by scanning electron microscopy (SEM) and X-ray diffraction (XRD) and PVA composites, prepared through film casting method, by XRD, thermal analysis and mechanical characterization.

2. State of the art

Cellulose was rediscovered at the end of the last century due to nanoscience and improved technologies available for the disintegration of cellulose in submicron and nanosize fibers. The native cellulose molecule consists of linear glucan chains with repeating (1-4)-β-glucopyranose units. The supramolecular structure of cellulose is very complex and it has been the subject of many studies (Eichhorn et al., 2010).

In living plants cellulose occurs in fibers, the cell wall of wood fibers consisting of repeated crystalline structures resulted from the aggregation of cellulose chains, termed microfibrils. These microfibrils, with high aspect ratio, are surrounded by an amorphous matrix of hemicelluloses and lignin. Besides wood, bast fibers (Cao et al., 2008; Li and Pickering, 2008; Shaikh et al. 2009; Belhassen et al., 2009), agricultural residues (El-Sakhawy and Hassan, 2007; Fama et al., 2009), leaf fibers (de Rodriguez et al., 2006; Zuluagaet al., 2009), bacterial cellulose (Nakagaito et al., 2005; Yano et al., 2008), the shell of some fruits and vegetables (Chen et al., 2009; Rosa et al., 2010) or tunicates (Petersson and Oksman, 2006) were used as sources for cellulose nanofibers preparation. Among the methods of nanofiber isolation, the most used are: mechanical disintegration, acid hydrolysis and biological treatments. Mechanical treatments involve conventional refining procedures (Hubbe et al., 2008), crushing and cryocrushing (Bhatnagar and Sain, 2005; Panaitescu et al., 2007a; Alemdar and Sain, 2008) or high pressure homogenization of cellulose source suspended in water (Lee et al., 2009a). Chakraborty isolated cellulose fibers with submicron diameters by combining the severe shearing in a refiner with high-impact crushing under liquid nitrogen (Chakraborty et al., 2005).

Mechanical treatments have some disadvantages related to the high energy required in the process and to the high degree of inhomogenity, the resulted material containing larger fibrils in addition to microfibrils (Nakagaito and Yano, 2004; Nakagaito and Yano, 2005; Andresen et al., 2006; Andresen and Stenius, 2007; Stenstad et al. 2008).

Chemical treatments involve mainly acid and alkaline hydrolysis. Acid hydrolysis leads to the isolation of micro and nanofibers with a high degree of crystallinity by removing the amorphous regions of the raw cellulosic material. Using sulfuric acid, a negatively charged surface of the cellulose fibers can be obtained, through the esterification of hydroxyl groups by the sulfate ions, which prevents the agglomeration of fibers.

Many researchers have successfully used this method, alone or in combination with others methods, to obtain cellulose nanofibers (Zhang et al., 2007; El-Sakhawy et al., 2007; Moran et al., 2008; Chen et al., 2009): Bondenson et al. (2006) treated microcrystalline cellulose (MCC) with sulfuric acid in concentration of 63.5% (w/w) and isolated cellulose whiskers with an aspect ratio between 20 and 40, Lee et al. (2009b) obtained nanocellulose fibers by hydrobromic acid hydrolysis of MCC, an increasing in MCC crystallinity being reported, Rosa et al. (2010) obtained cellulose whiskers with diameters as low as 5 nm and aspect ratio of up to 60 by sulfuric acid hydrolysis of coconut husk fibers.

Degradation of the cellulosic substrate may also occur in the presence of microorganisms (fungi, bacteria) or, directly, with enzymes. Enzymatic treatment of cellulose was performed by Henriksson et al. (2007) who reported the obtaining of cellulose nanofibers from two different commercial bleached wood sulphite pulps. Li et al. (2008) reported that removal of non-cellulosic components from cellulose fibers by enzyme treatment can increase the crystallinity, thermal stability and the amount of -OH groups of the treated fibers.

New and environmental friendly methods for cellulose nanofibers isolation were also tested in laboratory conditions. Ultrasonication had been used alone or in combination with other methods (e.g. acid hydrolysis) to obtain cellulose fibers. Filson and Dawson-Andoh (2009) obtained nanofibers with an average diameter between 21 and 23 nm applying this method.

Application of cellulose nanofibers in polymer reinforcement is a relatively new research field. Two directions could be detected: research studies which explore the development of nano-bio-plastics as fully biodegradable nanocomposites and studies aimed at dispersing cellulose nanofibers in non-biodegradable, petroleum derived polymers. Polymer composites containing cellulose nanofibers were prepared with polyvinyl alcohol (PVA), polylactide (PLA), starch and polycaprolactone but also with polyethylene or polypropylene. Improvement in term of brittleness, thermal stability or barrier properties were signaled in case of biodegradable polymers reinforced with cellulose nanofibers (Iwatake et al., 2008; Chen et al., 2009; Fama et al., 2009; Pandey et al., 2009; Suryanegara et al., 2009; Bledzki et al., 2009; Chang et al., 2010). Cellulose fibers reinforcement could be a good solution for starch, which is a low cost source of biodegradable composites but a material with very poor mechanical properties. Dufresne and Vignon (1998, 2000) reported an improvement by a factor of 3.5 of the tensile modulus of starch at 50 wt% cellulose fibers addition. PLA is a commercially available biopolymer with similar properties to petroleum derived thermoplastics. Some drawbacks like brittleness and low thermal stability restrict its applications. Using cellulose fibers (10 wt%) and a solvent exchange method, Iwatake et al. (2008) succeeded in fabricating a composite sheet with uniform filler distribution showing an increase of Young's modulus and tensile strength of 40 and 25% respectively. Chemical modification of cellulose has been explored as a route for improving filler dispersion in hydrophobic polymers such as polyethylene or polypropylene (Panaitescu et al., 2007b; Rahman et al., 2009; Yanjun et al., 2010).

PVA is a water-soluble and biodegradable polymer with excellent chemical resistance and an interesting material for biomedical applications. PVA has no toxic action on the human body being used to manufacture medicines cachets, yarn for surgery, controlled drug delivery systems (Tang et al., 2009). New fields of application regard cardiovascular devices (Millon and Wan, 2006), dialysis membrane, artificial cartilage, tissue engineering scaffold (Zhou et al., 2010). Development of ecofriendly packaging materials is still a challenging area and many studies are focused on the improvement of PVA mechanical and barrier properties by combination with other polymers or fillers in order to use it in the packaging industry (Sedlarik et al., 2006). For many other applications, mechanical properties of PVA should be substantially improved without damaging its valuable properties. Low cellulose fibers addition could be an appropriate solution. Many studies emphasized the effectiveness of large amount of cellulose fibers in improving mechanical properties of PVA. Zimmermann et al. (2004) reported an improvement of the elastic modulus and tensile strength of up to five times and three times, respectively, in the case of dispersing 20 wt% cellulose fibers in PVA. An increase of about five times of the tensile strength relative to the reference polymer was reported by Bruce et al. (2005) at 50 wt% cellulose fibers in PVA and an increase by a factor of 3.5 at the same concentration of fibers in the work of Leitner et al. (2007). Nevertheless, no increase in tensile strength and modulus was observed by Lu et al. (2008) above 10 wt% cellulose fibers in PVA.

Despite the important publication activity dealing with cellulose nanofibers and related nanocomposites the application of such materials is quite limited. There are several reasons for this situation. One of them is the difficulty of the separation of plant fibers into smaller constituents with uniform and reproducible characteristics. Another reason is the high energy demands for most isolation processes. Problems concerning interfacial adhesion and uniform dispersion of the cellulose fibers in the polymer matrix delayed also the wide application of these materials.

Moreover, results obtained in different laboratories on polymer composite containing cellulose fibers are often contradictory because of many factors controlling the process like filler size and content, interface adhesion, fiber aspect ratio and orientation, fiber dispersion in the matrix and stress transfer efficiency through the interface (Dufresne et al., 2003). However, the role of nanofiber characteristics and aggregation in influencing macroscale properties of polymer matrix is not yet well understood.

Important research work was focused on studying the changes induced in PVA characteristics by high concentration of cellulose fibers prepared by different processes. These results were obtained in different laboratories and different conditions were applied in order to prepare PVA composite films. PVA is a polymer sensitive to preparation conditions so that significant changes may occur depending on the working atmosphere and heat treatments applied. Knowing the high influence of preparing and characterization conditions on PVA properties, it is difficult to select the proper system PVA/cellulose fibers for a target application based on existing information. Moreover, achieving improved mechanical properties with the less concentration of cellulose fibers is desirable for preserving transparency and flexibility of PVA films. For these reasons, we further describe research results obtained in our laboratory for monitoring the influence of cellulose fibers obtained by different treatments on the microscopic and macroscopic properties of PVA, all the samples being prepared in the same conditions. PVA composites were prepared by solution casting using low concentration (≤5 wt.%) of cellulose fibers obtained by mechanical treatment (L), acid hydrolysis (H) and ultrasound treatment (U). The isolated cellulose fibers were characterized by scanning electron microscopy (SEM) and X-ray diffraction (XRD). PVA composites containing these fibers were characterized with the aim of determining thermal and mechanical behavior.

3. Experimental part

3.1 Materials

Polyvinyl alcohol, PVA 120-99, 1200 polymerization degree and 99% hydrolysis degree was used as matrix. Microcrystalline cellulose (MCC) with a mean particle size of 20 μm and an aspect ratio of 2 - 4, from Sigma-Aldrich, was used as raw material for the preparation of nanofibers and cellulose microfibers obtained from regenerated wood fibers (L) through mechanical treatment were donated by National Institute of Wood.

3.2 Nanofibers isolation

(a) By ultrasonic treatment

MCC with a mean particle size of 20 μm, as revealed by SEM image (Fig. 1), was dispersed under continuous stirring in distilled water (1/500) and sonicated at 80% power for 20 minutes using an ultrasonicator type Vibra Cell VC505 (500 W, 20 KHz). In order to control the process temperature, the beaker with the cellulose suspension was put in a water bath with controlled temperature. The appearance of turbidity in the supernatant was a definite indication of the presence of cellulose nanofibers. Cellulose fibers suspension was obtained by decanting the supernatant into other vessel. Cellulose fibers - sample U was characterized by X ray diffraction after drying 4 hours at 40^{0}C.

(b) By acid treatment

MCC was dispersed in distillated water (1/10). The resulted suspension was placed in an ice bath and stirred while concentrated sulfuric acid was added by drop until the desired acid concentration of 63% was reached. The suspension was heated at 40^{0}C while stirred for 5 h. The obtained mixture was washed with distillated water using repeated centrifuge cycles (20 min at 7000 rpm, Universal 320R Ultracentrifuge) until a pH of 4 was reached. Further, the obtained sample of cellulose fibers (H) was ultrasonicated (Elmasonic S40H, Elma) for five hours in an ice bath in order to avoid the sample overheating. Cellulose fibers - sample H was characterized by X ray diffraction after drying 4 hours at 40^{0}C.

(c) Cellulose fibers obtained through mechanical treatment (L) were used as received.

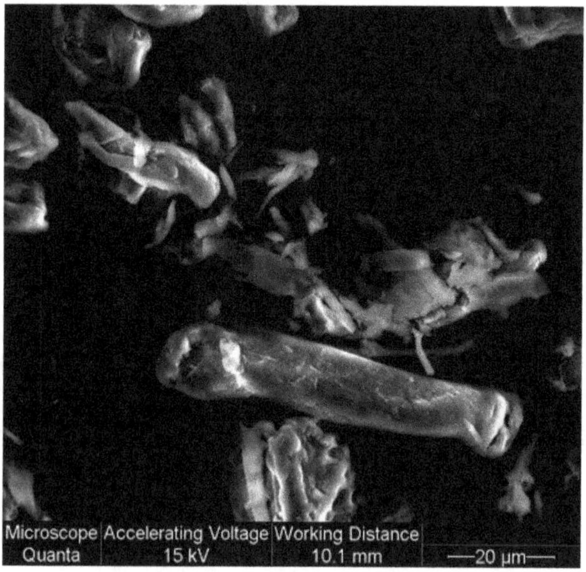

Fig. 1. SEM image of commercial microcrystalline cellulose (MCC)

3.3 Nanocomposite films preparation

In order to obtain PVA/cellulose fibers composite films with different filler concentrations, the required amount of PVA corresponding to a final concentration of 10% in the aqueous medium and the calculated amount of L, U and H cellulose fibers suspensions to achieve 1, 3 and 5 wt% filler concentration in the final composite were mixed using a high speed stirrer (500 min^{-1}). The stirring was performed at 80^{0}C for 3 hours, and the resulting mixture was degassed for approximately 15 minutes in an Elmasonic S40 H ultrasonic bath. The films were cast on a PET plate and were kept at room temperature for 2 days until they were completely dried and then removed from the PET plates and placed in a desiccator for two weeks before mechanical characterization. Neat PVA films were obtained under similar conditions.

3.4 Characterization methods

SEM micrographs of cellulose fibers were obtained with an environmental Quanta Scanning Electron Microscope 200, with a tungsten electron source in the following conditions: accelerating voltage 10 – 30 kV, no coating. The crystalline structure of cellulose fibers was determined by XRD with a DRON-UM diffractometer (horizontal goniometer Bragg-Brentano) using Co Kα radiation (wavelength λ = 1.79021 Å), scanning from the 2θ value of 5^0 to 36^0 at a scanning rate of 0.05^0/5 sec. Samples were analyzed in reflection mode.
Mechanical behavior of composite films at room temperature was tested in tensile mode according to ISO 527-1:1993 Part 1 and ISO 527-3:1995 Part 3 by an Instron 3382. The Universal Testing Machine was equipped with a video extensometer for strain measurement and with a load cell of 1 kN capacity. The following data: video axial strain at break (elongation at break), tensile stress at break (tensile strength), Young's modulus (tensile modulus) and other optional features were automatically display using the software of Instron 3382, Bluehill 2. Young's modulus on the strain channel with a start value of 0.05% and an end value of 0.25% was calculated by the system applying the least square fit algorithm to each region between the start and end value selecting the highest slope. The load cell of 1 kN capacity, used for PVA composites tensile properties measurements has a sensitivity of 0.001 N. Test specimens were cut to 110 mm (length) x 10 mm (width) x 0.035-0.045 mm (thickness) from the films maintained in desiccators, 7 specimens being tested for each sample. Tensile modulus was determined at a crosshead speed of 2 mm/min and tensile strength and elongation at 10 mm/min. Lower crosshead speed was necessary for correct determination of tensile modulus as stipulated by ISO 527 Part 1. Differential scanning calorimetry (DSC) and thermogravimetric analysis (TGA) were performed on a SDT Q600 V20.9 from TA Instruments under Heliom flow (100 mL/min). The samples weighing between 5 and 10 mg were packed in aluminium pans and placed in the DSC cell. The samples were tested from the ambient temperature to 600^0C at a heating rate of 10^0C/min. The DSC glass transition temperature (Tg) was taken at the onset of the glass transition region. The PVA composite samples were conditioned for 24 h at 40 ^{0}C in a vacuum oven prior to thermal analysis.

4. Results

4.1 Cellulose fibers characterization

The three samples of cellulose fibers, L, H and U, were investigated by scanning electron microscopy (Fig. 2, 3a-c and 4). SEM image obtained for sample L (Fig. 2) revealed the distribution of fiber diameters ranging from 5 to 20 μm. The surface of L is smoother and the aspect ratio l/d of 4 to 14, greater then the l/d of MCC sample, ranging from 2 to 4.
The acid hydrolysis of MCC is a heterogeneous process which involves the diffusion of acid into the cellulose fibers, and the subsequent cleavage of glycosidic bonds. Figures 3a-c are SEM micrograph of acid treated MCC after different time of acid hydrolysis (2, 4 and 5 hours). Fig. 3a shows how the acid broke the MCC particle, attacking amorphous areas and releasing aggregates of cellulose microfibrils. Fig. 3b shows that a big fraction of amorphous areas were entirely destroyed and cellulose crystallites in needle form released. Portions with agglomerated microfibrils could also be observed.

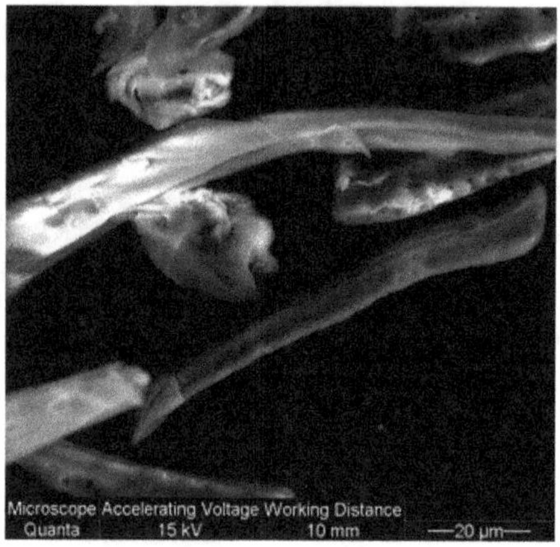

Fig. 2. SEM image of L type cellulose fibers

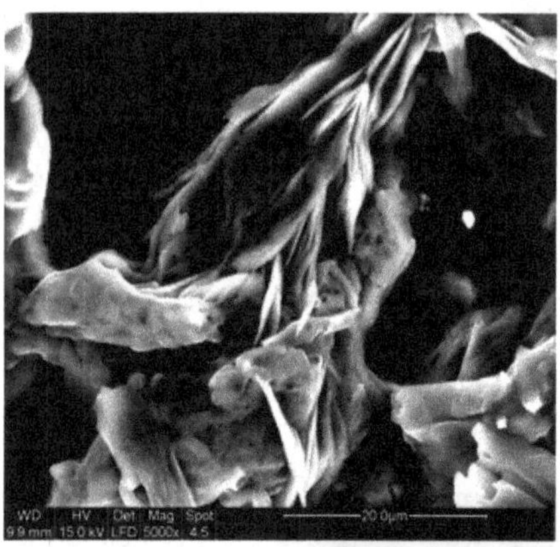

(a)

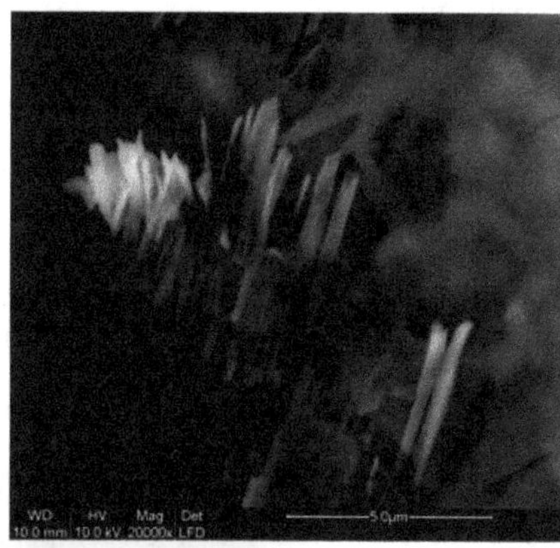

(b)

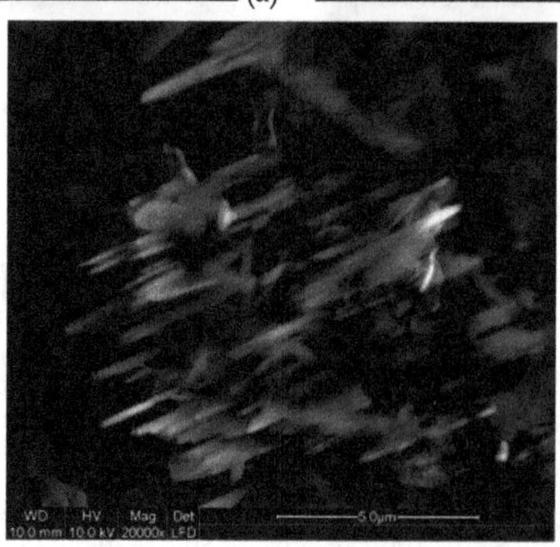

(c)

Fig. 3. SEM micrographs of (a) the first stages of acid attack of MCC, (b) releasing of cellulose microfibrils aggregates and (c) cellulose fibers type H

In Fig. 3c rod like cellulose microfibrils with diameter of 100 – 400 nm and aspect ratio of 10 to 20 are observed. Cellulose fibers with smaller diameters can not be detected in SEM images, at this magnitude. It is presumed that the further ultrasonic treatment destroyed the agglomerates.

SEM micrograph of U type cellulose fibers (Fig. 4) shows that MCC particles are entirely disintegrated into fragments with different shapes and size by sonication treatment. The effectiveness of MCC defibrillation process is very high, particles and fibers with a diameter of 20 - 800 nm being observed. Individual particles but also large aggregates of cellulose fragments can be easily observed. It is possible that a re-aggregation to take place, promoted by the strong hydrogen bonding between the individual cellulose whiskers prepared by ultrasonication. The heterogeneity of shapes and size is very great in this sample because of the high energy involved in ultrasonication, so that o separation process was imperative for removing higher size aggregates of cellulose fibers.

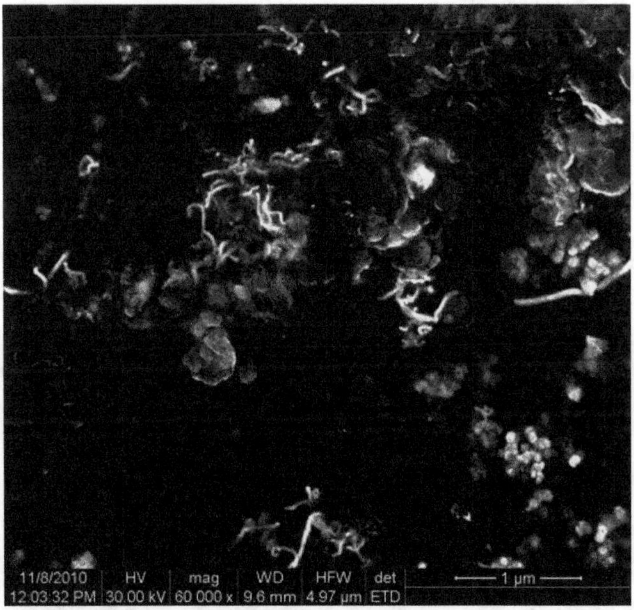

Fig. 4. SEM image of cellulose fibers prepared by ultrasonication

X-ray diffraction patterns of MCC and cellulose fibers (L, H and U) are shown in Fig. 5. The characteristics peaks originated from cellulose I are visible for all the samples at the same values of 2θ (17.3°, 18.9°, 26.2°). These values correspond to the interplanar spacings of the three principal planes (101), $(10\bar{1})$ and (002) of the monoclinic unit cell of cellulose: d_{101} = 5.99 Å, $d_{10\bar{1}}$ = 5.40 Å and d_{002} = 3.95 Å, respectively. Very close values for 2θ were reported by Lee et al. (2009b) taking into account the different wavelength of CuKα radiation used. In the case of sample L the two first peaks are not separated forming a broad peak, probably because of the more complex structure of sample L which contain beside celluloses, hemicelluloses and lignin.

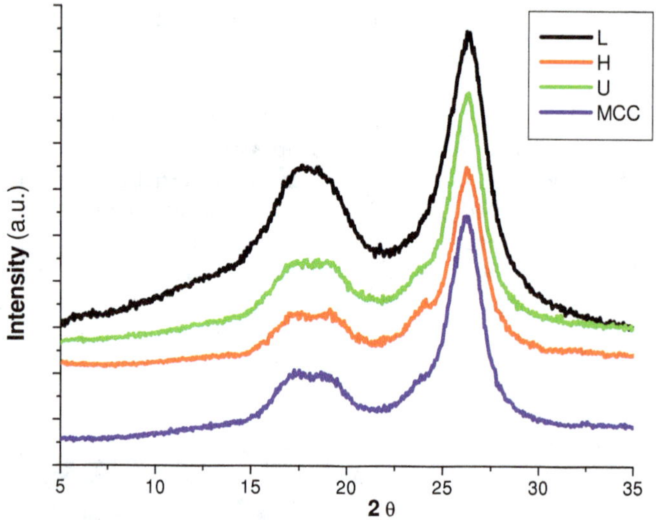

Fig. 5. X-ray diffraction patterns of MCC and cellulose fibers (samples L, H and U)

4.2 Polymer composites characterization

a. Mechanical characterization

The tensile strength, elongation at break, and elastic modulus of the PVA/cellulose fibers composites compared with those of unreinforced PVA are shown in Figs. 6 - 8.

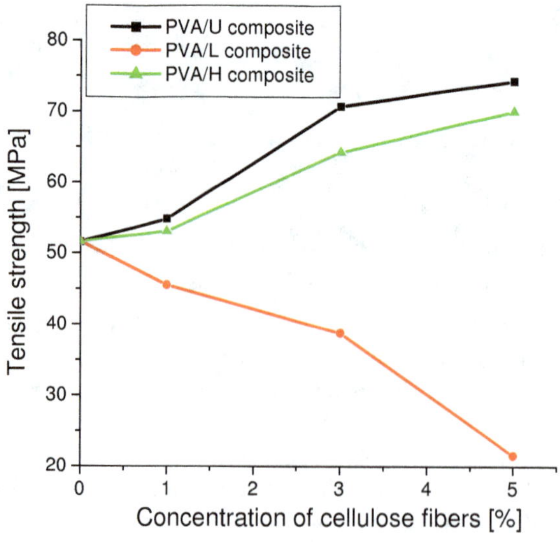

Fig. 6. Tensile strength of neat PVA and PVA reinforced with cellulose fibers U, L and H

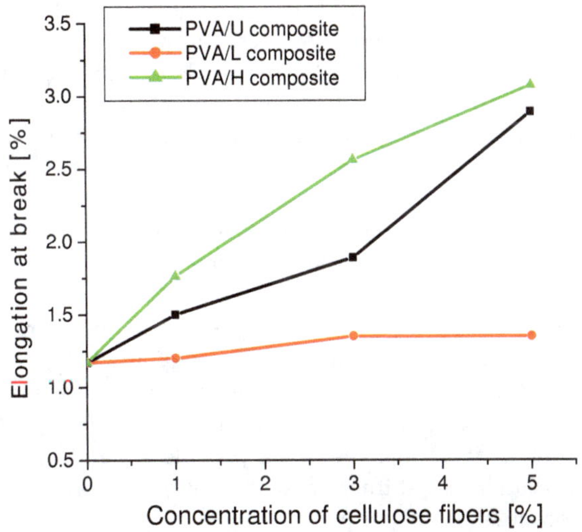

Fig. 7. Elongation at break of neat PVA and PVA reinforced with cellulose fibers U, L and H

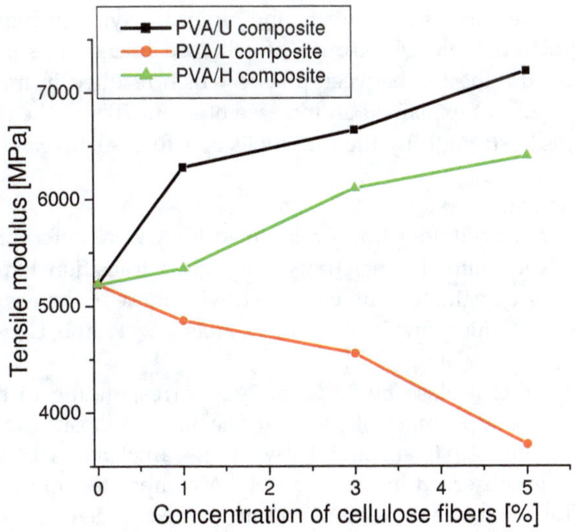

Fig. 8. Tensile modulus of neat PVA and PVA reinforced with cellulose fibers U, L and H

It is apparent that the three types of fibers differently influence the mechanical properties of PVA matrix. U and H fibers lead to an improvement of mechanical properties, both the strength and modulus and elongation at break being enhanced and the type L to a decrease in strength and stiffness. The observed differences are mainly due to the different size of fibers, U and H fibers having diameters on the nanoscale while L being micron sized fibers.

It is known that an important increase in strength and stiffness of polymer matrix can be obtained at concentrations above 30% of micron sized cellulose fibers, lower concentration of 20 - 25% being effective only in the case of a good treatment of fibers or coupling agent adding. In these composites, elongation at break knows significant decreases, polymer matrix ductility being severely altered by micron sized fibers (Bledzki et al., 2003; Ganster et al., 2006). Hence, it was not expected that the low concentration of L fibers could improve PVA mechanical properties.

In contrast to conventional filled polymer systems where the increase of strength and stiffness is accompanied by an important decrease of ductility, in nanocomposites a simultaneous increase of Young's modulus, tensile strength and elongation could appear, especially in the case of strong filler/matrix interface (Lin et al., 2009). Therefore, the incorporation of U and H fibers in PVA which provide a significant improvement of all mechanical properties (strength, stiffness and ductility) is a proof of a good interfacial adhesion in our composites.

The highest improvement of tensile strength and modulus in regard to PVA (Fig. 6 and 8) was obtained for PVA/U composites, 5 wt% U in PVA: almost 50% and 40%, respectively. The different behavior of U and H fibers as reinforcements in PVA could be a result of their different size (in the case of U fibers, the nanosized fraction were separated from the coarse fraction) and the higher tendency to form agglomerates in the case of higher aspect ratio fibers (H). The enhancement of mechanical properties of PVA/U and PVA/H composite films by comparison with neat PVA resulted from the good adhesion at the filler/matrix interface, favored by the small size of fibers and, accordingly, their high surface area. The hydrogen bonding between the OH groups of cellulose fibers and similar group of PVA matrix led to improved adhesion between phases which resulted in improved mechanical properties. Lee et al. (2009b) explained an increase of about 70% of the elastic modulus and up to 55% of the tensile strength by the intermolecular forces between cellulose fibers and PVA matrix.

b. Thermal characterization

The thermal properties of cellulose fibers reinforced PVA were determined from DSC and DTG thermograms. The main thermal transitions, glass transition temperature (Tg) and melting point (Tm), were evaluated and compared with those of the neat PVA. DSC results of PVA composites containing 5wt% of cellulose fibers, L, H and U types, are shown in Fig. 9.

PVA exhibits an endotherm close by 57°C (57.2^{0}C) corresponding to the glass transition temperature of PVA. The appearance of one Tg in the nanocomposite samples highlights the good interaction of cellulose fibers and PVA in the amorphous phase. No significant changes in Tg value are observed in the case of PVA composites containing 5 wt% U or H fibers as shown in Table 1, but a decrease with four degrees is detected in PVA filled with 5 wt% L fibers as regard to neat PVA. This behavior is caused by the decrease of the cohesive energy density in the amorphous phase of the PVA/L composites in comparison with neat PVA and can be explained by the rupture of the hydrogen bonds in PVA because of larger size of L type fibers (tens of microns in diameter and hundred of microns in length). As a result, the rubbery properties and the increase of the mobility of the macromolecules will appear from lower temperatures. Nanometer sized fibers, H and U, will not cause massive breaks of hydrogen bonds in PVA and, the expected interactions between these cellulose fibers and PVA will not alter the supramolecular structure of the amorphous phase. Nevertheless it seemed that these new formed interactions are not strong enough to slow the chain mobility associated with glass transition since the Tg values measured on PVA/H and PVA/U composite samples are close to that of neat PVA.

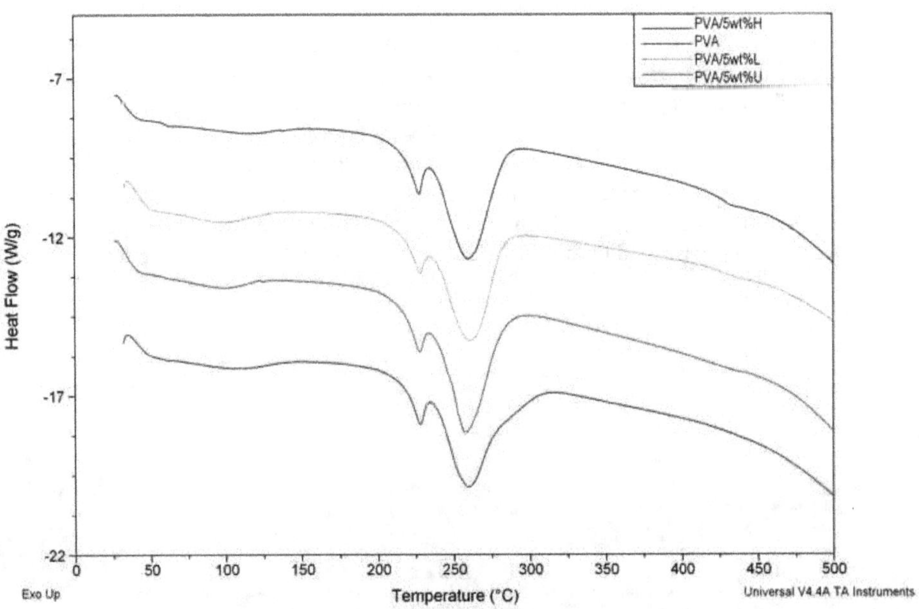

Fig. 9. DSC diagrams of neat PVA and PVA composites containing U, L and H fibers

Samples	Tg ^{0}C	Tm ^{0}C	ΔHm J/g	Xc %
PVA	57.2	227.1	79.3	48.7
PVA/5%U	56.8	226.9	70.5	45.5
PVA/5%L	53.0	227.2	60.0	38.7
PVA/5%H	57.0	227.3	78.1	50.4

Table 1. DSC data of neat PVA and PVA composites containing 5 wt.% fibers

PVA exhibits a sharp endothermic curve with a peak at 227.1^{0}C, as shown in Fig. 9, corresponding to the melting of the crystalline phase of PVA. Tm values remains roughly constant for all PVA composites whatever the type of fiber used but the heat of fusion is different depending on the filler characteristics. The degree of crystallinity (Xc) of the PVA component in the composite was obtained as follows:

$$X_c = \frac{\Delta H_f}{w * \Delta H_0} * 100 \tag{1}$$

where ΔH_f and ΔH_o are the heats of fusion for PVA and 100% crystalline PVA, respectively and w is the mass fraction of PVA in the composite. ΔH_o was taken 163 J/g (Ramaraj et al., 2010).

A highest degree of crystallinity is observed in PVA/H composite and the lowest in PVA/L composite. PVA filled with H type fibers show higher crystallinity than neat PVA. This could be ascribed to stronger interactions between cellulose fiber surface and adjacent PVA chains. The nucleating effect of cellulose fibers could also explain this increase of crystallinity. PVA/H composite shows a higher crystallinity as regard to PVA/U composite but the mechanical behavior of the latter is better. This is due to the fact that, besides the matrix crystallinity, other factors influence the tensile properties of the composite. The nanoscale dispersion of the filler and its orientation in the matrix are among these factors.

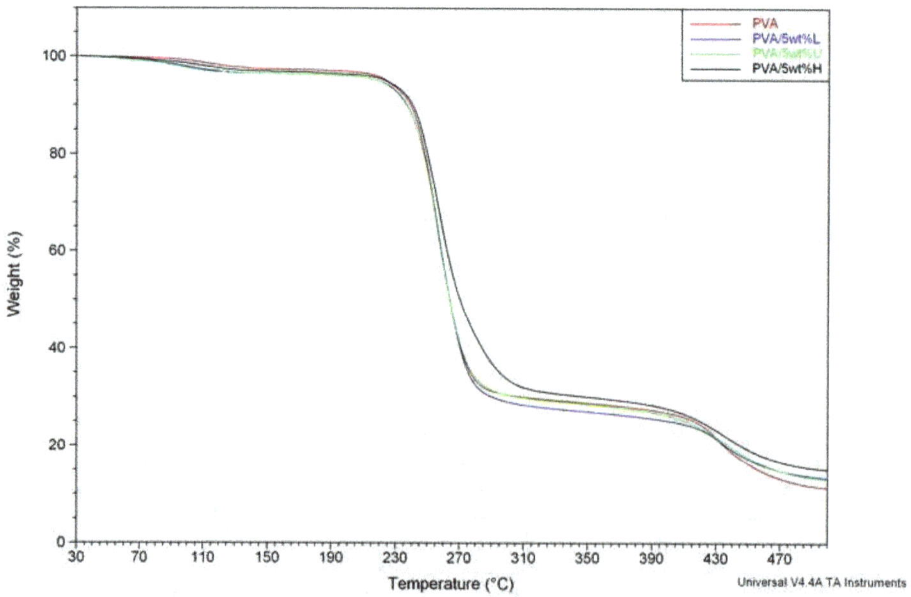

Fig. 10. TGA thermograms of neat PVA and PVA composites containing U, L and H fibers

Thermogravimetric analysis (TGA) was used to investigate the effect of cellulose fibers on the thermal stability of the composites. In the obtained PVA thermograms (Fig. 10) three main weight loss regions can be observed (Lee et al., 2009b; Qua et al., 2009). All the samples show an initial weight loss in the region 75 - 150 ^{0}C caused by the evaporation of water. Between 2.5 and 3.5 wt.% physically and chemically bound water was detected in neat PVA and PVA composites thermograms. Figure shows that the second degradation region is located between 220 and 300 ^{0}C and is due to the pyrolysis of cellulose fibers and to the degradation of PVA films, the weight loss being around 70% for all the samples. As reported by Qua et al. (2009), the second stage of degradation mainly involves dehydration reactions and the formation of volatile products.

The third stage weight loss occurrs above 400 ^{0}C and consists of decomposition of carbonaceous matter (Lee et al., 2009b).

It can be observed in Fig. 10 that the differences in TGA curves are negligible for neat PVA and PVA composites. However, some differences can be detected in the case of PVA/H composite. TGA's first order derivative (Fig. 11) show a broadening of the main decomposition peak and a shift of the onset temperature for the third decomposition stage to higher temperatures. This can be due to sulfonic groups bound to the cellulose fibers by acid hydrolysis which influence the degradation process.

The onset degradation temperature (Td) and enthalpy (ΔHd) could be easily determinated from DTG curves (Table 2). Slightly higher onset degradation temperatures were obtained for PVA composites as regard to neat PVA, showing a marginal increase of the thermal stability caused by the cellulose fibers. Lee showed that the thermal stability of PVA composites was improved with the increase of the nanocellulose loading (Lee et al., 2009b).

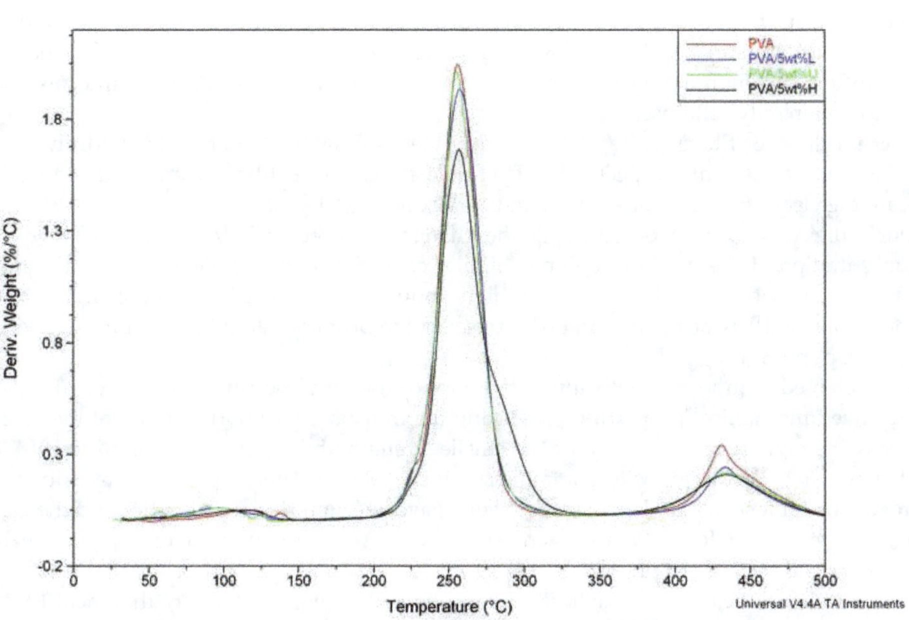

Fig. 11. DTG thermograms of neat PVA and PVA composites containing U, L and H fibers

Samples	Td ^0C	ΔHd J/g
PVA	238.7	513.8
PVA/5%U	242.9	509.2
PVA/5%L	239.7	480.5
PVA/5%H	239.8	525.3

Table 2. DSC data of neat PVA and PVA composites containing 5 wt.% fibers

The lowered PVA/L decomposition enthalpy was thought to be related to the more complex structure of these fibers containing beside cellulose, hemicellulose and lignin and to the more porous fiber structure.

5. Conclusion

The application of cellulose nanofibers in polymer reinforcement is a relatively new research field. The development of fully biodegradable nanocomposites is still a challenging area.
In this chapter discussion is focused on the physical and mechanical properties of polyvinyl alcohol as a biodegradable matrix reinforced with cellulose fibers prepared by different methods. Three types of fibers with different characteristics in terms of composition, size and aspect ratio were tested as reinforcements in PVA: L fibers of micron size and high aspect ratio contain beside cellulose, hemicellulose and lignin, H fibers with a nanosize dimension and high aspect ratio and U fibers with a nanosize dimension and low aspect ratio contain only cellulose.
These cellulose fibers were obtained by mechanical treatment, acid hydrolysis and ultrasound treatment, respectively. The isolated cellulose fibers were characterized by scanning electron microscopy (SEM) and X-ray diffraction (XRD).
Our work was focused on studying the changes induced in PVA characteristics by low concentration of these cellulose fibers. Similar conditions were applied in order to prepare PVA composite films, knowing the high influence of preparing and characterization conditions on PVA properties, in order to select the proper system PVA/cellulose fibers for a target application.
The obtained composites with low concentration of cellulose fibers (1 - 5 wt.%) showed improved mechanical properties, preserving the transparency and flexibility of the original films. The highest improvement of the tensile strength and modulus in regard to PVA was obtained for PVA composite containing 5 wt% cellulose fibers prepared by ultrasound treatment: almost 50% and 40%, respectively. Favorable interfacial properties and the lack of agglomerations at low fibers concentration in PVA were supposed to explain the high values of the mechanical properties.
PVA filled with acid treated cellulose fibers showed higher crystallinity than neat PVA, as resulted from DSC analysis. Stronger interactions between cellulose fiber surface and adjacent PVA chains and the nucleating effect of cellulose fibers were proposed to explain

this increase of crystallinity. Slightly higher onset degradation temperatures were obtained for PVA composites as regard to neat PVA, showing an increase of the thermal stability caused by the addition of cellulose fibers.

The biocomposites presented in this chapter are advanced materials with improved mechanical and thermal properties, high transparency and flexibility and large possibilities of application in packaging and other fields.

Future work will be focus on the study concerning the processing behavior of these materials and on achieving new preparation methods with industrial application.

6. References

Alemdar, A., and Sain, M. (2008). Biocomposites from wheat straw nanofibers: Morphology, thermal and mechanical properties. *Composites Sci. Technol,* Vol. 68, No. 2, pp. 557-565

Andresen, M., Johansson, L. S., Tanem, B. S., and Stenius P (2006). Properties and characterization of hydrophobized microfibrillated cellulose. *Cellulose,* Vol. 13, pp. 665–677

Andresen, M., and Stenius, P. (2007). Water-in-oil emulsions stabilized by hydrophobized microfibrillated cellulose. *J Dispers Sci Technol,* Vol. 28, pp. 837–844

Belhassen, R., Mendez, J. A., Boufi, S., Lopez, J. P., Puig, J., Pelach, A., and Mutje, P. (2009). Preparation and Properties of Biocomposites Based on Jute Fibers and Blend of Plasticized Starch and Poly(*b*-hydroxybutyrate). *Journal of Applied Polymer Science,* Vol. 114, pp. 313–321

Bhatnagar, A., and Sain, M. (2005). Processing of cellulose nanofiber-reinforced composites. *J Reinforced Plastics Composites,* Vol. 24, No. 12, pp. 1259-1268

Bledzki, A. K., and Faruk, O. (2003). Wood Fibre Reinforced Polypropylene Composites: Effect of Fibre Geometry and Coupling Agent on Physico-Mechanical Properties. *Applied Composite Materials,* Vol. 10, pp. 365–379

Bledzki, A. K., Jaszkiewicz A., and Scherzer, D. (2009). Mechanical properties of PLA composites with man-made cellulose and abaca fibres. *Applied Science and Manufacturing,* Vol. 40, pp. 404-412

Bondeson, D., Mathew, A., and Oksman, K. (2006). Optimization of the isolation of nanocrystals from microcrystalline cellulose by acid hydrolysis. *Cellulose,* Vol.13, No.2, pp. 171-180

Bruce, D. M., Hobson, R. N., Farrent, J. W., Hepworth, and D. G. (2005). High-performance composites from low-cost plant primary cell walls. *Appl Sci Manufact,* Vol. 36, pp. 1486–1493

Cao, X., Chen, Y., Chang, P. R., Stumborg M. and Huneault, M. A. (2008). Green Composites Reinforced with Hemp Nanocrystals in Plasticized Starch. *J. of Appl. Polym. Sci.,* Vol. 109, No. 6, pp. 3804–3810

Chakraborty, A., Sain, M. and Kortschot, M (2005). Cellulose microfibrils: A novel method of preparation using high shear refining and cryocrushing *Holzforschung,* Vol. 59, pp. 102–107

Chang, P. R., Jian, R., Zheng, P., Yu, J., and Ma, X. (2010). Preparation and properties of glycerol plasticized-starch (GPS)/cellulose nanoparticle (CN) composites. *Carbohydrate Polymers,* Vol. 79, pp. 301–305

Chen, Y., Liu, C., Chang, P. R., Cao, X., and Anderson, D. P. (2009). Bionanocomposites based on pea starch and cellulose nanowhiskers hydrolyzed from pea hull fiber: Effect of hydrolysis time. *Carbohydrate Polymers*, Vol. 76, No. 4, pp. 607-615

de Rodriguez, N. L. G., Thielemans, W., and Dufresne, A. (2006). Sisal cellulose whiskers reinforced polyvinyl acetate nanocomposites. *Cellulose*, Vol. 13, No. 3, pp. 261–270

Dufresne, A, and Vignon, M R (1998). Improvement of starch film performances using cellulose microfibrils. *Macromolecules*, Vol. 31, pp. 2693–2696

Dufresne, A., Dupeyre, D., and Vignon M. R. (2000). Cellulose microfibrils from potato tuber cells: processing and characterization of starch-cellulose microfibril composites. *J Appl Polym Sci*, Vol. 76, pp. 2080–2092

Dufresne, A., Dupeyre, D., and Paillet, M. (2003). Lignocellulosicflour-reinforced poly(hydroxybutyrate-co-valerate) composites. *J Appl Polym Sci*, vol. 87, pp. 1302–1315

Eichhorn, S. J., Dufresne, A., Aranguren, M., Marcovich, N. E., Capadona, J. R., Rowan, S. J., Weder, C., Thielemans, W., Roman, M., Renneckar, S., Gindl, W., Veigel, S., Keckes, J., Yano, H., Abe, K., Nogi, M., Nakagaito, A. N., Mangalam, A., Simonsen, J., Benight, A. S., Bismarck, A., Berglund L. A., and Peijs, T. (2010). Review: current international research into cellulose nanofibres and nanocomposites. *Journal of Materials Science*, Vol. 45, No. 1., pp. 1–33

El-Sakhawy, M., and Hassan, M. L. (2007). Physical and mechanical properties of microcrystalline cellulose prepared from agricultural residues. *Carbohydrate Polymers*, Vol. 67, pp. 1–10

Famá, L., Gerschenson, L., and Goyanes, S. (2009). Starch-vegetable fiber composites to protect food products. *Carbohydrate Polymers*, Vol. 75, No. 2, pp. 230–235

Filson, P. B., and Dawson-Andoh, B. E. (2009). Sono-chemical preparation of cellulose nanocrystals from lignocellulose derived materials. *Bioresource Technology*, Vol. 100, pp. 2259–2264

Ganster, J., and Fink, H. P. (2006). Novel cellulose fibre reinforced thermoplastic materials. *Cellulose*, Vol. 13, pp. 271 –280

Henriksson, M., Henriksson, G., Berglund, L.A., and Lindstrom, T. (2007). An environmentally friendly method for enzyme-assisted preparation of microfibrillated cellulose (MFC) nanofibers. *European Polymer Journal*, Vol. 43, No. 8, pp. 3434–3441

Hubbe, M. A., Rojas, O. J., Lucia, L. A. and Sain, M. (2008). Cellulosic Nanocomposites: A review. *BioResources*, Vol. 3, No. 3, pp. 929-980

Iwatake, A., Nogi, M., and Yano, H. (2008). Cellulose nanofiber-reinforced polylactic acid. *Composites Science and Technology*, Vol. 68, pp. 2103–2106

Lee, S. Y., Chun, S. J., Kang, I. A., and Park, J. Y. (2009a). Preparation of cellulose nanofibers by high-pressure homogenizer and cellulose-based composite films. *Journal of Industrial and Engineering Chemistry*, Vol. 15, pp. 50–55

Lee, S. Y., Mohan, D. J., Kang, I. A., Doh, G. H., Lee, S., and Han, S. O. (2009b). Nanocellulose reinforced PVA composite films: Effects of acid treatment and filler loading. *Fibers and Polymers*, Vol. 10, No. 1, pp. 77-82

Leitner, J., Hinterstoisser, B., Wastyn, M., Keckes, J., and Gindl, W. (2007). Sugar beet cellulose nanofibril-reinforced composites. *Cellulose*, Vol. 14, pp. 419–425

Li, Y. and Pickering, K.L. (2008). Hemp fibre reinforced composites using chelator and enzyme treatments. *Composites Science and Technology*, Vol. 68, No. 15/16, pp. 3293–3298

Lin, N., Chen, G., Huang, J., Dufresne, A., and Chang, P. R. (2009). Effects of Polymer-Grafted Natural Nanocrystals on the Structure and Mechanical Properties of Poly(lactic acid): A Case of Cellulose Whisker-graft-Polycaprolactone. *J Appl Polym Sci*, Vol. 113, pp. 3417–3425

Lu, J., Wang, T., and Drzal, L. T. (2008). Preparation and properties of microfibrillated cellulose polyvinyl alcohol composite materials. *Appl Sci Manufact*, Vol. 39, pp. 738–746

Millon, L. E., and Wan, W. K. (2006). The Polyvinyl Alcohol–Bacterial Cellulose System As A New Nanocomposite for Biomedical Applications. *J Biomed Mater Res Part B: Appl Biomater* , Vol. 79B, pp. 245–253

Moran, J. I., Alvarez, V. A, Cyras, V. P., and Vazquez, A. (2008). Extraction of cellulose and preparation of nanocellulose from sisal fibers. *Cellulose*, Vol. 15, No. 1,pp. 149-159

Nakagaito, A. N., and Yano, H. (2004). The effect of morphological chances from pulp fibers towards nano-scale fibrillated cellulose on the mechanical properties of high strength plant fiber based composites. *Applied Physics A*, Vol. 78, No. 4, pp. 547-552

Nakagaito, A.N., Iwamoto, S. and Yano, H. (2005). Bacterial cellulose: the ultimate nano-scalar cellulose morphology for the production of high-strength composites. *Appl. Phys* A, Vol. 80, No. 1, pp. 93–97

Nakagaito, A. N., and Yano, H. (2006). Novel high-strength biocomposites based on microfibrillated cellulose having nano-order-unit web-like network structure. *Appl. Phys. A*, Vol. 80, pp. 155–159

Panaitescu, D. M., Matasaru, I., Iovu, H., Ghiurea, M., Iorga, M. D., and Stanescu, P. (2007a). Thermoplastic Composites with Cellulose Microfibrils Obtained by Mechanical Treatments. *Materiale Plastice*, Vol. 44, No. 2, pp. 144-147

Panaitescu, D. M., Donescu, D., Bercu, C., Vuluga, D. M., Iorga, M. and Ghiurea, M. (2007b). Polymer Composites with Cellulose Microfibrils. *Polymer Eng. Sci.*, Vol. 47, No. 8, pp. 1228-1234

Pandey, J. K., Chu, W. S., Kim, C.S., Lee, C. S., and Ahn, S.H. (2009). Bio-nano reinforcement of environmentally degradable polymer matrix by cellulose whiskers from grass. *Composites: Part B*, Vol. 40, pp. 676–680

Petersson, L., and Oksman, K. (2006). Biopolymer based nanocomposites: Comparing layered silicates and microcrystalline cellulose as nanoreinforcement. *Composites Science and Technology*, Vol. 66, pp. 2187–2196

Qua, E. H., Hornsby, P. R., Sharma, H. S., Lyons, G. and McCall, R. D. (2009). Preparation and Characterization of Poly(vinyl alcohol) Nanocomposites Made from Cellulose Nanofibers. *Journal of Applied Polymer Science*, Vol. 113, pp. 2238–2247

Rahman, W.A.W. A., Adenan, N., Rasit Ali, R. and Sulaiman, H. (2009). Effect of Silane Crosslinker on the Thermal Properties of Rice Straw/HDPE Biocomposite. *Journal of Applied Polymer Science*, Vol. 9, pp. 3041-3047

Ramaraj, B., Nayak, S. K., and Yoon, K. R. (2010). Poly(vinyl alcohol) and Layered Double Hydroxide Composites: Thermal and Mechanical Properties. *J Appl Polym Sci,* Vol. 116, No. 3, pp. 1671–1677

Rosa, M. F., Medeiros, E.S., Malmonge, J. A., Gregorski, K. S., Wood, D. F., Mattoso L. H. C., Glenn, G., Orts, W. J., and Imam, S. H. (2010). Cellulose nanowhiskers from coconut husk fibers: Effect of preparation conditions on their thermal and morphological behavior. *Carbohydrate Polymers,* Vol. 81 , pp. 83–92

Sedlarık, V., Saha, N., Kuritka, I., and Saha, P. (2006). Characterization of Polymeric Biocomposite Based on Poly(vinyl alcohol) and Poly(vinyl pyrrolidone). Polym Compos, Vol. 27, pp. 147–152

Shaikh, H., Pandare, M. K. V., Nair, G., and Varma, A. J. (2009). Utilization of sugarcane bagasse cellulose for producing cellulose acetates: Novel use of residual hemicellulose as plasticizer. *Carbohydrate Polymers,* Vol. 76, No. 1, pp. 23–29

Siro, I., and Plackett, D. (2010). Microfibrillated cellulose and new nanocomposite materials: a review. *Cellulose,* Vol. 17, pp. 459–494

Stenstad P, Andresen M, Tanem BS, Stenius P (2008) Chemical surface modifications of microfibrillated cellulose. *Cellulose,* Vol. 15, pp. 35–45

Suryanegara, L., Nakagaito, A. N., and Yano, H. (2009). The effect of crysatllization of PLA on the thermal and mechanical properties of microfibrillated cellulose-reinforced PLA composites. *Composites Science and Technology,* Vol. 69, pp. 1187-1192

Tang, Y., Du, Y., Li, Y., Wang, X., and Hu, X. (2009). A thermosensitive chitosan/poly(vinyl alcohol) hydrogel containing hydroxyapatite for protein delivery. *J Biomed Mater Res* , Vol. 91A, pp. 953–963

Taniguchi, T. and Okamura, K. (1998). New films produced from microfibrillated natural fibres. *Polym Int,* Vol. 47, pp. 291–294

Yanjun, X., Hill, C. A. S., Xiao, Z., Militz, H., and Mai, C. (2010). Silane coupling agents used for natural fiber/polymer composites: A review. *Applied Science and Manufacturing,* Vol. 41, pp. 806-819

Yano, S., Maeda, H., Nakajima, M., Hagiwara, T., and Sawaguchi, T. (2008). Preparation and mechanical properties of bacterial cellulose nanocomposites loaded with silica nanoparticles. *Cellulose,* Vol. 15, No. 1, pp. 111-120

Zhang, J., Elder, T. J., Pu, Y., and Ragauskas, A. J. (2007). Facile synthesis of spherical cellulose nanoparticles. *Carbohydrate Polymers,* Vol. 69, pp. 607–611

Zhou, W. Y., Guo, B., Liu, M., Liao, R., Rabie, B. A. M., and Jia, D. (2010). Poly(vinyl alcohol)/Halloysite nanotubes bionanocomposite films: Properties and in vitro osteoblasts and fibroblasts response. *J Biomed Mater Res* , Poly(vinyl alcohol)/Halloysite nanotubes bionanocomposite films: Properties and in vitro osteoblasts and fibroblasts response , Vol. 93A, pp. 1574–1587

Zimmermann, T., Pöhler, E. and Geiger, T. (2004). Cellulose Fibrils for Polymer Reinforcement. *Advanced Engineering materials,* Vol. 6, no. 9, pp. 754-761

Zuluaga, R., Putaux, J. L., Cruz, J., Vélez, J., Mondragon, I., and Gañán, P. (2009). Cellulose microfibrils from banana rachis: Effect of alkaline treatments on structural and morphological features. *Carbohydrate Polymers,* Vol. 76, pp. 51–59

6

Buckling and Post-buckling Analysis of Composite Plates

Elena-Felicia Beznea and Ionel Chirica
University Dunarea de Jos of Galati
Romania

1. Introduction

Thin walled stiffened composite panels are among the most utilized structural elements in ship structures. The composite layered panels with fibers are the most usually used in shipbuilding, aerospace industry and in engineering constructions as well. These structures possess the unfortunate property of being highly sensitive to geometrical and mechanical imperfections. These panels, unfortunately, have one important characteristic connected to big sensitivity on geometrical imperfections (different dimensions comparative with the design ones). The defects are of following types: different directions of fibers design, variations in thickness, inclusions, delaminations or initial transversal deformations.

Ship structure plates are subjected to any combination of in plane, out of plane and shear loads during application. Due to the geometry and general load of the ship hull, buckling is one of the most important failure criteria of these structures.

This is why it is necessary to develop the appropriate methodologies able to correctly predict the behavior of a laminated composite plate in the deep postbuckling region, at the collapse load, which is characterized by separation between the skin and the stiffeners, delaminations, crack propagations and matrix failure, as well as to understand its behavior under repeated buckling.

During its normal service life, a ship hull, which is composed of many curved laminated composite stringer stiffened panels, may experience a few hundreds of buckling-postbuckling cycles. Although it is well recognized that CFRP stiffened structures are capable of withstanding very deep post-buckling, yielding collapse loads equal to three - four times their buckling load (Bisagni & Cordisco, 2004, Knight & Starnes, 1988), there exists scarce knowledge in the literature about the effects of repeated buckling on the global behavior of the laminated composite panels under combined loading influences.

According to the studies, it is possible to predict on how far into the post-buckling region it is possible to increase loading without loosing structural safety.

Buckling failure mode of a stiffened plate can further be subdivided into global buckling, local skin buckling and stiffener crippling. Global buckling is collapse of the whole structure, i.e. collapse of the stiffeners and the shell as one unit.

Local plate buckling and stiffeners crippling on the other hand are localized failure modes involving local failure of only the skin in the first case and the stiffener in the second case. A grid stiffened panel will fail in any of these failure modes depending on the stiffener configuration, plate thickness, shell winding angle and type of applied load.

Over the past four decades, a lot of research has been focused on the buckling, collapse and post buckling behavior of composite shells. The simplest stiffened panel consists of only orthogonal stiffeners (stiffened orthogrid) such as longitudinal and transversal girders.
Another type of stiffener arrangement is the transversal framing system.
Different analytical tools have so far been developed by researches to successfully predict the three buckling failure modes associated with stiffened panels subjected to different loading conditions.
The use of finite-elements analysis for investigation of buckling problem of composite panels is becoming popular due to the improvement in computational hardware and emergence of highly specialized software. Depending on the degree of accuracy desired and limit of computational cost, three types of buckling analysis can be carried out. Linear bifurcation analysis is the basic analysis type which does not take into consideration the pre-buckling deformation and stresses. This analysis can accurately predict the buckling load of a geometrically perfect compression loaded panel, and the pre-buckling deformation and stress in the panel have an insignificant effect on the predicted bifurcation buckling load of the shell. The second kind of bifurcation analysis takes into consideration the nonlinear pre-buckling deformation and stresses and results in a much more accurate buckling load.
The third analysis, the nonlinear buckling analysis, allows for large nonlinear geometric deflections. Unlike the previous two bifurcation analyses that are eigenvalue problems, the nonlinear analysis is iterative in nature. In this analysis the load is steadily increased until the solution starts to diverge.
In this chapter, layered composite plates with imperfections are analysed.

1.1 Plates with initial imperfection

The composite layered panels with fibers are the most usually used in shipbuilding industry and in engineering constructions as well. Taking into account that fabrication technologies of composite materials are hand made based, the probabilistic occurrence of defects is quite too high. These panels, unfortunately, have one important characteristic connected to big sensitivity on geometrical imperfections (initial transversal deformation).
In this chapter is analyzing the buckling behavior of the plates placed between two pairs of stiffeners of the ship hull structure. Objective is to present the results obtained after buckling analysis of ship hull plates made of composite materials taking into account the transversal imperfection (spatial cosine form) due to fabrication. Due to the special behavior of the layered composite plates, the nonlinear analysis of the buckling behavior of the plates is to do. In certain cases, to determine the buckling load (ultimate strength), the failure criterion is applied. The buckling load is determined when the first failure occures in an element, based on the Tsai-Wu failure criterion, who provides the mathematical relation for the strength under combined stresses.

1.2 Plates with delaminations

Delamination in composite structures can be a serious threat to the safety of the structure.
Delamination leads to loss of stiffness and strength of laminates under some conditions. This is particularly so in the case of compressively loaded structures as the loss of stiffness may lead to buckling, the consequences of which can be catastrophic.
Causes of delamination are many. In shipbuilding and aerospace applications, this includes manufacturing defects, as well as operationally induced defects such as bird strikes hits due

to runway debris and tool drops (aerospace), or cargo operating and slamming loading (shipbuilding).
The type of delamination that is dealt with in this report is the one that is already initiated by one of the above causes.
When a laminate is subjected to in-plane compression, the effects of delamination on the stiffness and strength may be characterised by three sets of analytical results:

a. Buckling load;
b. Post-buckling solutions under increased load;
c. Results concerning the onset of delamination growth and its subsequent development.

Many of the analytical treatments deal with a thin near surface delamination. Such approaches are known as "thin-film" analysis in the literature. The thin-film analytical approach may involve significant errors in the post-buckling solutions.

2. Theory, finite element and experimental modeling of perfect composite plates

The buckling of composite plates is treated with specific methods applied to each particular case. This part deals with classical bifurcation and FEM buckling analysis, discusses the relevant plate equations and their solutions and edge conditions for perfect plates. The buckling phenomena mean collapse of the structure at the maximum point in a load versus deflection curve and bifurcation buckling. The way in which buckling occurs depends on how the structure is loaded and on its geometrical and material properties. The prebuckling process is often nonlinear if there is a large bending energy being stored in the structure throughout the loading history. According to the level of bending energy, the buckling of plates can occur in two ways: the first is bifurcation buckling and the second is limit point buckling. Bifurcation buckling is an instability in which there is a sudden change of shape of the structure. A bifurcation point is a point in a load-deflection space where two equilibrium paths intersect. On a load-frequency curve, a bifurcation point can be characterized by the load-frequency curve passing through a frequency of zero with a non-zero slope.
Limit point buckling is an instability in which the load-displacement curve reaches a maximum and then exhibits negative stiffness and releases strain energy. During limit point buckling there are no sudden changes in the equilibrium path; however, if load is continuously increased then the structure may jump or "snap" to another point on the load-deflection curve. For this reason, this type of instability is often called "snap-through" buckling, because the structure snaps to a new equilibrium position. A limit point is characterized by the load-frequency curve passing through a frequency of zero with a zero slope. The load-deflection curve also has a zero slope at the point of maximum load (limit point).
Buckling analysis of a plate may be divided into three parts:

- Classical buckling analysis;
- Difficult classical effects;
- Non classical phenomena.

The classical buckling theory may be described by the curve 1-3, from the Fig. AA1 where is plotted the in-plane loading force (N) versus the transversal displacement (w) of a representative point of the plate. By supposing that the loading is applied in the median plane of the plate, no transversal displacement will occur, and in the conditions of perfect symmetry the loading may increase up to the yield point, according to the curve 1-2. At a value over the N_{cr}, the instability may occur. At $N=N_{cr}$ one bifurcation point does exist. This means that on the curve loading-displacement, on the path 4, other way, much stable for buckling occurring does exist. At the value N_{cr} a little perturbation will generate a transversal displacement.

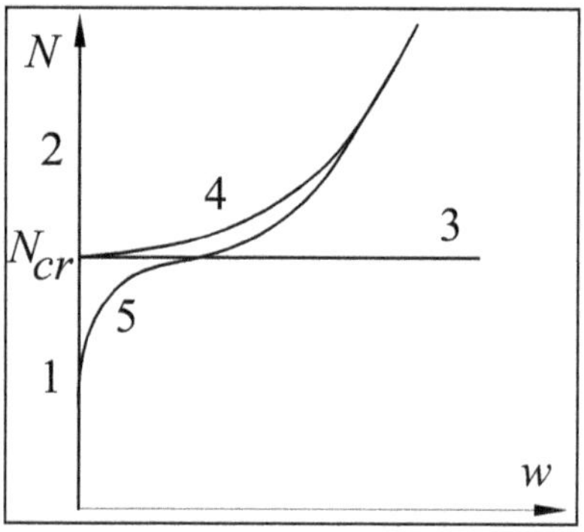

Fig. 1. Variation of transversal displacement, w, of a plate's point versus in-plane load N

The classical linear analysis, which is a generalization of the Euler buckling for beams, should indicate the fact that w increases to infinite at $N=N_{cr}$ (curve 3). Really, the nonlinear effects started to act and after an initial finite displacement, the in-plane loading N is increasing since the displacement is increasing. In this case the plate is able to carry loads far in excess of N_{cr} before it collapse. The latest curve (4) is so called a "postbuckling curve" because it depicts the behaviour of the plate after the buckling load N_{cr} is reached.

The difficult effects in classical buckling analysis are in connection with vibrations, shear deformations, springs, non homogeneities and variable thicknesses, nonlinear relations between stresses and strains. Non classical buckling analysis involves considerations such as imperfections, non-elastic behaviour of the material, dynamic effects of the loading, the fact that the in-plane loading is not in the initial point of the plate.

Finally, it may point out that no plate is initially perfect (perfectly flat or perfectly symmetry) and if initial deviation (from flatness or symmetry) exists, the behaviour of the plate will follow the path similar with curve 1-5. In this case, no clear buckling phenomenon may be identifying. The deviations of the plate from the flatness and symmetry are usually called imperfections (initial transversal imperfections, delaminations) as it will be treated in the following chapters.

The following cases (in numerical and experimental ways) are presented: compressive buckling, shear buckling, mixed compressive and shear buckling. The results (for linear and nonlinear model) are presented as variation of the buckling loads function of maximum transversal displacement (buckling and post-buckling behaviour).

The state of equilibrium of a plate deformed by forces acting in the plane of the middle surface is unique and the equilibrium is stable if the forces are sufficiently small. If, while maintaining the distribution of forces constant at the edge of the plate, the forces are increased in magnitude, there may arise a time when the basic form of equilibrium ceases to be unique and stable and other forms become possible, which are characterized by the curvatures of the middle surface.

The equation of the deflected surface of symmetrically laminated plates is

$$D_{11}\frac{\partial^4 w}{\partial x^4}+4D_{16}\frac{\partial^4 w}{\partial x^3\partial y}+2(D_{12}+D_{66})\frac{\partial^4 w}{\partial x^2\partial y^2}+4D_{26}\frac{\partial^4 w}{\partial x\partial y^3}+D_{22}\frac{\partial^4 w}{\partial y^4}-$$
$$-N_x\frac{\partial^2 w}{\partial x^2}-2N_{xy}\frac{\partial^2 w}{\partial x\partial y}-N_y\frac{\partial^2 w}{\partial y^2}=0 \tag{AA1}$$

For symmetrically laminated cross-ply plates there is no coupling between bending and twisting. So, $D_{16}=D_{26}=0$. In this case, the equation (AA1) will have the form as the buckling equation for a homogeneous, orthotropic plate

$$D_{11}\frac{\partial^4 w}{\partial x^4}+2(D_{12}+D_{66})\frac{\partial^4 w}{\partial x^2\partial y^2}+D_{22}\frac{\partial^4 w}{\partial y^4}-N_x\frac{\partial^2 w}{\partial x^2}-2N_{xy}\frac{\partial^2 w}{\partial x\partial y}-N_y\frac{\partial^2 w}{\partial y^2}=0 \tag{AA2}$$

where D_{11}, D_{22}, D_{66}, D_{16}, D_{26} are the orthotropic plate stiffnesses, calculated according to the equation

$$D_{ij}=\frac{1}{3}\sum_{k=1}^{N}\bar{Q}_{ij}^{k}\left(\bar{z}_k^3-\bar{z}_{k-1}^3\right) \tag{AA3}$$

The thickness and position of every ply can be calculated from the equation

$$t_k=z_k-z_{k-1} \tag{AA4}$$

and

$$\bar{z}_k=z_{k-1}+\frac{t_k}{2} \tag{AA5}$$

The second and fourth terms from equation (AA2) are the measure of the orthotropic coupling, resulting from the fact that the principal orthotropic axes are not orthogonal with the plate geometry axes.

Linear buckling of beams, membranes and plates has since been studied extensively. A linearized stability analysis is convenient from a mathematical viewpoint but quite restrictive in practical applications. What is needed is a capability for determining the nonlinear load-deflection behaviour of a structure. Considerable effort has also been expended on this problem and two approaches have evolved: class-I methods, which are incremental in nature and do not necessarily satisfy equilibrium; and class-II methods, which are self-correcting and tend to stay on the true equilibrium path (Thurley & Marshal, 1995).

Historically, class-I was the first finite element approach to solving geometrically non-linear problems (Ambarcumyan, 1991). In this method the load is applied as a sequence of sufficiently small increments so that the structure can be assumed to respond linearly during each increment.

To solving of geometrically and material nonlinear problems, the load is applied as a sequence of sufficiently small increments so that the structure can be assumed to respond linearly during each increment.

For each increment of load, increments of displacements and corresponding increments of stress and strain are computed. These incremental quantities are used to compute various corrective stiffness matrices (variously termed geometric, initial stress, and initial strain matrices) which serve to take into account the deformed geometry of the structure. A subsequent increment of load is applied and the process is continued until the desired number of load increments has been applied. The net effect is to solve a sequence of linear problems wherein the stiffness properties are recomputed based on the current geometry prior to each load increment. The solution procedure takes the following mathematical form

$$(K + K_1)_{i-1} \Delta d_i = \Delta Q \tag{AA6}$$

where
K is the linear stiffness matrix,
K_I is an incremental stiffness matrix based upon displacements at load step i-1,
Δd_i is the increment of displacement due to the i–th load increment,
ΔQ is the increment of load applied.
The correct form of the incremental stiffness matrix has been a point of some controversy. The incremental approach is quite popular (this is the procedure applied in all studies in this chapter). This is due to the ease with which the procedure may be applied and the almost guaranteed convergences if small enough load increments are used.
The plate material is damaged according to a specific criterion.
For various materials classes three dimensional failure criteria are developed. These include both isotropic and anisotropic material symmetries, and are applicable for macroscopic homogeneity. In the isotropic materials form, the properly calibrated failure criteria can distinguish ductile from brittle failure for specific stress states. Although most of the results are relevant to quasi-static failure, some are for time dependent failure conditions as well as for fatigue conditions.
The buckling load determination may use the Tsai-Wu failure criterion in the case if the general buckling does not occurred till the first-ply failure occurring. In this case, the buckling load is considered as the in-plane load corresponding to the first-ply failure occurring.
The Tsai-Wu failure criterion provides the mathematical relation for strength under combined stresses. Unlike the conventional isotropic materials where one constant will suffice for failure stress level and location, laminated composite materials require more elaborate methods to establish failure stresses. The strength of the laminated composite can be based on the strength of individual plies within the laminate. In addition, the failure of plies can be successive as the applied load increases. There may be a first ply failure followed by other ply failures until the last ply fails, denoting the ultimate failure of the laminate. Progressive failure description is therefore quite complex for laminated composite structures. A simpler approach for establishing failure consists of determining the structural integrity which depends on the definition of an allowable stress field. This stress field is usually characterized by a set of allowable stresses in the material principal directions.
The failure criterion is used to calculate a failure index (F.I.) from the computed stresses and user-supplied material strengths. A failure index of 1 denotes the onset of failure, and a value less than 1 denotes no failure. The failure indices are computed for all layers in each element of your model. During postprocessing, it is possible to plot failure indices of the mesh for any layer.
The Tsai-Wu failure criterion (also known as the Tsai-Wu tensor polynomial theory) is commonly used for orthotropic materials with unequal tensile and compressive strengths. The failure index according to this theory is computed using the following equation (Altenbach & all, 2004, Ambarcumyan, 1991).

$$F.I. = F_1 \cdot \sigma_1 + F_2 \cdot \sigma_2 + F_{11} \cdot \sigma_1^2 + F_{22} \cdot \sigma_2^2 + F_{66} \cdot \sigma_6^2 + 2F_{12} \cdot \sigma_1 \cdot \sigma_2 \quad \text{(AA7)}$$

where

$$F_1 = \frac{1}{R_1^T} - \frac{1}{R_1^C};\ F_{11} = \frac{1}{R_1^T \cdot R_1^C};\ F_2 = \frac{1}{R_2^T} - \frac{1}{R_2^C};\ F_{22} = \frac{1}{R_2^T \cdot R_2^C};\ F_{66} = \frac{1}{R_{12}^2}. \quad \text{(AA8)}$$

The coefficient F_{12}, which represents the parameter of interaction between σ_1 and σ_2, is to be obtained by a mechanical biaxial test. In the equations (AA8), the parameters R_i^C, R_i^T are the compressive strength and tensile strength in the material in longitudinal direction (i=1) and transversal direction (i=2). The parameter R_{12} is in-plane shear strength in the material 1-2 plane.

According to the Tsai-Wu failure criterion, the failure of a lamina occurs if

$$F.I. > 1. \quad \text{(AA9)}$$

The failure index in calculated in each ply of each element. In the ply where failure index is greater than 1, the first-ply failure occurs, according to the Tsai-Wu criterion. In the next steps, the tensile and compressive properties of this element are reduced by the failure index. If the buckling did not appeared until the moment of the first-ply failure occurring, the in-plane load corresponding to this moment is considered as the buckling load.

3. Finite element and experimental modeling of composite plates with initial transversal imperfection

3.1 Presentation

In this part, the buckling behavior of the plates with initial transversal deformation, placed between two pairs of stiffeners of the ship hull structure is analyzing. The results obtained on buckling analysis of ship hull plates made of composite materials taking into account the imperfection due to fabrication, are presented. Due to the special behavior of the layered composite plates, the nonlinear analysis of the buckling behavior of the plates is done. The buckling load is determined according to the first failure occurring in an element, based on the Tsai-Wu failure criterion.

The most used framing construction types for the ship deck plates made of composite materials are transversal or mixed. So, the following case of plate placed in the ship deck structure is analyzed: the plate placed between two pairs of parallel stiffeners (two transversal web frames and two longitudinal frames - longitudinal framing construction system).

The geometry of the plate is square one, having the side length of 320mm, and total thickness of 4.96mm. The layers were grouped into the macro-layers (group of layers having the same characteristics: thickness (t), direction of fibers (α) and type of material).

The imperfections of plates are considered the initial transversal deformations. The shape of the deformation is just the first form of the buckling of perfect plate clamped on the sides. The analysis presented in this part is done for the most usual magnitudes of the transversal deformation of the imperfections (versus side length of the plate) occurred in the ship deck plates after fabrication.

The following cases (in numerical and experimental ways) are presented: compressive buckling, shear buckling, mixed compressive and shear buckling. The results (for linear and nonlinear model) are presented as variation of the buckling loads function of maximum transversal displacement (buckling and post-buckling behaviour).

The simplest stiffened plate consists of only orthogonal stiffeners (stiffened orthogrid) such as longitudinals and transversal girders. Other type of stiffener arrangement is the transversal framing system.

The characteristics of the material used in this chapter are:

Ex=46 GPa, Ey=13 GPa, Ez=13 GPa, Gxy=5 GPa, Gxz=5 GPa, Gyz = 4.6 GPa, μ_{xy}=0.3, μ_{yz}=0.42, μ_{xz}=0.3,

- traction strengths R_x=1.062 GPa, R_y=0.031 GPa,
- compression strength R_y=0.118 GPa,
- shear strength R_{xy}=0.72 GPa.

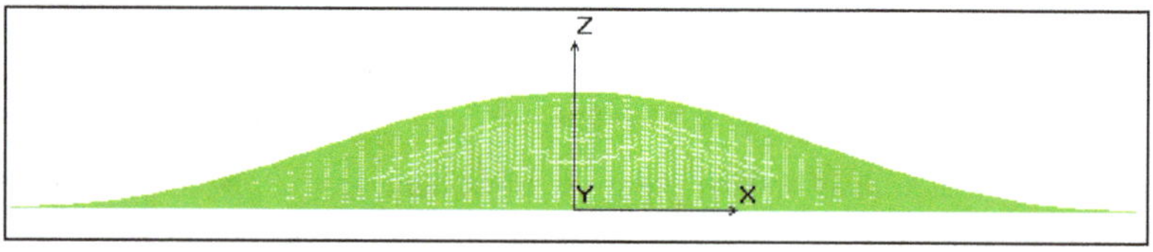

Fig. 2. Plate cross section in plane xz

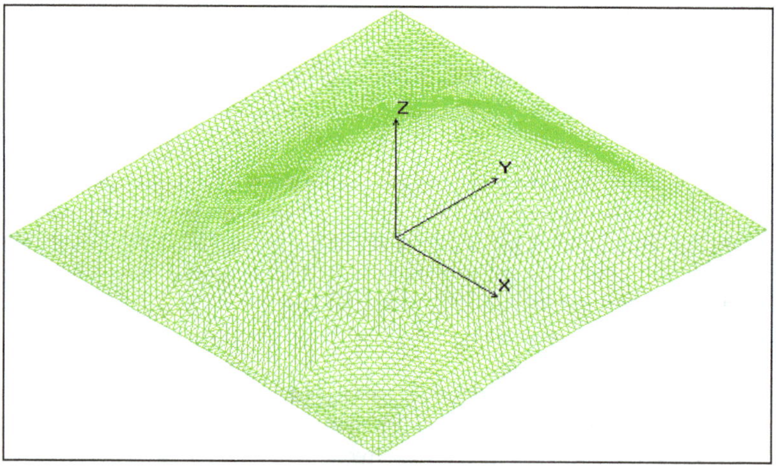

Fig. 3. Imperfect plate

Different analytical tools have so far been developed by researches to successfully predict the three buckling failure modes associated with stiffened panels subjected to different loading conditions. These analytical tools developed are divided into three major categories. The use of finite-elements analysis for investigation of buckling problem of composite panels is becoming popular due to the improvement in computational hardware and emergence of highly specialized software. Depending on the degree of accuracy desired and limit of computational cost, three types of buckling analysis can be carried out. Linear bifurcation analysis is the basic analysis type which does not take into consideration the pre-buckling deformation and stresses. This analysis can accurately predict the buckling load of a geometrically perfect compression loaded plate, and the pre-buckling deformation and

stress in the plate have an insignificant effect on the predicted bifurcation buckling load of the shell (Ambarcumyan, 1991, Thurley & Marshall, 1995). The second kind of bifurcation analysis takes into consideration the nonlinear pre-buckling deformation and stresses and results in a much more accurate buckling load (Adams & all, 2003, Chirica & all, 2008).
The third analysis, the nonlinear buckling analysis, allows for large nonlinear geometric deflections. Unlike the previous two bifurcation analyses that are eigenvalue problems, the nonlinear analysis is iterative in nature. In this analysis the load is steadily increased until the solution starts to diverge (Adams & all, 2003). A lot of work has been done in finite elements analysis pertaining to the investigation of buckling of stiffened panels (Beznea, 2008, Chirica & all, 2008). One of the major drawbacks associated with this tool is the tedious model-building phase involved and the subsequent inconvenient parametric study.
Following to these considerations, in this chapter the results of the buckling behavior analysing of the plates placed between two pairs of stiffeners of the ship hull structure are analysed. In the following pages only the results obtained after buckling analysis of ship hull plates made of composite materials taking into account the imperfection due to fabrication will be presented. Due to the special behavior of the layered composite plates, the nonlinear analysis of the buckling behavior of the plates is to do (Altenbach, 2004, Hilburger, 2001).

3.2 Numerical studies on compression buckling

The most used framing construction types for the ship deck panels made of composite materials are transversal or mixed. So, in this chapter the following case is presented (case of plates placed in the ship deck structure): plate placed between two pairs of parallel stiffeners (2 transversal web frames and 2 longitudinal frames, in the case of longitudinal framing construction system);
The geometry of the plates is square one, having the side length of 320mm, and total thickness of 4.96mm respectivelly. The orthotropic directions (α), thickness of the macro-layers (t) and plate lay-up are presented in Table. The layers were grouped into the macro-layers (group of layers having the same characteristics: thickness (t), direction of fibers (α) and type of material).
The constrains are considered according to each plate. The degrees of freedom (displacements u and rotations r), considered to be zero are:

- on the sides parallel with x axis: u_z, r_x, r_y;
- on the sides parallel with y axis: all d.o.f. (clamped side); u_y, u_z, r_x, r_y, r_z (charged side).

The shape of the initial transversal deformation is considered as just the first form of the buckling of perfect plate clamped on the sides.
The analysis presented in this chapter is done for the most usual magnitudes (w_0) of the transversal deformation of the imperfections (versus side length of the plate) occurred in the ship deck plates after fabrication. The following three cases are considered: w_0=1.06mm (symbol 12); w_0=3.2mm (symbol 32); w_0=9.6mm (symbol 92).
In figures the FEM model using shell composite elements is presented. In figure, the variation of the transversal displacement of the middle point of the plate versus in-plane loading magnitude for the plate is plotted.
Since the plates have initial transversal deformation, as it is seen in the figure the increasing of the transversal deformation is starting from the beginning, which is so named buckling is starting as the in-plane load is starting to increase from 0.

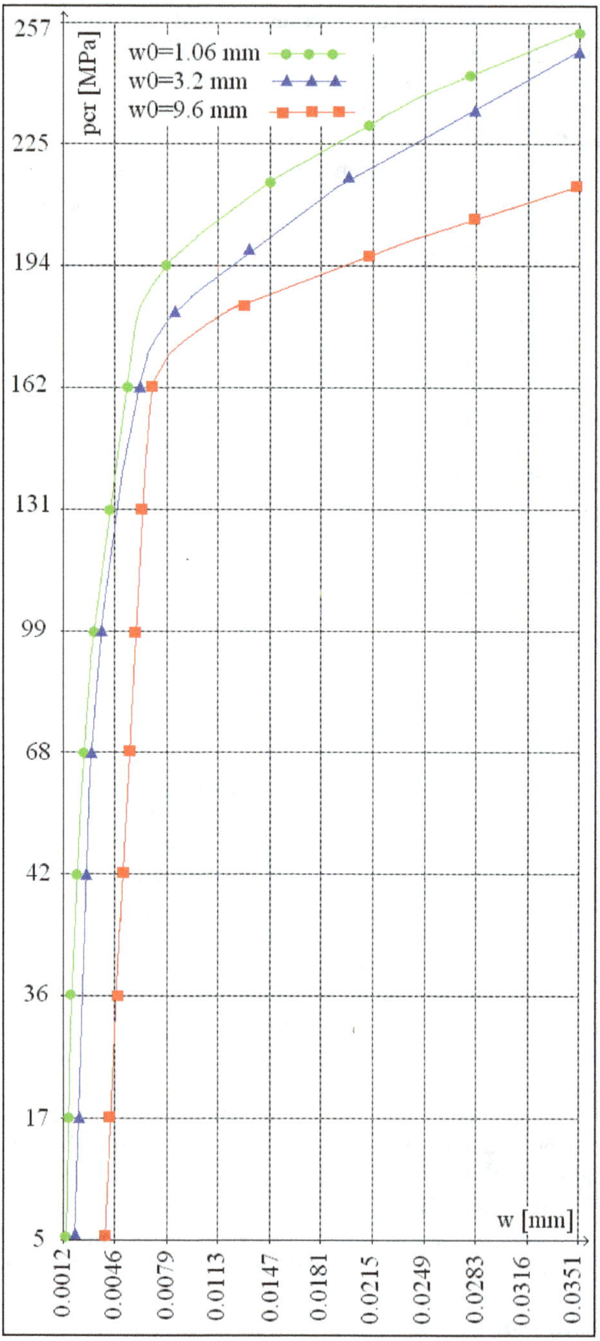

Fig. 4. Buckling and post buckling behaviour of compression buckling of plate with transversal imperfection

Macro-layer	α	t [mm]
	2x0^0	0.62
	45^0	0.31
1	2×90^0	0.62
	45^0	0.31
	2x0^0	0.62
	2x0^0	0.62
	45^0	0.31
2	2×90^0	0.62
	45^0	0.31
	2x0^0	0.62

Table 1. Plate lay-up

The explanation is that due to the initial deformation, the in-plane loading produces the compression in the plate and also bending in the area of imperfection. Therefore it is difficult to determine the buckling load by numerical way.

This is why we have chosen the graphical method, by drawing the asymptote to the curve in the area where the slope is changing almost suddenly.

The intersection of the asymptote with the loading axis can be considered as buckling load. Anyway the asymptote is not an unique one and we may determine the buckling loading in a range of values. Also, according to the curves the buckling load is decreasing since the amplitude of the imperfection is increasing.

So, as it is seen in figure, according to the plotted asymptotes, the buckling loads of the pates are placed in the domain

140 MPa $< p_{cr} <$ 175 MPa.

On the other part, we can determine the behaviour of the plate after the buckling (so name post-buckling behaviour), to estimate the ultimate capacity (ultimate strength) using one of the failure criteria. So, the in-plane loading value of first fail occurring in the material may be considered as the ultimate strength of the plate.

In the studies described in this chapter, the occurring of the first fail was done according to the Tsai-Wu failure criterion. Due to the elasticity of the thin plate, no any fail occurs for the in-plane pressure increasing up to the 260 MPa.

3.3 Experimental studies on compression buckling

The experiments were performed on plates having an initial transversal deformation with magnitude of 9.6 mm to determine the results obtained by numerical tests, such as:

- variation of the transversal displacement (w) of the point placed in the middle of the plate, on the surface, function of in-plane loading magnitude (p);
- variation of the strains in the certain point placed so on the concave face and on the convex face of the plate.

The measurements were done with the stretching machine, displacement transducer, strain gauge measurement system. In figure, the plate and equipment used for experiments are shown.

Fig. 5. Test rig and imperfect composite plate

To assure the boundary condition on the plate sides a special very rigid frame was made. So, the plate is clamped on three sides and simple supported on the side where the pressure is acting.

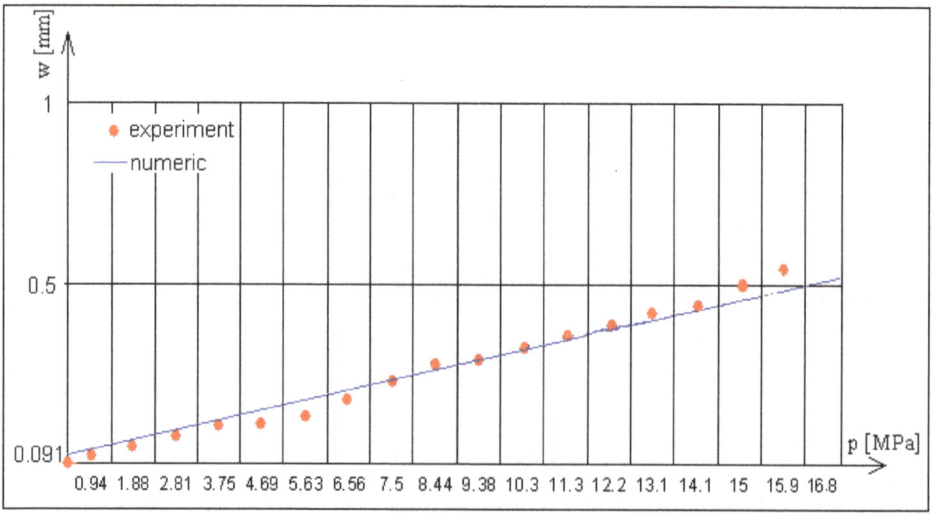

Fig. 6. Variation of compression load versus displacement of the point placed in the middle of the plate

In the figure, the results obtained from experiments are plotted with dots. Acceptable deviations between numerical and experimental results, for the transversal displacement of the point placed in the middle of the plate, are observed. In the figures only the elastic range of the curve is presented, because the experiment was made in this domain. In the abscise, a gap for pression corresponds to a force $\Delta F = 1$ kN.

3.4 Numerical studies on shearing buckling

The geometry of the plates is square one, having the side length of 320mm, and total thickness of 4.96mm. The orthotropic directions (α), thickness of the macro-layers (t) and plate lay-up are presented in Table. The layers were grouped into the macro-layers (group of layers having the same characteristics: thickness (t), direction of fibers (α) and type of material).

The constrains are considered according to plate presented in the previous chapter. The loading acting on the plate is according to figure.

The all types of amplitude of initial transversal deformation have been analysis.

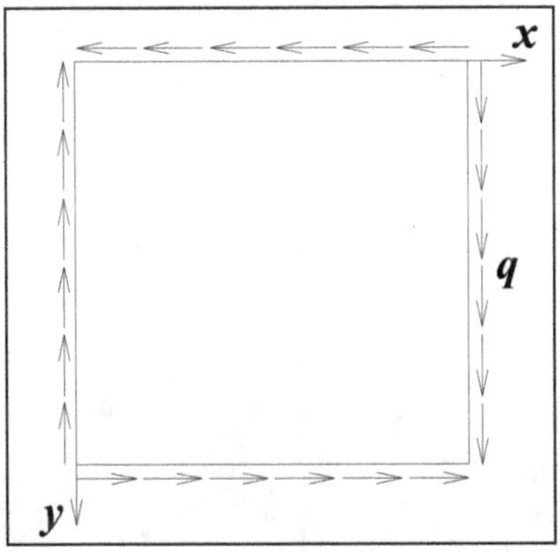

Fig. 7. Plate in shearing

In the table the buckling loads so for perfect and for imperfect plates are presented.

As it is seen, according to the graphical method, the buckling loads for the all three imperfect plates are in the range 104 MPa < p_{cr} < 275.78 MPa. As it is seen in figure, the buckling load is decreasing since the magnitude of the transversal imperfection is increasing. In this case, the load capacity of the plate is decreasing since the transversal deformation of the plate is increasing.

But, after the nonlinear calculus according to Tsai-Wu criterion, the ultimate strength, presented in table, may be considered as buckling load, due to the fact the values of the ultimate strength are almost constant for all plates (15 MPa).

Failure type	Perfect plate	w_0=1.06 mm	w_0=3.2 mm	w_0=9.6mm
FAIL 1 (tension)	20	15	15	15

Table 2. Shear buckling load (ultimate strength) in [MPa]

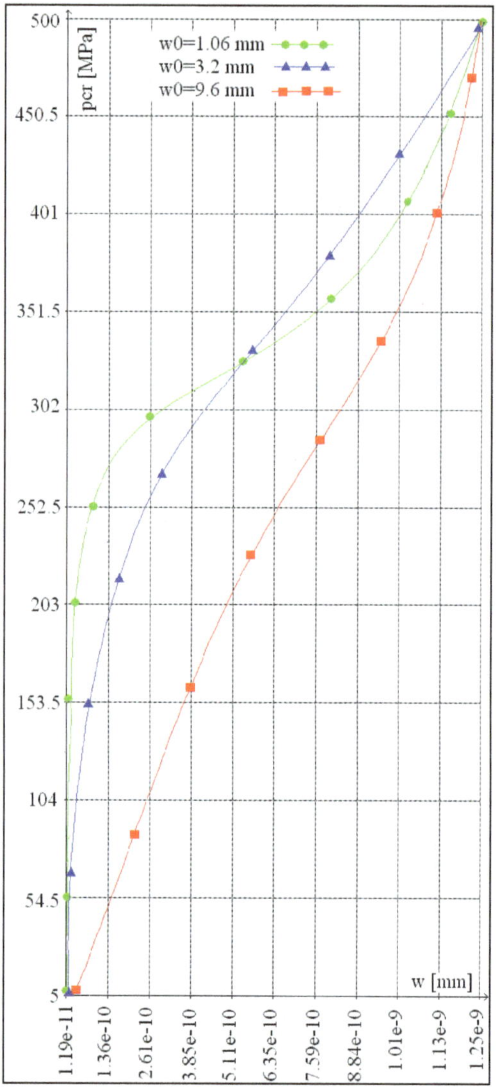

Fig. 8. Shear buckling and post buckling behaviour of the imperfect plate

3.5 Numerical studies on combined (compression and shear) buckling

The compression and shear buckling analysis is made only for thin plates (with material characteristics presented in previous chapter). The geometry of the plates is square one, having the side length of 320mm, and total thickness of 4.96mm. The orthotropic directions (α), thickness of the macro-layers (t) and plate lay-up are presented in Table.

The constrains are considered according to plates type b from the previous chapter. The loading acting on the plate is according to figure. For analysis the parametric calculus was done for various ratios $a=q/p$, that is {0.2; 0.4; 0.6; 0.8; 1}.

The all types of amplitude of initial transversal deformation are analysed.
In the table the buckling loads so for perfect and for imperfect plates are presented.
Only the cases of a=0.2 and a=0.4 are presented in figure. As it is seen in figure, according to the graphical method, the buckling loads for the all three imperfect plates are in the range 0.134MPa $< p_{cr} <$ 0.288MPa.

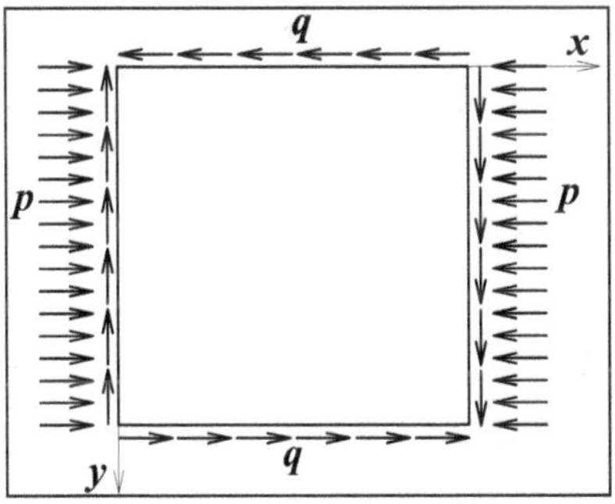

Fig. 9. Plate in shearing and compression

$a=q/p$	w_0=0	w_0=1.06 mm	w_0=3.2 mm	w_0=9.6mm
0.2	41.765	0.288	0.222	0.134
0.4	41.561	0.266	0.111	0.087
0.6	41.007	0.288	0.088	0.071
0.8	40.167	0.277	0.111	0.071
1	39.119	0.305	0.124	0.081

Table 3. Buckling load in [MPa] for shearing and compression loading of plate

$a=q/p$	Failure mode	Perfect plate	w_0=1.06 mm	w_0=3.2 mm	w_0=9.6mm
0.2	Tension	2.499	0.706	0.605	0.403
	Compression	-	1.413	1.312	1.211
0.4	Tension	2.499	0.706	0.504	0.403
	Compression	-	1.311	1.211	0.999
0.6	Tension	1.874	0.504	0.504	0.302
	Compression	-	0.908	0.876	0.802
0.8	Tension	1.562	0.403	0.403	0.302
	Compression	-	0.806	0.706	0.706
1	Tension	1.249	0.302	0.302	0.302
	Compression	-	0.706	0.563	0.563

Table 4. Buckling load (ultimate strength) in [MPa] for shearing and compression loading of plate

As it is seen in figure, the buckling load is decreasing since the magnitude of the transversal imperfection is increasing for all cases of loading ratios.

In this case, the load capacity of the plate is decreasing since the transversal deformation of the plate is increasing. According to the nonlinear calculus (according to Tsai-Wu criterion), the ultimate strength is presented in table.

The ultimate strength (pression corresponding to the first fail occurring) for perfect plates is greater than the ultimate strength of the plate with transversal imperfection.

In the central area, in the layers the compression fails occurs at an in-plane pression greater than the pression corresponding to tension fails.

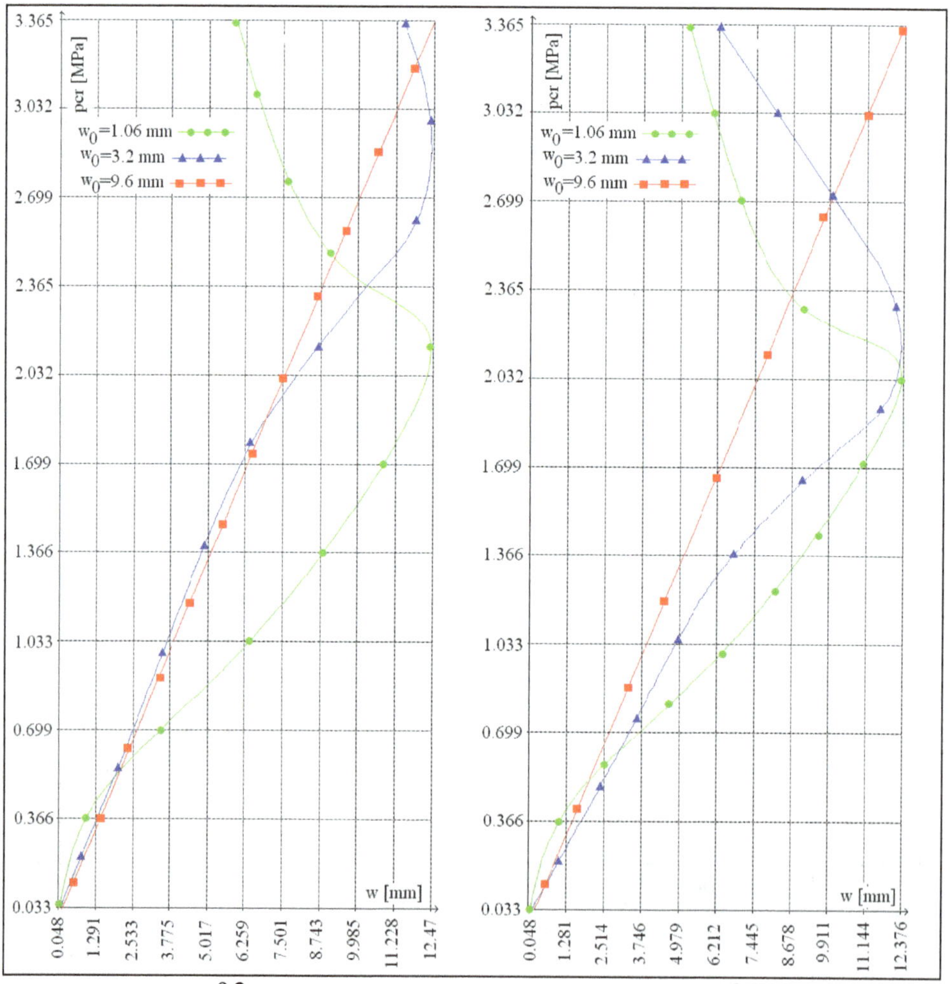

$a = 0.2$ $a = 0.4$

Fig. 10. Shear and compression behaviour of the plate

The ultimate strength decreases since the transversal imperfection is increasing.

The tension fails occur in a corner of the plate. On the other part, the compression fails occur in the central part (imperfection area) of the plate.

3.6 Conclusions for buckling of plates with transversal imperfections

The FEM based methodology and the experimental test program were successfully developed for the investigation of buckling problems of composite plates. The FEM model is robust in that it can be used to predict the global buckling loads of composite plates either on one side or both sides. Finite-elements analysis and experimentation were carried out to assess the reliability of the methodology. The buckling behaviour of the orthotropic composite plates is different than the buckling behaviour of the isotropic plates. A special remark is to do for the plates having initial transversal deformation: the transversal deformation is increasing since the in-plane load starts to increase.

The buckling load determination is too difficult without applying a graphical method, or applying the Tsai-Wu failure criterion.

So, for a plate with transversal imperfection, since the magnitude of the initial transversal deformation is increasing, the value of the force for compression failure or tension failure of an element is decreasing.

The lack of the criterion is referring to the anticipation of the real mode to occurring the cracking.

Taking into account the mathematical formulation, the Tsai-Wu failure criterion is easy to be applied. Additionally, this criterion offers advantages concerning the real prediction of the strength at variable loadings. It is to remark that by applying linear terms, it is possible to take into account the differences between the tension and compression strengths of the material.

A good correlation between the numerical and experimental results is concluded.

4. Finite element modelling of delaminated composite plates

4.1 Presentation

Due to the anisotropy of composite laminates and non-uniform distribution of stresses in lamina under flexural bending as well as other types of static/dynamic loading, the failure process of laminates is very complex.

Large differences in strength and stiffness values of the fiber and the matrix lead to various forms of defect/damage caused during manufacturing process as well as service conditions. In shipbuilding, many structures made of composite laminates are situated such that they are susceptible to foreign object impacts which can result in barely visible impact damage. Often, in the form of a complicated array of matrix cracks and interlaminar delaminations, these barely visible impact damages can be quite extensive and can significantly reduce a structure's load bearing capability.

Delamination or separation of two adjacent plies in a composite laminate is one of the most common modes of damage. The presence of delamination may reduce the overall stiffness as well as the residual strength leading to structural failure. A clear understanding of the influence of delamination on the performance of the laminates is very essential to use them efficiently in structural design applications.

Since such damage is in general difficult to detect, structures must be able to function safely with delamination present.

Although several studies are available in the literature in the field of delamination prediction and growth, effect of delamination on buckling, post-buckling deformation and delamination propagation under fatigue loading, etc. the work on the effect of delamination on the first ply failure of the laminate is scarce.

However, there clearly exists the need to be able to predict the tolerance of structures to damage forms which are not readily detectable (Chirica & al, 2006).

(Ambarcumyan, 1991, Adams & al, 2003) have analysed experimental characterization of advanced composite materials. When a laminate is subjected to in-plane compression, the effects of delamination on the stiffness and strength may be characterized by three sets of results, (Finn & Springer, 1993):

a. Buckling load;
b. Postbuckling solutions under increased load;
c. Results concerning the onset of delamination growth and its subsequent development.

Many of the analytical treatments deal with a thin near surface delamination. Such approaches are known as "thin-film" analysis in the literature (Kim & al,1999, Thurley & al, 1995). The thin-film analytical approach may involve significant errors in the post-buckling solutions.

(Naganarayana & Atluri, 1995) have analysed the buckling behaviour of laminated composite plates with elliptical delaminations at the centre of the plates using finite element method. They propose a multi-plate model using 3-noded quasi-conforming shell element, and use J-integral technique for computing point wise energy release rate along the delamination crack front.

(Pietropaoli & al, 2008) studied delamination growth phenomena in composite plates under compression by taking into account also the matrix and fibers breakages until the structural collapse condition is reached.

The aim of the work presented in this chapter is to present the studies on the influence of elliptical delamination on the changes in the buckling behaviour of ship deck plates made of composite materials. This problem has been solved by using the finite element method, in (Beznea, 2008). An orthotropic delamination model, describing mixed mode delaminating, by using FEM analysis, was applied. So, the damaged part of the structures and the undamaged part have been represented by well-known finite elements (layered shell elements). The influence of the position and the ellipse's diameters ratio of delaminated zone on the critical buckling force was investigated.

If an initial delamination exists, this delamination may close under the applied load. To prevent the two adjacent plies from penetrating, a simple numerical contact model is used.

Taking into account the thickness symmetry of the plates, only cases of position of delamination on one side of symmetry axis are presented. The variations of the transversal displacement of the point placed in the middle of the plate versus the in-plane applied pressure are plotted for each position of delamination. Buckling load determination for the general buckling of the plate has been done by graphical method. The post-buckling calculus has been performed to explain the complete behaviour of the plate.

Only cases with one delamination placed between two laminas is presented here.

There are several ways in which the panel can be modeled for the delamination analysis. For the present study, a 3-D model with 4-node shell composite elements is used. The plate is divided into two sub-laminates by a hypothetical plane containing the delamination. For this reason, the present finite element model would be referred to as two sub-laminates model. The two sub-laminates are modeled separately with shell composite elements, and then joined face to face with appropriate interfacial constraint conditions for the corresponding nodes on the sub-laminates, depending on whether the nodes lie in the delaminated or undelaminated region.

The delamination model has been developed by using the surface-to-surface contact option (Fig.). In case of surface-to-surface contact, the FE meshes of adjacent plies do no need to be identically. The contact algorithm used in the FEM analysis has possibility to determine which node of the so-called master surface is in contact with a given node on the slave surface. Hence, the user can define the interaction between the two surfaces.

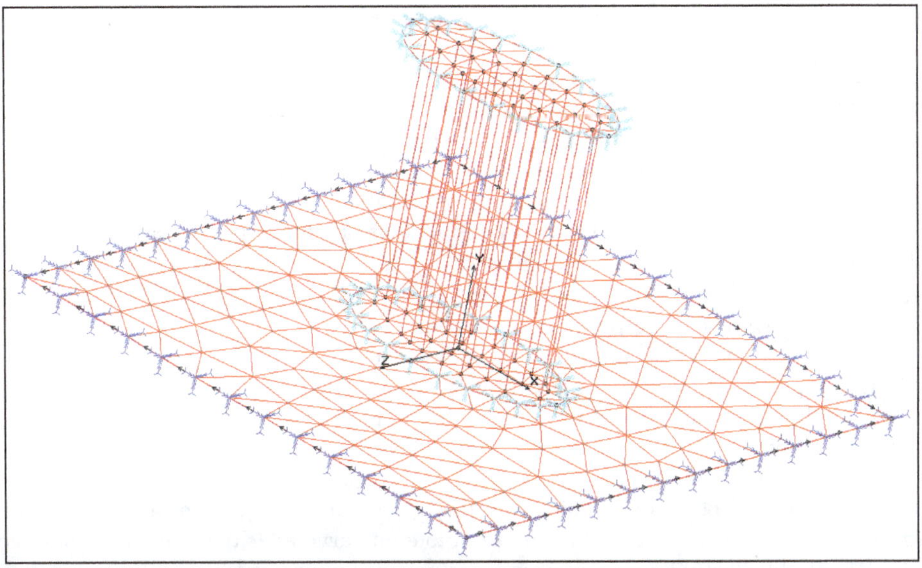

Fig. 11. The FEM delamination model

The condition is that the delaminated region does not grow. These regions were modeled by two layers of elements with coincident but separate nodes and section definitions to model offsets from the common reference plane. Thus their deformations are independent. At the boundary of the delamination zones the nodes of one row are connected to the corresponding nodes of the regular region by master slave node system.

Typically, a node in the underlaminated region of bottom sub-laminate and a corresponding node on the top sub-laminate are declared to be coupled nodes using master–slave nodes facility. The nodes in the delaminated region, whether in the top or bottom laminate, are connected by contact elements. This would mean that the two sublaminates are free to move away from each other in the delaminated region, and constrained to move as a single laminate in the undelaminated region.

The material characteristics presented in previous chapter are used.

Two material models were used: quasi-nonlinear model and non linear model.

A quasi-nonlinear model means that the material behaviour is not according to a failure criterion.

The non-linear model is the material fulfilling Tsai-Wu failure Criterion.

The ellipse's diameters of the delamination area placed in the middle of the plate are considered from the condition of the same area for all cases. In the parametric calculus, the following diameters ratios were considered:

- Case 1 (Dx/Dy=0.5): transversal diameter Dy=141mm, longitudinal diameter Dx= 70.5mm;
- Case 2 (Dx/Dy=1): transversal diameter Dy= 100 mm; longitudinal diameter Dx = 100mm;
- Case 3 (Dx/Dy=2): transversal diameter Dy= 70.5 mm; longitudinal diameter Dx = 141mm.

In this chapter the following cases (in numerical and experimental ways) are presented: compressive buckling, shear buckling, mixed compressive and shear buckling. The results (for linear, and nonlinear model) are presented as variation of the buckling loads function of maximum transversal displacement (buckling and post-buckling behaviour).

The buckling analysis of delaminated plates was done on square plates (320x320mm), made of E-glass/epoxy (biaxial layers having the thickness t=0.32mm).

These layers are grouped into macro-layers as are presented in table. The position of the delamination is considered between two neighbors layers i and i+1, (i=1,10). The calculus was done for the all 9 cases. Only results obtained for a specific case of position of delamination (i=4) are presented in the chapter.

4.2 Numerical studies on compression buckling

The plate is considered as clamped on the sides. The in-plane loading was applied as an uniform compressive pressure in the x direction (Fig.). The force was increased step by step (with and certain increment). In the figure, the variation of the transversal displacements of the central point, versus applied loading is drawn. Each curve corresponds to one diameters ratio.

Buckling load has been determined by the graphical method. On each curve that corresponds to a delamination type, an asymptote on curve after the bifurcation has been plotted. Critical value for the buckling load was obtained in the range:

$43.41 \text{ MPa} < p_{cr} < 59.5 \text{ MPa}$

In the case of nonlinear model of the material behaviour the buckling load (ultimate strength) was determined by Tsai Wu criterion. So, the degradation index (failure index) for tension and compression of the delaminated plates was determined. Only the plate global buckling was examined.

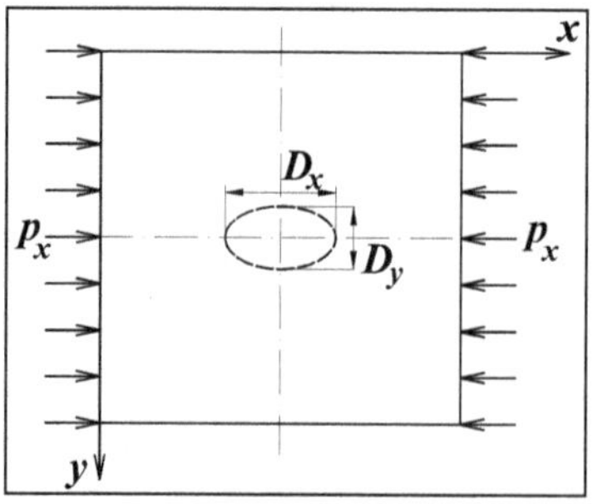

Fig. 12. Delaminated plate with in-plane loading

Failure mode	Fail 1 (Tension)	Fail 2 (Compression)
Dx/Dy=0.5	35	74
Dx/Dy=1	37	78
Dx/Dy=2	39	82

Table 5. Buckling load (ultimate strength) in [MPa] for compression loading

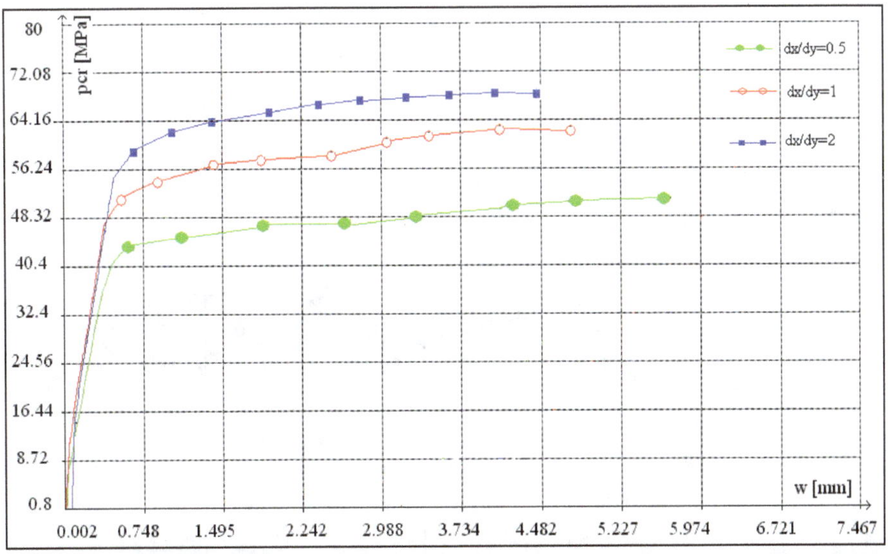

Fig. 13. Compression buckling and post buckling behaviour of the plates with central delamination placed between macro-layers 4 and 5

4.3 Numerical studies on shearing buckling

4.3.1 Shearing buckling of thin plates

In this chapter, the thickness and the lay-up of the plate are according to Table. The plate, considered as clamped on the sides, was loaded with an uniform shear pressure on the sides (Fig. AA14). The force was increased step by step (with and certain increment).

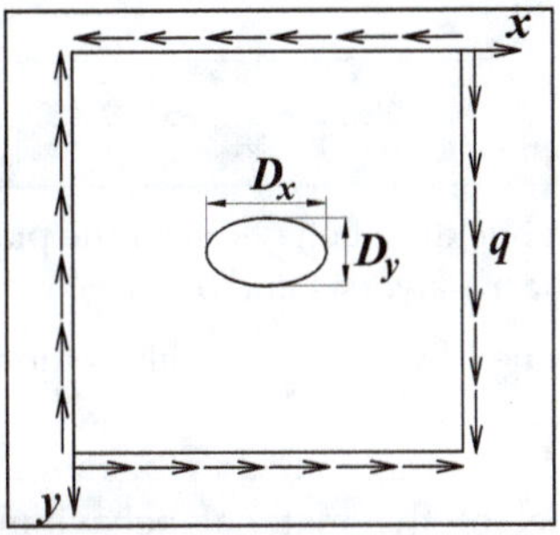

Fig. 14. Shear loaded plate with central delamination

In the figure, the variation of the transversal displacements of the central point, versus applied loading is drawn, for delamination placed between macro-layers 4 and 5. Each curve corresponds to one diameters ratio.

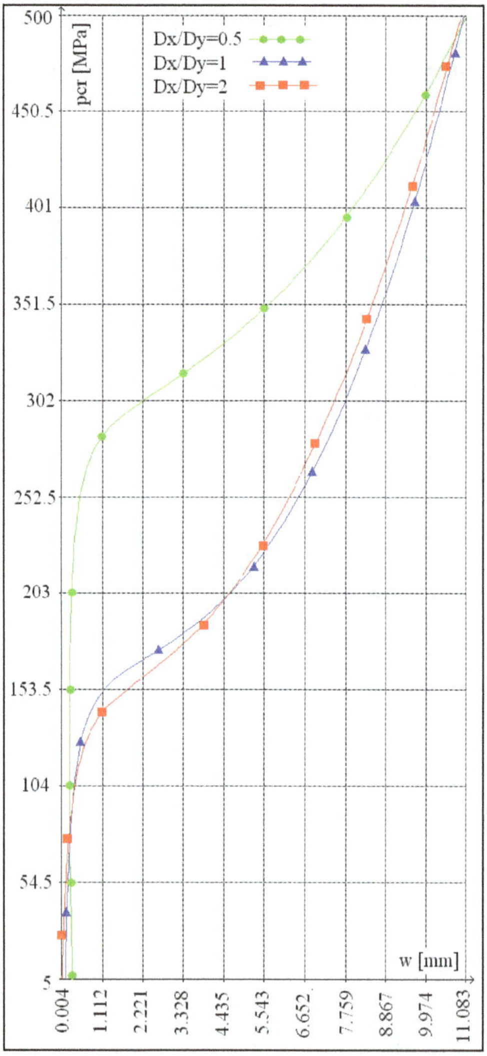

Fig. 15. Shear buckling and post buckling behaviour of the plates with central delamination placed between macro-layers 4 and 5

Buckling load has been determined by the graphical method (an asymptote on curve after the bifurcation has been plotted).

Critical value for the buckling load was obtained in the range 109 MPa $< p_{cr} <$ 289 MPa.

In the case of nonlinear model of the material behaviour the buckling load (ultimate strength) was determined by Tsai Wu criterion. So, the degradation index (failure index) for tension and compression of the delaminated plates was determined (Table). Only the plate global buckling was examined.

Dx/Dy		0.5	1	2
Position of delamination	Fail type			
Macro-layer 1 Macro-layer 2	Traction	25	20	20
	Compression	90	-	-
Macro-layer 2 Macro-layer 3	Traction	25	20	20
	Compression	90	-	-
Macro-layer 3 Macro-layer 4	Traction	25	20	20
	Compression	90	-	-
Macro-layer 4 Macro-layer 5	Traction	25	20	20
	Compression	90	-	-

Table 6. Buckling load (ultimate strength) in [MPa] for shear loading of the plates

4.4 Numerical studies on combined (compression and shear) buckling

The discussions and results on compression and shear buckling analysis will be done only for plate presented in figure, having the same geometry and the orthotropic directions (α), thickness of the macro-layers (t) and plate lay-up according to Table.

For analysis the calculus was done for various ratios $a=q/p$, that is {0.2; 0.4; 0.6; 0.8; 1}.

In figure the combined loading (shear and compression) buckling and post buckling behaviour of the plates with central delamination placed between macro-layers 1 and 2, for a=0.2 is presented.

$a=q/p$	Failure mode	Perfect plate	Dx/Dy=0.5	Dx/Dy=1	Dx/Dy=2
0.2	Tension	3962	3.962	3.962	4.239
	Compression	8.879	8.059	7.649	6.147
0.4	Tension	3.688	3.688	3.688	4.098
	Compression	7.649	6.966	6.966	5.874
0.6	Tension	3.142	3.005	2.868	2.868
	Compression	6.966	6.693	5.601	5.191
0.8	Tension	2.868	2.322	2.049	2.049
	Compression	6.693	6.425	6.425	5.327
1	Tension	1.912	1.912	1.639	1.639
	Compression	6.147	4.508	4.508	4.508

Table 7. Buckling load in [MPa] for shearing and compression loading

According to the curves presented in figure the buckling load is decreasing since the value of the diameters ratio is increasing.

So, as it is seen in figure, according to the plotted asymptotes, the buckling loads of the thin plates are placed in the domain

0.368 MPa $< p_{cr} <$ 2.634 MPa.

In the case of nonlinear model of the material behaviour, the buckling load (ultimate strength) was determined by Tsai Wu criterion, so that the degradation index (failure index) for tension and compression of the delaminated plates may be shown in Table.

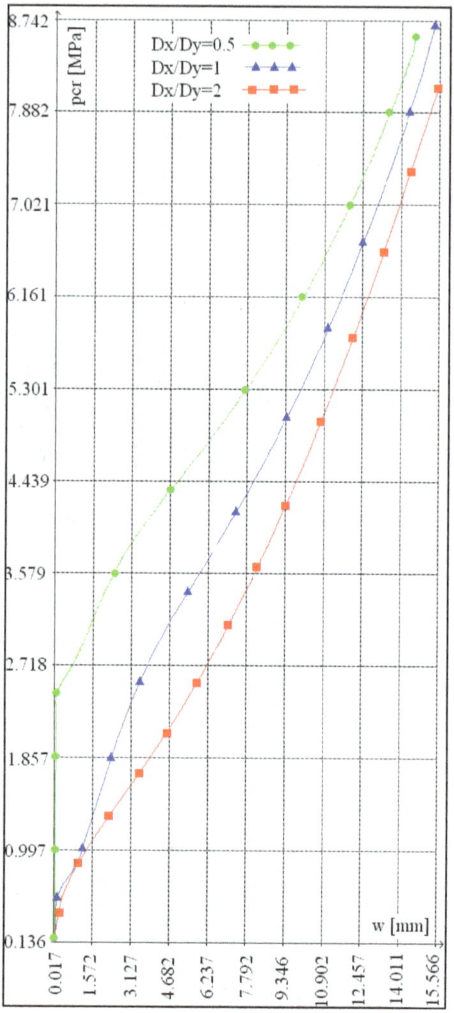

Fig.16. Buckling and post buckling behaviour of the plate

4.5 Conclusions for buckling of plates with delaminations

Delamination is a phenomenon of critical importance for the composite industry. It involves a breakdown in the bond between the reinforcement and the matrix material of the composite. Understanding delamination is essential for preventing catastrophic failures. Due to the geometry and general load of the ship deck, buckling is one of the most important failure criteria. The FEM based methodology was successfully developed for the investigation of buckling problems of composite plates having a central delamination. Two hypotheses regarding the type of material modeling were used: linear and nonlinear model. In the chapter, a convenient methodology to model delaminated composite plates was presented. The FEM modeling can be used to predict the global buckling loads of composite plates with delaminations.

In the case of linear model used for material, the following conclusions can be formulated:

- For the values of in-plane loads lower than 40 MPa, the displacement values are increasing since the diameters ratio is increasing. This trend is due to the fact the transversal diameter is decreasing (since the delaminated area remains constant). Smaller transversal diameter means increasing of shear stiffness.
- In the case of the in-plane loading values bigger than 40 MPa, the displacement values are decreasing since the diameters ratio is increasing. This trend is due to the contact pressure between the layers in contact in the delamination, which is increasing since the loading force is increasing.

In the case of nonlinear material model (Figure) the trend of the curves is so that the transversal displacement is increasing since the diameters ratio is increasing for the same in-plane loading, for a in-plane loading value under 25-40 MPa. For an in-plane loading value between 40 and 65 MPa in each case, a small instant jumping of transversal displacement is observed. This means that what is recover in plate stiffness after the increasing of contact pression in the delamination area, is lost due to the lamina damage occurring.

The buckling load determination is too difficult without the applying a graphical method, or the applying the Tsai-Wu failure criterion in the case of the general buckling not occurred till the first-ply failure occurring.

The first failure occurring in an element is based on the Tsai-Wu failure criterion, which provides the mathematical relation for strength under combined stresses was used.

The failure index is calculated in each ply of each element. In the ply where failure index is greater than 1, the first-ply failure occurs, according to the Tsai-Wu criterion. In the next steps, the tensile and compressive properties of this element are reduced by the failure index. If the buckling did not appeared until the moment of the first-ply failure occurring, the in-plane load corresponding to this moment is considered as the buckling load.

The lack of the criterion is referring to the anticipation of the real mode to occurring the cracking.

Taking into account the mathematical formulation, the Tsai-Wu failure criterion is easy to be applied. Additionally, this criterion offers advantages concerning the real prediction of the strength at variable loadings. In the case of shear buckling of thin plates, the ultimate strength of the delaminated plate is not depending on the position of delamination. For diameters ratio greater than 1, the ultimate strength in shear loading occurs only for traction fail.

Final remarks may be done. A very good postbuckling load carrying ability specific to light weight structures can be exploited to obtain high performance. Finite element analysis can offer more accurate analyses with a high degree of fidelity.

Since the plates included in the ship hull structure (so perfect or imperfect) are subjected to in-plane forces combined with transverse pressure, the buckling response of such plates is a basic concern in the design process. Furthermore, the presence of imperfections may weaken the load-carrying capacity of structures due to stiffness loss and cause an increase in stress concentrations in the imperfection area.

After the buckling, the composite plates, so perfect or imperfect, have a great capacity to be overloaded, so that the plate and finally the structure is not collapsed immediately.

5. Acknowledgement

The results presented in this chapter are based on the work supported by CNCSIS-UEFISCSU, project number PNII - IDEI, code 512/2008.

6. References

Adams, D.F., Carlsson, L.A. & Pipes, R.B. (2003). *Experimental Characterization of Advanced Composite materials*, Ed. Taylor & Francis Group.

Altenbach, H., Altenbach, J. & Kissing, W. (2004). *Mechanics of Composite Structural Elements*, Ed. Springer, Berlin.

Ambarcumyan, S.A: (1991). *Theory of Anisotropic Plates: Strength, Stability, and Vibrations*, Hemispher Publishing,Washington.

Beznea, E.F., (2008). *Studies and Researches on the Buckling Behaviour of the Composite Panels*, Doctoral Thesis, University Dunarea de Jos of Galati.

Beznea, E.F., & Chirica, I.,(2009). Studies on buckling behaviour of the composite plates with delamination, *Proceedings of the 26-th DAS-2009: Danubia-Adria Symposium on Development in Experimental Mechanics*, Leoben, Austria, pp.13-14, ISBN 978-3-902544-02-5

Bisagni, C. & Cordisco, P., (2004). Testing of Stiffened Composite Cylindrical Shells in the Postbuckling Range until Failure, *AIAA Journal*, Vol. 42, No. 9, pp. 1808- 1817

Chirica, I., Beznea, E.F., Chirica, R., Boazu, D., Chirica, A. & Berggreen C.C., (2008). *Journal of Material Testers Magazine*, www.anyagvizsgaloklapja.hu, Vol. 18/1, p. 24.

Chirica, I., Beznea, E.F.,& Chirica, R., (2009). Studies on buckling of the ship deck panels with imperfections made of composite materials, *Materiale Plastice*, vol. 46, nr.3, , pp. 243-248, ISSN0025/5289

Engelstad, S. P., Reddy, J. N. & Knight, N. F., Jr. (1992). Postbuckling Response and Failure Prediction of Graphite-Epoxy Plates Loaded in Compression, *AIAA Journal*, 30(8), 2106-2113

Finn, S.C. & Springer, G.S. (1993). Delamination in Composites Plates under Transverse Static or Impact Loads–a model, *Composite Structures*, vol. 23

Hilburger, M.F. (2001). Nonlinear and Buckling Behavior of Compression-loaded Composite Shells, *Proceedings of the 6th Annual Technical Conference of the American Society for Composites*, Virginia.

Kawai, T. & Ohtsubo, H. (1968). A Method of Solution for the Complicated Buckling Problems of ElasticPlates With Combined Use of Rayleigh-Ritz's Procedure in the Finite Element Method, *AFFDLTR*-68-150

Kim, H. & Kedward, K.T., (1999). A Method for Modeling the Local and Global Buckling of Delaminated Composite Plates. *Composite Structures* 44: 43-53

Knight N. F. Jr. & Starnes J. H. Jr., (1988). Postbuckling Behavior of Selected Curved Stiffened Graphite Epoxy Panels Loaded in Axial Compression, *AIAA Journal*, Vol. 26, No. 3, pp.344-352.

Naganarayana, B.P.& Atluri, S.N. (1995). Strength Reduction and Delamination Growth in Thin and Thick Composite Plates under Compressive Loading, *Computational Mechanics*, 16: 170-189.

Pietropaoli, E., Riccio, A. & Zarrelli, M., (2008). Delamination Growth and Fibre/Matrix Progresive Damage in Composite Plates under Compression. ECCM13, *The 13-th European Conference on Composite Materials*, June 2-5, Stockholm, Sweden

Schlack, A.L., Jr. (1964). Elastic Stability of Pierced Square Plates. *Experimental Mechanics*, June 167–172.

Thurley, G.J. & Marshall, I.H. (1995). *Buckling and Postbuckling of Composite Plates*, Ed. Chapman & Hall, London.

7

Thermoplastic Polymer based Modified Red Mud Composites Materials

A.H. Bhat[1], H.P.S. Abdul Khalil and A. K. Banthia[2]
[1]School of Industrial Technology, Universiti Sains Malaysia, Penang-11800,
[2]Materials Science Centre, Indian Institute of Technology Kharagpur-721302,
[1]Malaysia
[2]India

1. Introduction

Soaring prices are a reminder of the essential role that affordable products play in sustainable economic growth and higher human development. Utilization of waste materials has become more pressing than ever. Red mud is accumulating at a rate of 30 million ton annually throughout the world. Under normal conditions when 1 ton of alumina is produced from bauxite, an equal amount of red mud is generated as a waste. A further aspect is their reuse as starting materials for other products (Grjotheim & Welch, 1998; Kasliwal, & Sai, 1999). Red mud has been suggested as filler for polymer reinforcement or as a cheap adsorbent for removal of toxic metals or an acid by several researchers. Chand and Hashmi, 1999, tried to improve the mechanical properties and abrasive wear properties of polymer blend filled with red mud. Pradhan et al. 1999 had reported that activated red mud as a good adsorbent was used for adsorption of phosphate or chromium. In addition, the mechanical and thermal properties of polymers are generally improved by the addition of inorganic fillers. The challenges in this area of high-performance organic–inorganic hybrid materials are to obtain significant improvements in the interfacial adhesion between the polymer matrix and the reinforcing material since the organic matrix is relatively incompatible with the inorganic phase. Generally, a better interfacial bonding will impart better properties to a polymer composite such as high modulus, strength, and stiffness (Agag et al., 2001; Jang, 1992).This reinforcement of polymer filled with inorganic material is largely dependent on the physical interfacial phenomenon between the filler and matrix. This can be determining the degree of adhesion by physical interaction, such as active functional groups, hydrogen bonding, Lewis acid–base interactions, surface energy, and crystallite faces of filler surface at the interface (Park & Kim, 2000; Park & Cho, 2000). To increase physical interaction, various surface treatment techniques are applied, such as oxidation in acid solutions, (Donnet & Bansal, 1990) dry oxidation in oxygen, (Yuan et al., 1991) anodic oxidation, (Ishikawa & Matsumoto, 2001; Park, S. J.; Kim) and plasma treatments (Dilsiz et al., 1995). Surface modification leads to development of surface functional activity on a filler surface, resulting in modification to achieve good interfacial adhesion between the reinforcement and the matrix.

In materials research, the development of polymer nanocomposites is rapidly emerging as a multidisciplinary research activity whose results could broaden the applications of polymers to the great benefit of many different industries. The details about worldwide generation of Red Mud, country wise and region wise is shown in fig. 1.

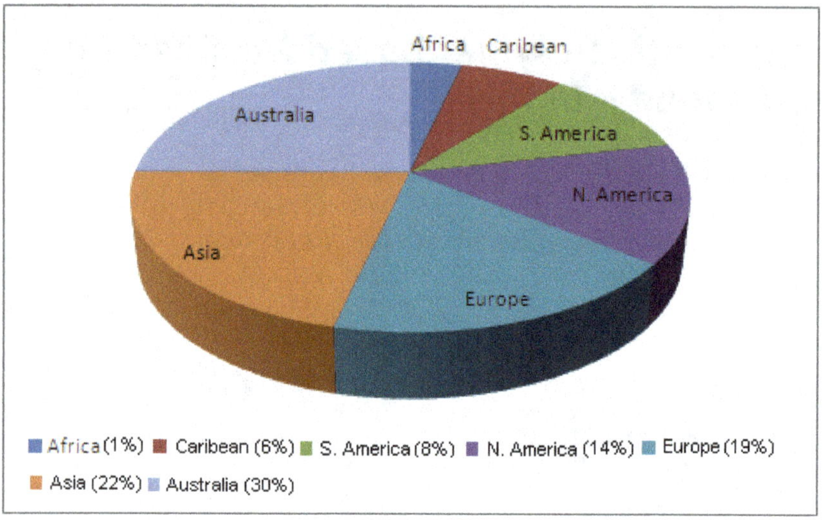

Fig. 1. Red Mud generation Continent/Region wise

The use of layered inorganic fillers has been a common practice in the plastics industry to improve the properties of thermoplastics. The effects of filler on the material properties of composite materials depend strongly on its particle size, shape, aggregate size, surface characteristics, and degree of dispersion. Polymer composites prepared in combination with an organic component as a matrix and an inorganic component as filler on the nanoscale (Huang et al., 1987; Suriyet et al., 1992; Novak & Ellsworth, 1993; Hajji et al., 1999) have additional advantages, such as the possibility of obtaining a material that has the advantages of both organic materials (e.g., light weight, flexibility, good moldability) and inorganic materials (e.g., high strength, heat stability, and chemical resistance)(Shao et al, 2003).

Low volume additions (≤5%) of nanoparticles of layered red mud provide property enhancements with respect to the neat resin that are comparable to those achieved by conventional loadings (15-40%) of traditional fillers. Red mud (nanofiller) is a major waste material obtained during the production of alumina from bauxite by the Bayer's process. It comprises of silicates and oxides of iron, aluminum, sodium, calcium and titanium, along with some other minor constituents. Based on economics as well as environmental related issues, enormous efforts have been directed worldwide towards red mud management issues i.e. of utilization, storage and disposal. Different avenues of red mud utilization are more or less known but none of them have so far proved to be economically viable or commercially feasible.

Thermoplastics have a big potential for applications in the industry as well as in construction, electrical applications and food packagings. One of the few disadvantages associated with the use of nanofillers, is their high cost. The present research work has been undertaken with an objective to explore the use of red mud as a reinforcing material as a low cost option. This is due to the fact that red mud alone contains all these reinforcement elements and is plentifully available.

In mechanical reinforcement major issues are the homogeneous dispersion of nanofillers in the polymeric matrix and the developments of chemical bonding or strong interaction at the nanofiller-matrix interface. In this study we have used acid modified and organically modified red mud for better homogeneous dispersion as well as enhanced material properties. The focus of this research was to experimentally characterize the two polymer nanocomposite systems and investigate the role of modification of filler in their behavior. Modified Red mud nanoparticles were dispersed in poly (vinyl alcohol) (PVA) and poly hydroxy ether of bisphenol-A (Ph) matrices.
The conventional solvent casting technique was employed to generate polymer nanocomposites. Red mud was treated with boric acid and phosphomolybdic acid to develop the acidic functional groups or active oxygen, resulting in the better dispersion of the red mud into the polymer matrices. Red mud was also organically modified with the oligomers of aniline formaldehyde, for better interaction between the filler and the polymer matrices. The particle size of the modified red mud was determined by field emission scanning electron microscopy (FESEM). The as-synthesized composite films were typically characterized by FTIR spectroscopy and X-Ray Diffraction. The morphological image of the composite materials was studied by scanning electron microscopy (SEM) and the dispersion of the modified fillers within the matrix was studied by transmission electron microscopy (TEM).The thermal properties measured by thermogravimetric analysis (TGA) showed enhanced thermal stability of a series of composite materials. The differential scanning calorimetry (DSC) showed increase in glass transition temperature and crystallization of the composite films.
The physical topography of the composite materials was studied by Atomic Force Microscopy (AFM).
Polyvinyl alcohol (PVA) is cornmercially available in dry granular or powdered form. It is a water-soluble and fully biodegradable polymer (Chen et al., 2000; Bachtsi & Kiparissides, 1996). PVA is having planar zigzag structure like polyethylene (Horii et al., 1992). All PVA grades are readily soluble in water. As a hydrophilic polymer, PVA exhibits excellent water retention properties. Conditions for dissolution are governed primarily by degree of hydrolysis, but they are influenced by other factors such as molecular weight, particle size distribution and particle crystallinity [Peppas & Merrill 1977]. Optimum solubility occurs at 87-89% hydrolysis. The partially hydrolyzed grades in this range exhibit a high degree of cold-water solubility. For total dissolution, however, they require water temperatures of about 185°F (85°C) with a hold time of 30 minutes. It is, in fact, a refinement of PVAcetate since the most common manufacturing process is to replace by hydrolysis (or alcoholysis) the acetate groups with hydroxyl groups. This is commonly achieved using the presence of catalytic quantities of alkali such as sodium hydroxide (which, since it acts only as a catalyst, should not in theory remain in the final product). The extent of hydrolysis will determine the amount of residual acetyl groups and this in turn apparently affect the viscosity characteristics.[Ray & Bousmina ,2005] PVA exists only as a polymer; a monomer has not yet been isolated, so the chemical structure is described in Fig.2.

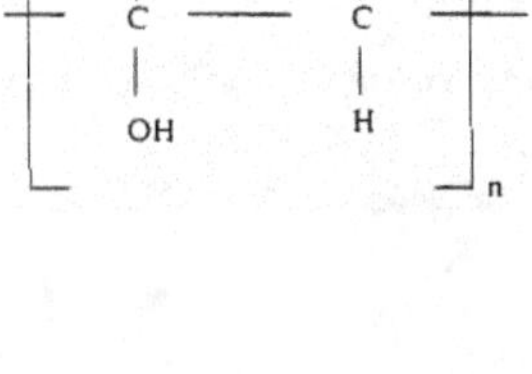

Fig. 2. Chemical formula of PVA

The PVA concentration in an aqueous solution is determined by the type of application. However, at concentrations greater than 10 wt%, the viscosity of the aqueous solution at room temperature is such that pouring becomes difficult. In addition to its solubility, PVA is also appreciated for its good mechanical properties in the dry state, resistance to common solvents, barrier effect in dry atmospheres, possibility of food contact for suitable grades, biodegradability. Some of the physical properties of PVA are as presented in table 1.

Properties	Values
Appearance	White to cream granular powder
Density	1.23-1.30 g/cm^3
Thermal stability	Gradual discoloration above 100 °C; darkens rapidly above 150 °C; rapid decomposition above 200 °C.
Coefficients of thermal expansion	7-8 x 10^{-5}/ °C
Thermal conductivity	0.2 W/m.k
Yield Stress	40-50 MPa
Elongation at break	100-200%
Melting point	230 °C for fully hydrolysed grades 180-190 for partially hydrolysed grades.
Glass transition temperature	75-85 °C

Table 1. Physical properties of PVA

PVA can be plasticized and processed by casting, dipping, injection and extrusion. (Biron, 2007).

The main engineering applications, possibly in combination with other polymers, are:

- Films for packing chemicals, fertilizers, herbicides, disinfectants, dyes, colorants, scalers, cosmetics etc.
- Release films for composite moulding.
- Solvent resistant tubes and pipes.
- Membranes for pumps carrying petroleum or chemical products

Trade names: Elvanol, Polyviol, Mowiol, Rhodoviol.

The commercial name of poly (hydroxy ether) of bisphenol A is Phenoxy, and as a thermoplastic polymer it possesses many excellent properties such as (Zhang et al., 2002)

- Good chemical stability
- Excellent matrix material for producing polymer nanocomposites
- Thermal stability
- Tractability
- Transparency

The poly (hydroxy ether of bisphenol A) (phenoxy (Ph)) has been revealed as a polymeric matrix able to intercalate in, and partially exfoliate a commercial organically modified montmorillonite. Dispersion was attributed to chemical interactions between the Ph and the inorganic clay (Fornes et al., 2004).
Poly (hydroxy ether of bisphenol A) (Ph) based polymer nanocomposites (PN) reinforced with a layered red mud with acidic and organic modifications were prepared by conventional solvent casting technique. The best dispersion occurred in the PN where the interactions between the functional groups of the polymer matrix and those of the organic substitution of the red mud appeared to be the highest. The modulus increase is an indirect but quantitative measurement of the attained dispersion level.
The structure of Poly (hydroxy ether) of bisphenol A is shown in fig.3.

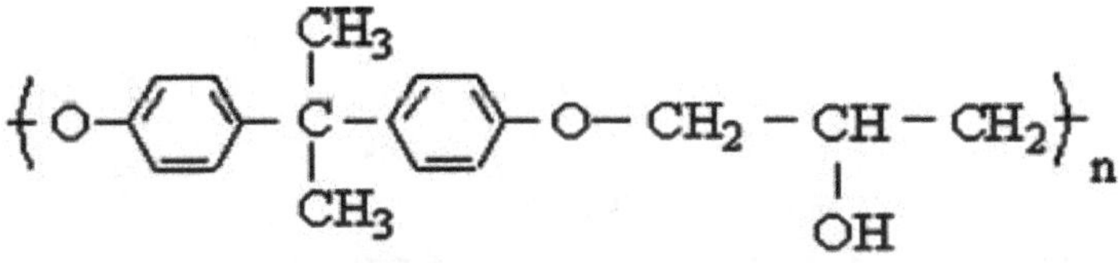

Fig. 3. Poly (hydroxy ether) of bisphenol A

2. Modification of red mud

2.1 Preparation of boric/phosphomolybdic acid modified-red mud

A known amount of raw red mud was first washed with distilled water two to three times and then dried in an oven. When the red mud dried completely; it was treated with a 5 M solution of boric acid/ phosphomolybdic acid for 24 hours. Then it was being filtered, dried, weighed and grinded in order to get its powdered form. The weight of the red mud increased by 24 %. The powder was then sieved through 53-micron mesh to remove the larger particles. The powder so obtained was fine and acidically modified. Thus, Layered silicates and double layered hydroxides, $Al_2 (OH)_7$ (AH) and $MgAl (OH)_5$ (MAH), of red mud were modified with inorganic acids. The inorganic acid modifier compatibilizes the silicate and hydroxide surface to polymer matrices and spaces the crystalline layers apart to minimize the energy needed for exfoliation process.

2.2 Preparation of organically modified-red mud

The organic modification of red mud was done by the following two steps:

i. Freshly prepared Aniline Hydrochloride is mixed with red mud with a magnetic stirrer.
ii. Formaldehyde is then added drop by drop to the mixture with intense stirring action.

The addition of aniline hydrochloride to red mud replaces the cation present in the octahedral sites of the silicate with aniline occupying the same. The formaldehyde added would form a condensation oligomer as the product with the pendent group as formaldehyde which is compatible with the polymer. The ratio of Aniline to formaldehyde is kept 1:1 so as to stop any further condensation of the aniline and formaldehyde as this could lead to the polymer blend type nanocomposite, which could have the problems associated with miscibility of polymer blends and this filler would not be universal filler for the polymer matrix. The figure 4 depicts the experimental setup for the organic modification of red mud.

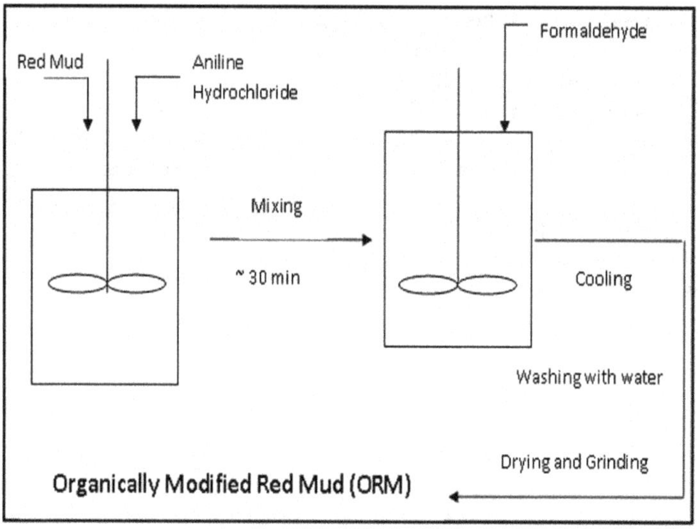

Fig. 4. Experimental setup of organic modification of red mud

2.3 Mechanism:

Figure 5 showed the mechanism and condensation reaction between substituted aniline and formaldehyde. After substitution of the metal cation, condensation reaction occurs between the substituted amine group with extra hydrogen and the formaldehyde molecule to form water as a byproduct. Thus the organic entity enters the space between the silicate layers thus providing a suitable site for binding the polymer. When this filler is mixed with the polymer, the polymer chains are attracted due to the presence of the organic species at the interlayer spaces, and thus get intercalated in between the layers, which have about nanometer size openings.

The reaction mechanism can be best depicted by the following sequence

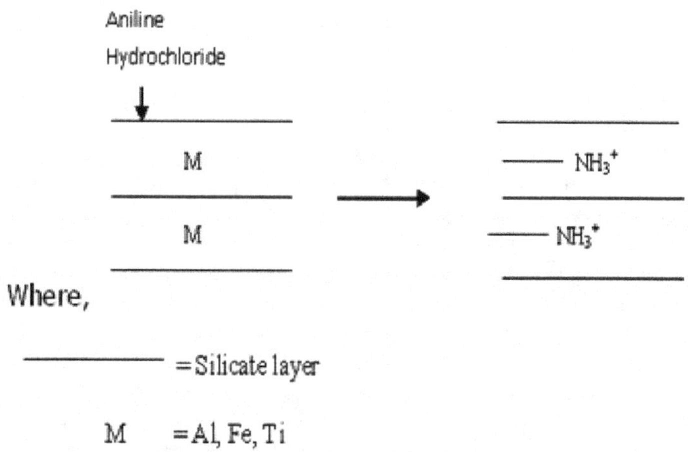

Mechanism for organic modification of red mud

Condensation reaction between substituted aniline and formaldehyde

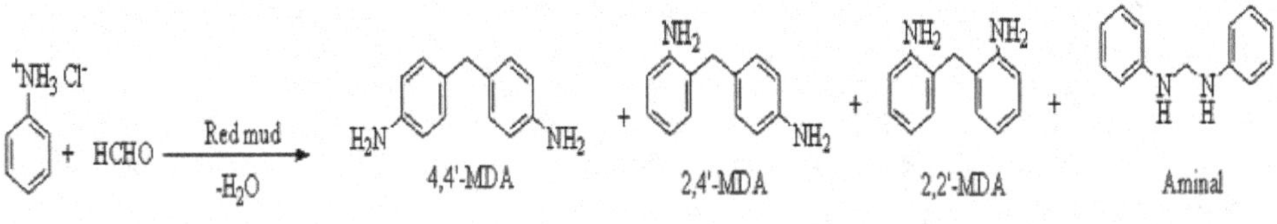

Fig. 5. Mechnaism and Condensation reaction showing Organic Modification of Red Mud

3. Experimental

3.1 Preparation of PVA-acid red mud nanocomposite materials and organically modified red mud nanocomposite materials

10% (w/v) aqueous solution of PVA was prepared by dissolving 10g of dried powdered PVA in 100ml of distilled water at a temperature of 100 °C with continuous stirring. Polymer modified red mud nanocomposite (PRM) films were made by mixing different loading percentage of acid modified red mud and organically modified red mud into the virgin PVA solution. The PVA-red mud solution thus obtained was sonicated for better dispersion and removal of any air bubble present in the solution. The solution was cast on the plane glass surface and dried at room temperature. The composition of the PRM materials was varied from 0 to 5 wt % of RM with respect to the PVA content.

3.2 Red mud loading levels

Sample index	PVA	SP1	SP2	SP3	SP4	SP5	SP6	SP7
% Loading of boric acid mod. red mud	0	0.5	1.0	1.5	2.0	3.0	4.0	5.0

Sample index	PVA	CP1	CP2	CP3	CP4	CP5	CP6	CP7
% Loading of organically mod. red mud	0	0.5	1.0	1.5	2.0	3.0	4.0	5.0

Sample index	PVA	PRM1	PRM2	PRM3	PRM4	PRM5	PRM6	PRM7
% Loading of Phosphomolybdic acid mod. red mud	0	1.0	1.5	2.0	2.5	3.0	4.0	5.0

Table 2. Sample Designation

3.3 FTIR spectroscopy

The nature of the chemical bonds in raw red mud and the modified red mud were characterized by FTIR spectroscopy. The FTIR spectra of raw red mud and acid modified red mud are shown in Fig. 5 and it clearly depicts the main characteristic peaks associated with them. The characteristic vibration band of raw red mud for hydroxyl stretching vibration (O-H) was found at 3417 cm^{-1} while for boric acid modified red mud, the hydroxyl stretching occurred at 3379 cm^{-1} and the same characteristic peak for phosphomolybdic acid was observed at 3129 cm^{-1}. A red shift in the hydroxyl stretching vibration (O-H) at 3379 cm-

1 in the boric acid modified red mud (BRM) and 3129 cm^{-1} in the phosphomolybdic acid modified red mud (PRM) arises due to disappearance of the hydroxyl groups upon cross linking reaction with the boric acid and the phosphomolybdic acid respectively. The peak at υ = 993 cm^{-1} in case of raw red mud could be attributed to Si-O stretching vibration and the same peak for BRM and PRM were observed at 1000 cm^{-1} and 996 cm^{-1} respectively. The blue shift observed in the peak position of Si-O occurred due to the interaction of red mud with the inorganic acids.

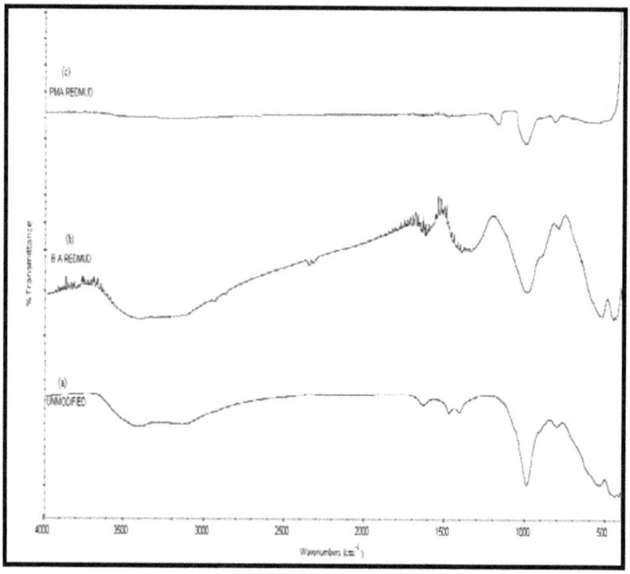

Fig. 5. FTIR Spectra of (a) Raw red mud (b) Boric acid modified red mud (BRM) (c) Phosphomolybdic acid modified red mud (PRM)

According to the Rocchiccioli - Deltcheff et al. [80], the four absorption bands in the spectrum of pure phosphomolybdic acid (PMA) shows the typical features of Keggin anions and the peak frequencies at 1064, 961, 868, 780 cm^{-1} are assigned to the□ v (P-O),□ v (Mo-O_t) (O_t refers to the terminal oxygen),□ v (Mo-O_c-Mo) (O_c refers to the corner oxygen) and□ v (Mo-O_e-Mo) (O_e refers to the edge oxygen) respectively. All these characteristic bands of PMA are not present in the spectra of the PRM, indicating that there is specific interaction between the keggin structure and red mud.

Groups	Raw red mud (cm^{-1})	BRM (cm^{-1})	PRM (cm^{-1})
OH	3417	3379	3129
OH bend.	1636	1631	-
B-O	-	1368	-
Si-O	993	1000	996
P-O	-	-	1074
Mo-O_c-Mo			872

Table 3. Assignment of FTIR spectral bands for raw red mud and the acid modified red mud

An upward shift was observed for the v (P-O) and v (Mo-O_c-Mo) stretching vibrations, whereas bands ascribed to the stretching vibration v (Mo-O_t) does not appear in the PRM.
The major vibration bands for organically modified red mud as shown in Fig. 6 occurs at OH/NH stretching (3320 cm^{-1}), NH_3^+ (1598 cm^{-1}) and Ar-N (1512 cm^{-1}). A red shift in the hydroxyl peak at 3320cm^{-1} for modified red mud arises due to the interaction of red mud platelets with the aniline hydrochloride moiety through physical interactions. The characteristic Si-O stretching vibration is not found in case of ORM due to formation of a coating of organic moiety around the silicate galleries.

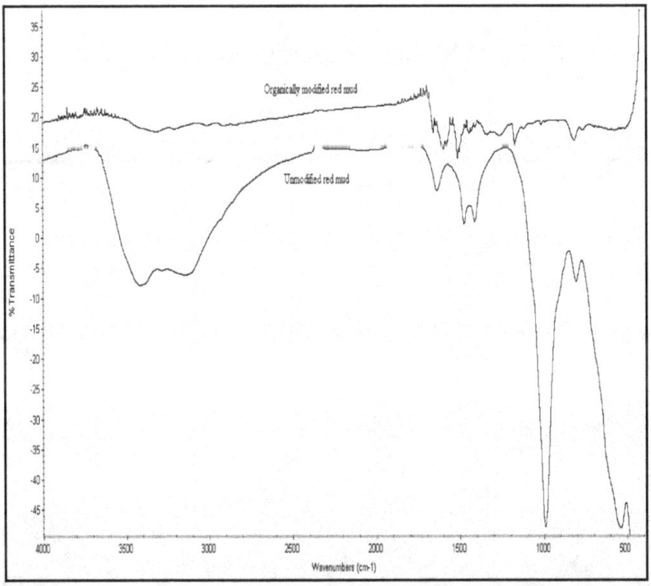

Fig. 6. FTIR Spectra of Raw red mud and organically modified red mud (ORM)

Groups	Raw red mud (cm^{-1})	ORM (cm^{-1})
OH	3417	3320
OH (bending)	1636	-
NH3+	-	1598
Ar-N	-	1512
Si-O	993	-

Table 4. Assignment of FTIR spectral bands for raw red mud and organically modified red mud

The representative FTIR spectra of the pure PVA and various polymer nanocomposite materials are shown in Fig.7.The characteristic vibration band of PVA for hydroxyl stretching vibration(O-H) having polymeric association is shown at 3292 cm^{-1}, While the alkyl stretching band of C-H is shown at 2942 cm^{-1}.The peak at υ=1736 cm^{-1} is due to C=O

stretching of saturated aliphatic esters (acetate), 1042cm^{-1}(O-H Bending), and 1424 cm^{-1} (C-H Bending).
The FTIR spectrum in Fig. 3.6 also shows the spectra associated with PRM2, SP2 and CP2. The intensity of the O-H stretching vibration peak (v = 3200-3400 cm^{-1}) decreased and peak frequency shifted from 3292 to 3250 cm^{-1}, i.e. to a lower frequency as compared to that of pure PVA. The red shift in the hydroxyl stretching frequency occurs due to the various degree of hydrogen bonding between the polymer and the filler which lengthens and weakens the O-H bond and hence lowers the vibrational frequency. Also, the peak frequency of C-H for PRM2, SP2 and CP2 occurs at 2916 cm^{-1}, 2945 cm^{-1} and 2960 cm^{-1} respectively as tabulated in Table 5.
By loading 1% ORM (CP2) in the polymer matrix, the alkyl stretching vibration of C-H increases from 2942-2960 cm^{-1} and occurs due to the interaction of polymer matrix with the organic moiety of the filler.

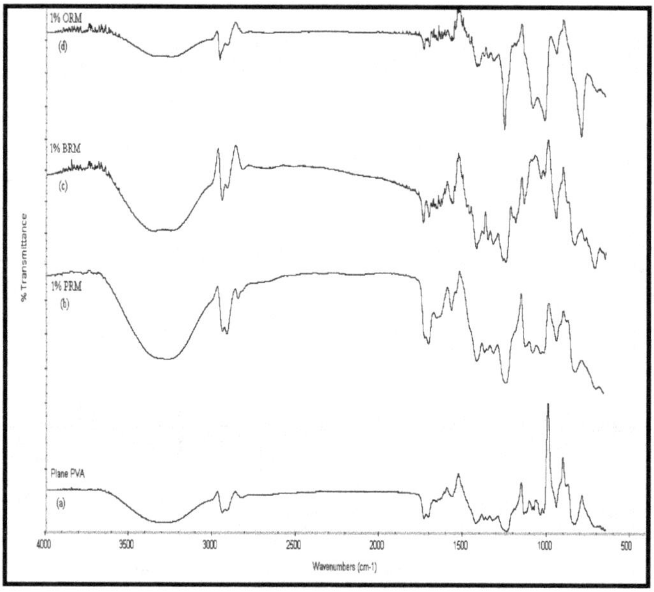

Fig. 7. FTIR Spectra of (a) Pure PVA (b) PRM2 (c) SP2 (d) CP2

Groups	PVA (cm^{-1})	PRM2 (cm^{-1})	SP2 (cm^{-1})	CP2 (cm^{-1})
OH	3292	3268	3267	3250
C-H	2942	2916	2945	2960
C=O	1736	1712	1743	1742
O-H $_{(bending)}$	1042	1043	1041	1019
CH_2 $_{(bending)}$	1424	1423	1424	1414

Table 5. Assignment of FTIR spectral bands for Pure PVA and the PVA/modified red mud nanocomposite membranes

The FTIR spectra in Fig. 8 shows the spectra associated with PRM4, CP4 and SP4.The intensity of the O-H stretching vibration peak (v = 3200-3400 cm^{-1}) decreased and the peak frequency shifted from 3292 in case of pure PVA to 3278 cm^{-1}, 3245 cm^{-1} and 3230 cm^{-1} for PRM4, CP4 and SP4 respectively i.e. to a lower frequency as compared to that of pure PVA. This is again due to the intermolecular hydrogen bond formation, thereby decreasing the O-H stretching frequency with the increase in the percent loading of the modified filler. The other characteristic peaks are given in table 6.

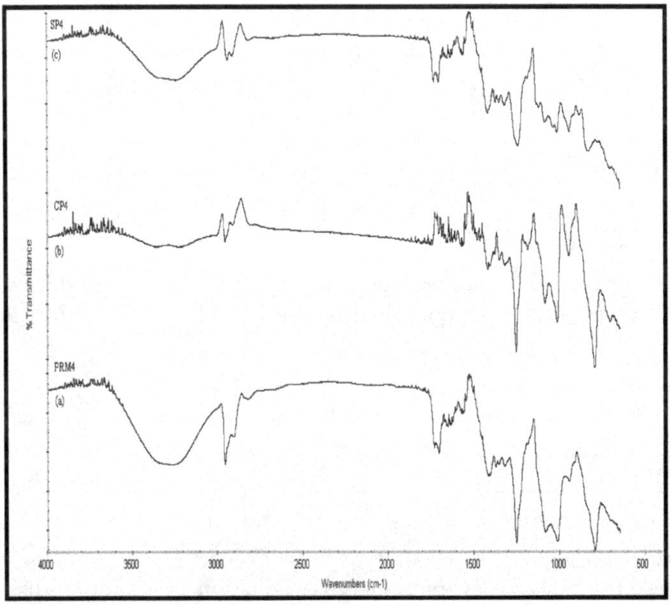

Fig. 8. FTIR Spectra of (a) PRM4 (b) CP4 (c) SP4

Groups	PRM4 (cm^{-1})	SP4	CP4
OH	3278	3245	3230
C-H	2961	2943	2959
C=O	1711	1737	1754
O-H $_{(bending)}$	1020	1021	1020
CH_2 $_{(bending)}$	1414	1424	1424

Table 6. Assignment of FTIR spectral bands for PVA/modified red mud nanocomposite membranes with different loading percentage of filler

3.4 X - ray diffraction study

Wide angle X-ray diffraction studies were performed on raw red mud and the modified red mud as shown in Fig. 9.

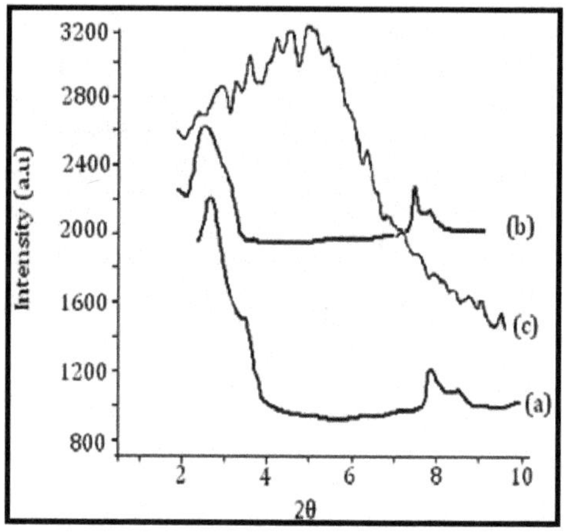

Fig. 9. Xrd pattern of (a) Raw red mud (b) Phosphomolybdic acid modified red mud (c) Organically modified red mud

The spacing, d_{001}, of the (001) basal reflections of the Raw red mud and PRM were measured and are shown in table 7. It was found that the basal spacing of the PRM is larger than the raw red mud.

Sample	2θ	d-spacing (A°)
Raw red mud	2.71	32.76
PRM	2.46	36.72

Table 7. XRD basal spacing results of raw red mud, and acid modified red mud

In the ORM, the case appears to be slightly different with the peak being lost. Thus, we could infer that the silicate layers undergo exfoliation or random dispersion at the modification stage itself. This suggests that the aniline formaldehyde type precursor have formed a coating on the partially exfoliated silicate layers leading to random dispersion thus rendering the absence of peaks in the XRD micrograph (Alexandre & Dubois, 2000).

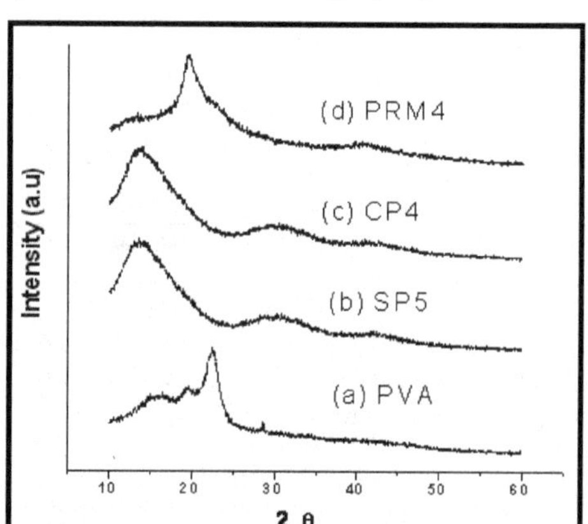

Fig. 10. X-ray Diffraction pattern of (a) PVA (b) SP5 (c) CP4 (d) PRM4 film

The XRD patterns for the Virgin PVA and the PVA nanocomposite films with different critical loading percentage of modified red mud were depicted in Fig. 10. The crystallinity of the nanocomposite membranes was found to be mainly due to PVA. The characteristic diffraction peak for pure PVA was observed at 2θ value of 22.56°. While the diffraction peak of the nanocomposite films were considerably broadened and displaced to lower 2θ values as shown in table 8, indicating reduction in the crystallinity of the nanocomposites.
The spacing, d_{001}, of the (001) basal reflections of the silicate galleries of the PVA nanocomposite films were measured as shown in table 3.9. It was found that the basal spacing for CP4 is larger than other nanocomposites materials, thereby giving evidence of more exfoliation in CP4 than the rest nanocomposites. Hence, it can be concluded that the interlayer spacing got increased by intercalation of polymer matrix. So, a lot of surface area of modified red mud is now available to the polymer for bonding.

Sample	2θ	d_{001}-spacing (A°)
SP5	13.47	6.56
CP4	13.11	6.74
PRM4	19.15	4.63

Table 8. XRD d_{001}-spacing Results

The change of crystalline behavior was further evidenced by the studies of the morphological image of as-synthesized materials. Individual silicate layers, along with two, three and four layer stacks, were found to intercalate and partially exfoliate in the PVA matrix.

3.5 SEM

On higher magnification (10,000 X) viewing of morphology of pristine PVA SP4 and SP5 type of nanocomposite films in Figure 11 (a) , (b) and (c) respectively, it was found that the morphology of polymer nanocomposite films becomes much smooth than pure PVA.

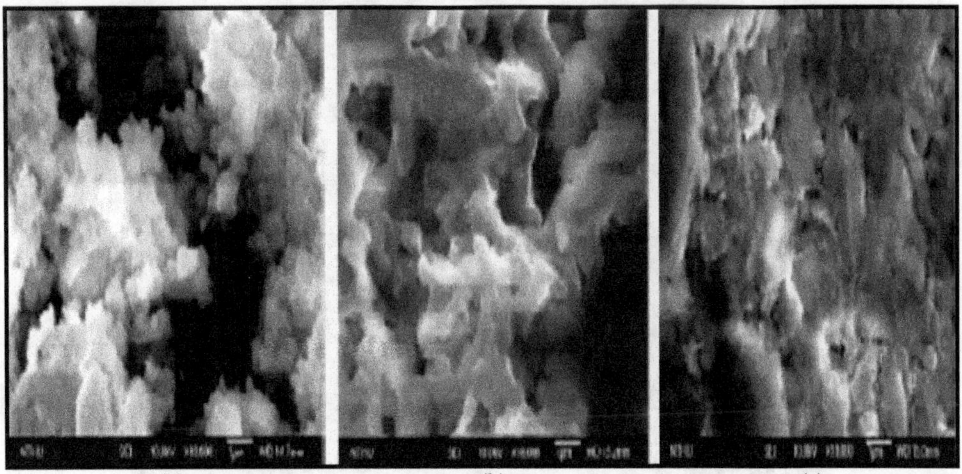

(a) (b) (c)

Fig. 11. SEM micrographs of (a) PVA (b) SP4 (c) SP5

The granules shape of PVA indicates some crystalline behavior occurring in pure polymer. The incorporation of red mud seems to destroy the orientation of semicrystalline polymer and to convert the polymer morphology approaching to amorphous state. [Bhat & Banthia, 2007] The SP5 showed much smooth morphology that is different to crystallites of PVA observation. This result is consistent with the wide-angle powder XRD patterns as described previously.

3.6 Transmission Electron Microscopy (TEM)

TEM complements XRD by observing a very small section of the material for the possibility of intercalation or exfoliation. It also provides information about the particle size and nanodispersion of particles. It however, supplies information on a very local scale. However, it is a valuable tool because it enables us to see the polymer and the filler on a nanometer scale.

The images, in Fig.12, 1 3 and Fig. 14, provide information regarding likely sizes of particles present in the matrix. Figure shows each of the polymer nanocomposite systems with critical loading of modified red mud. In Fig. 12 the silicate galleries of the organically modified red mud showed partial exfoliation and intercalation as depicted by the ridges in the image. While SP5 and PRM4 showed completely intercalated system. Nanoparticles showed agglomeration in some parts of the composite films due to the conformation of the polymer chains adhered to the nanoparticles.

Particle sizes on the TEM images are worth noting. In Figure 12, several ORM particles in PVA are ~8 nm , in Figure 13, BRM in PVA are as small as ~12 nm, and in Fig. 14, PRM in PVA ranges to a size of ~23 nm. Thus, it could be inferred, that there were numerous platelets that were expanded by the penetration of polymer chains.

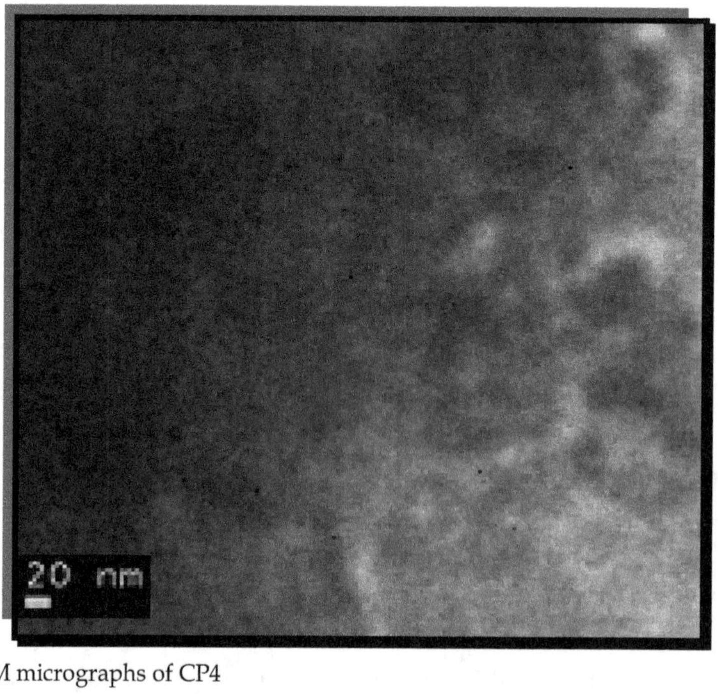

Fig. 12. TEM micrographs of CP4

Fig. 13. TEM micrographs of SP5

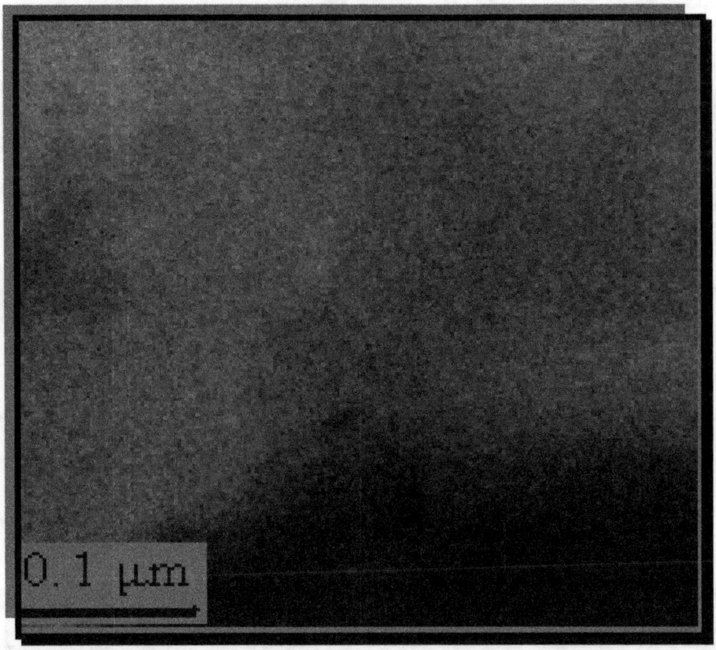

Fig. 14. TEM micrographs of PRM4

3.7 Atomic Force Microscopy (AFM)

The surface morphology of the pure PVA membrane and the PVA-modified nanocomposite membranes were analyzed by Tapping Mode - Atomic Force Microscopy (TM-AFM). Quantitatively, the differences in the morphology can be expressed in terms of various roughness parameters such as the mean roughness R_a, the root mean square (rms) of vertical data R_q, and the maximum height R_{max}. Here, the mean roughness is the mean value of surface relative to the center plane, the plane for which the volume enclosed by the image above and below this plane are equal; R_{max} the height difference between the highest and lowest points on the surface relative to the mean plane and R_q is the standard deviation of the Z values within the given area. The roughness parameters were calculated for pure PVA and PVA nanocomposite membrane surfaces with critical loading of filler and have been summarized in tabular format.

The TM-AFM images of the Pure PVA membrane is depicted in Fig. 15. The AFM images of PVA membrane shows randomly distributed hard crystalline regions (crystallites) of 100 - 250 nm in dimension on the membrane surface as evident from the observed morphology. The pure PVA membrane surface has a mean roughness of 1.026 nm.

The nanocomposite membrane with critical loading percentage of 3 wt. % BRM content (Fig. 16) shows a homogeneous and non-porous morphology. The mean roughness for the surface topography of the membrane was found to be 6.595nm. This shows that there is a good compatibility between the filler (BRM) and the PVA matrix. While critical loading surface topography of CP4 and PRM4 as shown in Fig. 17 and 18 respectively again showed homogenous and non-porous morphology. The mean roughness of CP4 and PRM4 were observed to be 2.016 nm and 9.22 nm respectively as shown in table 9.

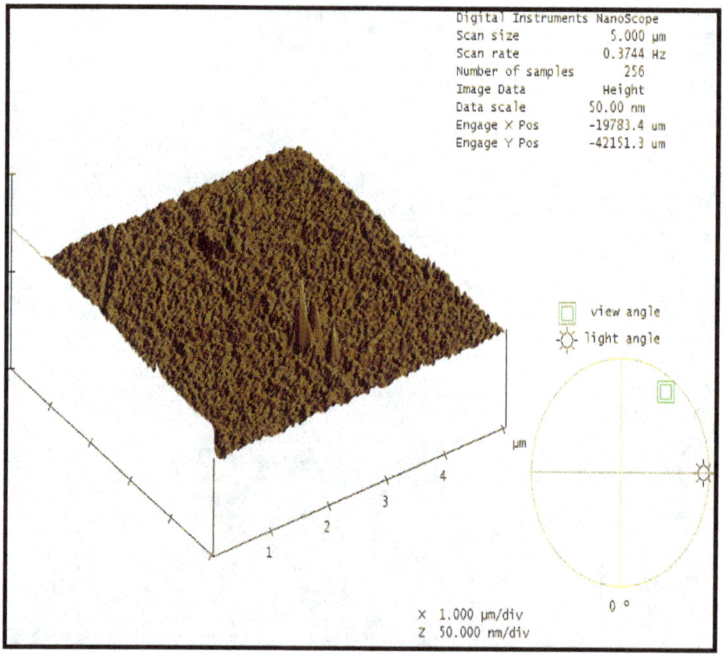

Fig. 15. Tapping Mode (3D) AFM images of Pure PVA membrane

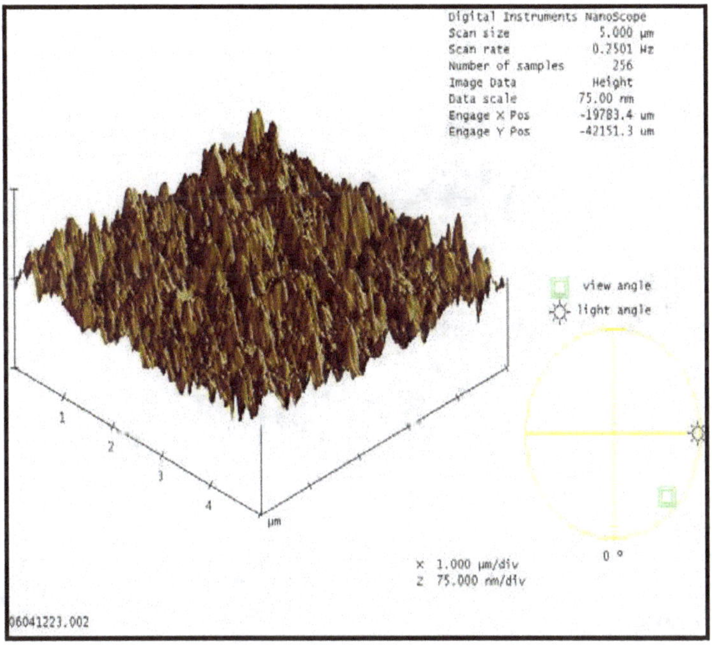

Fig. 16. Tapping Mode (3D) AFM images of SP5

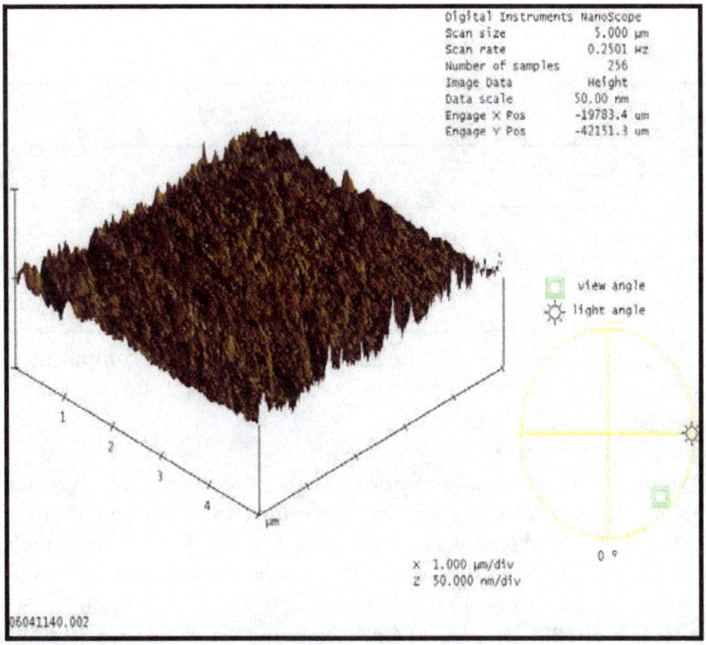

Fig. 17. Tapping Mode (3D) AFM images of CP4

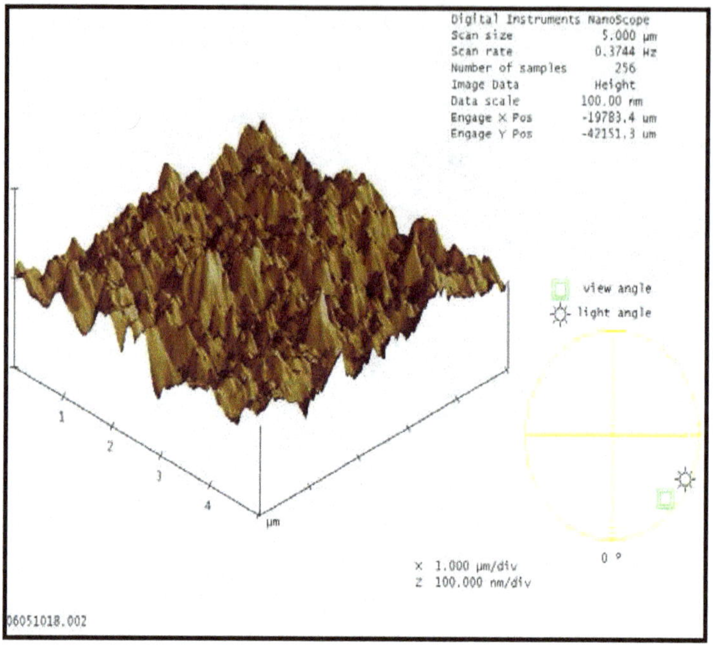

Fig. 18. Tapping Mode (3D) AFM images of PRM4

Membrane Details	R_a (nm)	R_{max} (nm)	R_q (nm)
Pure PVA	1.026	47.829	1.953
CP4	2.167	38.766	2.846
SP5	6.595	107.72	8.731
PRM4	9.22	89.803	11.572

Table 9. Roughness parameters for Pristine PVA and the PVA nanocomposite membrane surfaces

3.8 Thermogravimetric properties

The thermal stability of the raw red mud, ORM, pure PVA and PVA nanocomposite membranes with different filler content was investigated by thermogravimetric analysis. The TGA thermograms of the raw red mud, ORM, and PVA- nanocomposite films are shown in Fig. 19, Fig.20 and Fig. 21 respectively. The raw red mud is thermally stable up to 267 °C while ORM is stable upto 243 °C. Although ORM has 37% residue at 600°C, raw red mud is more stable, with 88% residue at 600°C.Raw red mud suffers an 11% weight loss at 600°C because of the probable loss of volatile impurities.

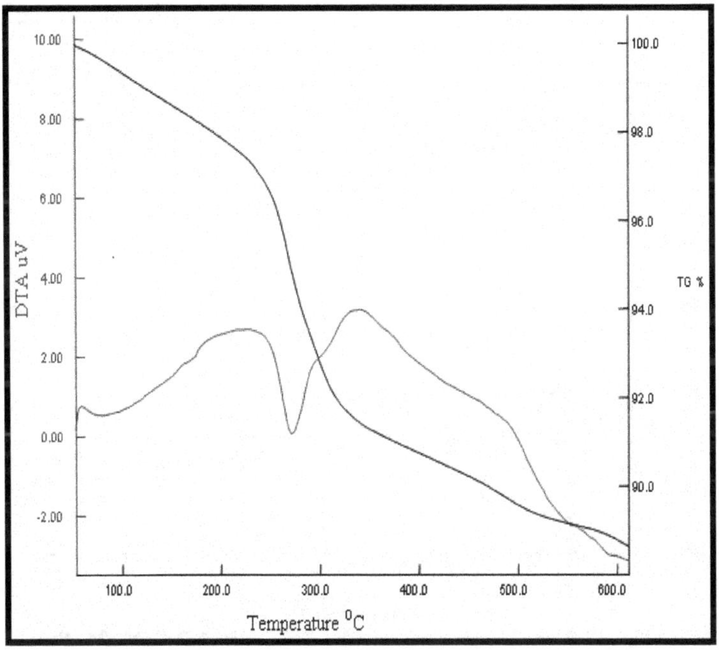

Fig. 19. DTA-TG Analysis of Raw red mud

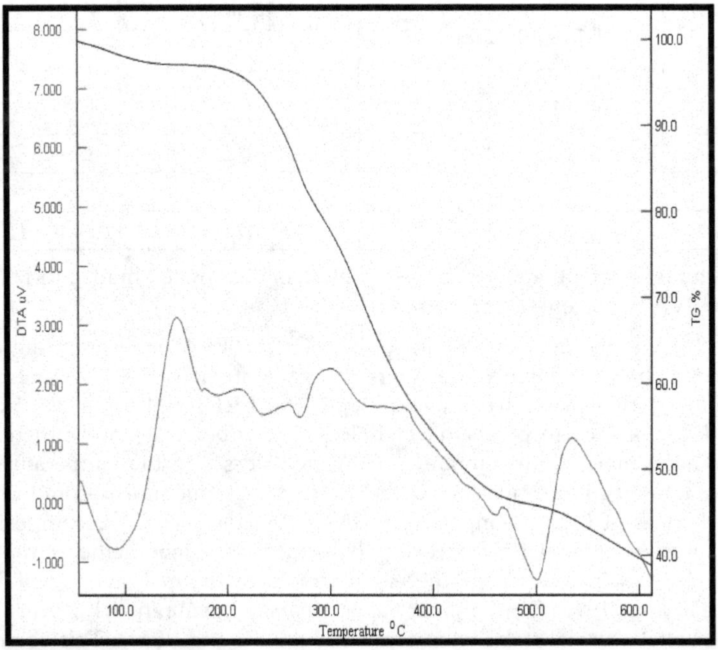

Fig. 20. DTA-TG Analysis of ORM

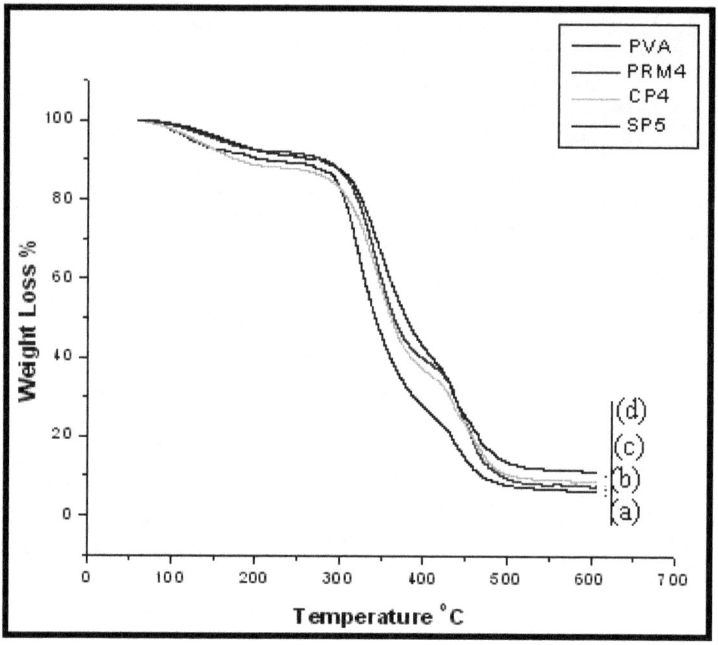

Fig. 21. Shows thermograms of (a) PVA (b) PRM4 (c) CP4 (d) SP5 film

Sample	T_i (°C)	T_f (°C)	Ash (%)
Raw red mud	267	356	88
ORM	243	482	37
Pure PVA	252	442	4.12
PRM4	274	462	7.23
CP4	277	467	10.13
SP5	281	474	12.43

Table 10. Degradation temperatures and ash content of the raw red mud, ORM, Pristine PVA and the PVA-nanocomposite membranes

The pristine PVA is thermally stable up to 250 °C. The PVA nanocomposite membranes show enhanced thermal stability; the thermal degradation of the PVA nanocomposites generally showed two major weight loss steps. The first weight loss of 15-20 wt % is centered on 150 °C and corresponds to weight loss of absorbed water and structural water of the polymer nanocomposite membranes. The weight losses in the temperature ranges of 353–483°C can be attributed mainly to the subsequent structural decomposition of the polymer backbones at higher temperatures. Evidently, the thermal decomposition of the polymer nanocomposite materials shift slightly towards the higher temperature range than that of pure PVA, which confirms the enhancement of thermal stability of intercalated polymer [Lee & Jang, 1996]. The Polymer nanocomposite membrane with critical loading of SP5, showed better thermal stability as compared to CP4 and PRM4. This could be due to better intercalation of polymer matrix within the silicate galleries of SP5 while in case of

CP4, inspite of partial exfoliation and intercalation, thermal stability is considerably lesser than SP5 .This could be due to presence of organic moiety in CP4 which causes a decrease in its thermal stability. After, 500°C, the curve all became flat and mainly the inorganic residue (i.e. Al2O3, MgO, and SiO2) remained.

3.9 Differential scanning calorimetry

DSC traces of PVA and polymer nanocomposite materials are shown in Fig.22. PVA exhibited an endotherm at 52.°C corresponding to the glass transition temperature (T_g) of PVA [90]. All the nanocomposite materials with different critical loading percentage of modified red mud were found to have a high T_g compared to the bulk PVA. This could tentatively be attributed to the confinement of the intercalated polymer chains within the red mud galleries that prevents the segmental motions of the polymer chains, thereby enhancing the glass transition temperature of the polymer matrix. The nanocomposite membrane SP5 showed maximum T_g and melting temperature (T_m) of 68 °C and 207 °C respectively. Thereby, revealing efficient dispersion of BRM within the polymer matrix.. The glass transition temperature (T_g) and melting temperature (T_m) of pure PVA and PVA nanocomposite membranes have been tabulated in table 11.

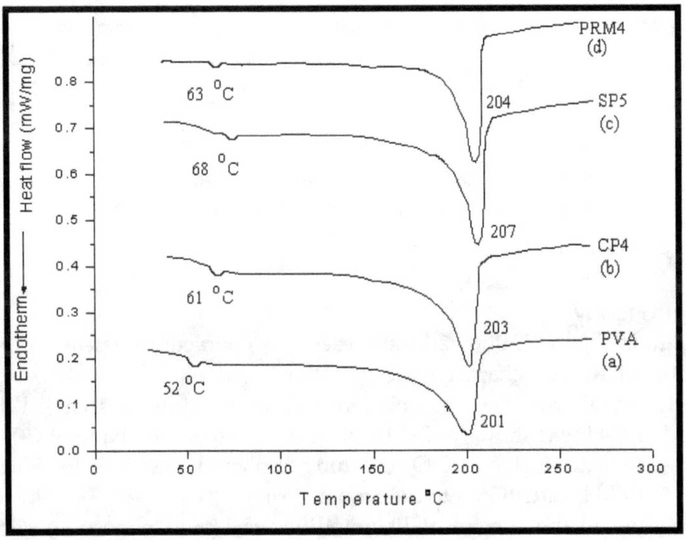

Fig. 22. DSC thermograms of (a) PVA (b) CP4 (c) SP5 (d) PRM4

Sample	T_g (°C)	T_m (°C)
Pure PVA	52	201
CP4	61	203
SP5	68	207
PRM4	63	204

Table 11. Glass Transition temperature (T_g) and melting temperature of pure PVA and the PVA- nanocomposite membranes

4. Preparation of Phe-phosphomolybdic acid modified red mud nanocomposite materials (PRC) and Phe-organically modified red mud nanocomposite materials (PNC)

10% (w/v) solution of Phe was prepared by dissolving dried Phe in THF with continuous stirring at 50 °C. Polymer nanocomposite (PN) films were made by mixing different loading percentage of phosphomolybdic acid modified red mud (PRM) and organically modified red mud (ORM) into the virgin Phe solution, stirred constantly and sonicated for half an hour for better dispersion. The composition of the PN materials was varied from 0 to 3wt % of modified RM with respect to the Phe content. The samples were designated as shown in Table 12.

Sample index	% Loading of filler
*Phe	0
**PRC1	1.0
PRC2	2.0
PRC3	3.0
***PNC1	1.0
PNC2	2.0
PNC3	3.0

*Phe: Phenoxy or poly (hydroxy ether) of Bisphenol A
** PRC: Phe-phosphomolybdic acid modified red mud nanocomposites
*** PNC: Phe-organically modified red mud nanocomposites

Table 12. Sample Designation

4.1 FTIR spectroscopy

The FTIR spectra in Figure 23 showed the presence of characteristic peaks of poly (hydroxy ether) of bisphenol A. The characteristic vibration band of Phe for hydroxyl stretching vibration (O-H) having polymeric association was observed at 3394 cm^{-1}, While the alkyl stretching band of C-H was shown at 2974 cm^{-1} and aryl stretching band of C-H at 3037 cm^{-1}. The peak at $v = 1183$ cm^{-1} is due to C-O stretching of phenyl ether linkage. Further, the peak at $v = 1607$ cm^{-1} could be attributed to the aromatic ring.

Figure 23 also showed the spectra of Phe-modified red mud nanocomposite membranes with different loading percentage of the filler. The characteristic vibration band of PRC3 for hydroxyl stretching vibration (O-H) was observed at 3270 cm^{-1}. The intensity of the O-H stretching vibration peak decreased and peak frequency shifted from 3394 to 3270 cm^{-1}, i.e. to a lower frequency as compared to that of pure phenoxy. The red shift in the hydroxyl stretching frequency occurs due to the various degree of hydrogen bonding between the polymer and the filler which lengthens and weakens the O-H bond and hence lowers the vibrational frequency. The alkyl stretching band of C-H in case of PRC3 also showed red shift from 2974 to 2861 cm^{-1} due to the interaction of modified red mud with the polymer matrix. Similarly, peak at $v = 3036$ cm^{-1}, was attributed to the stretching vibration of aryl C-H band, slightly lower frequency than the pure phenoxy. The other characteristic vibrational frequencies of PRC3 are presented in table 13.

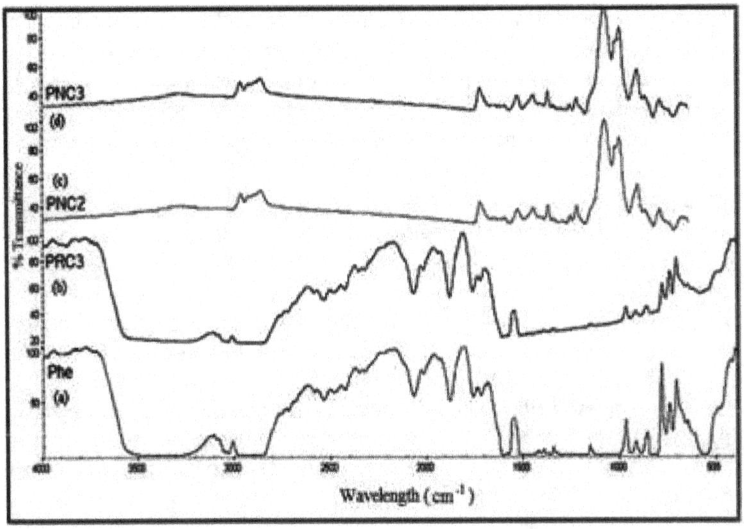

Fig. 23. FTIR Spectrum of (a) Pure Phe (b) PRC3 (c) PNC2 (d) PNC3

Groups	PHE (cm^{-1})	PRC3 (cm^{-1})	PNC2 (cm^{-1})	PNC3 (cm^{-1})
OH	3394	3270	-	-
C-H	2974	2861	2948	2947
C-H (Ar)	3037	3036	3016	3020
NH_3+	-	-	1571	1572
C-O	1183	1168	1182	1180
C=O	-	1766	1760	1760
Ar	1607	1613	1499	1503
Si-O	-	953	952	951
Mo-O_c-Mo	-	896	-	-

Table 13. Assignment of FTIR spectral bands for Pure Phenoxy and the Ph/modified red mud nanocomposite membranes

The spectra of PNC2 and PNC3 as shown in fig. 23, showed the disappearance of the hydroxyl stretching frequency (O-H). The disappearance of the hydroxyl peak depicted complete physical interaction of the O-H group with different groups of the organically modified red mud, thereby not leaving any free hydroxyl group. However, alkyl stretching band of C-H for PNC2 and PNC3 were observed at 2948 and 2947 cm^{-1} respectively. The red shift in the alkyl stretching frequency C-H of PNC2 and PNC3 as compared to the pure phenoxy showed interaction of the filler with the polymer matrix.

The characteristic vibrational frequencies of PNC2 and PNC3 at v = 1571 and 1572 cm^{-1} respectively, was attributed to the stretching frequency of ammonium ion (NH_3^+), thereby depicting formation of zwitter ion. The other vibrational frequencies associated with PNC2 and PNC3 have been tabulated in table 13. The organically modified red mud(ORM) showed better interaction with the poly (hydroxy ether) of bisphenol A matrix as compared to phosphomolybdic acid modified red mud (PRM), therefore, it can be inferred that ORM will show better dispersion into the polymer matrix as compared to PRM.

4.2 X - ray diffraction study

The wide angle diffraction pattern of PRC2 and PRC3 films are shown in Fig. 5. The crystallinity of the composite membranes is mainly due to Phe. The characteristic diffraction peak for PRC2 and PRC3 were observed at a 2θ value of 2.36° and 2.26°. The diffraction peak at 2θ = 2.26 ° is considerably broadened and the interplanner distance has also widened compared to the raw red mud and PRC2 nanocomposite, indicating intercalation of the phenoxy matrix within the nanofiller of the nanocomposite membranes.

The polymer nanocomposite films showed decrease in the intensity of the crystalline peak as the loading percentage of organically modified red mud increased. The decrease in the intensity of crystalline peaks showed an increase in the disorderness of the interlayers of modified red mud.

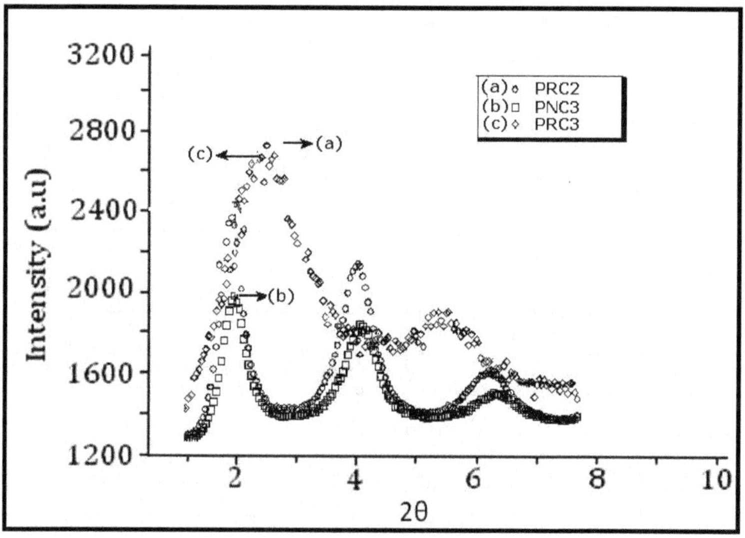

Fig. 24. XRD pattern of (a) PRC2 (b) PNC3 (c) PRC3 nanocomposites

From table 14, it is clear that the basal spacing for PRC3 is larger than the other PRM based nanocomposites materials, thereby giving evidence of better intercalation of polymer matrix within the silicate galleries of filler in case of PRC3 than the rest of nanocomposites.

Sample	2θ	d-spacing (A°)
PRC2	2.36	37.56
PRC3	2.26	39.48
PNC3	2.11	41.62

Table 14. XRD results of Phe-PRM and Phe-ORM based nanocomposite membranes

The X-ray diffraction pattern of phenoxy-organically modified red mud nanocomposite membranes are shown in fig. 24.The characteristic diffraction band of PNC3 was observed at 2 θ = 2.11°. The increase in the interlayer distance of silicate galleries of the organically modified red mud indicated that Phe has successfully intercalated into the silicate layers.

4.3 Field Emission Scanning electron microscopy (FESEM)

Field emission scanning electron microscope has evolved in which the electron beam cross section has become smaller and smaller increasing magnification several fold. Raw red mud and modified red mud were characterized by Field Emission Scanning Electron Microscopy (FESEM), as shown in Figure25. The FESEM analysis showed that the raw red mud consisted of aggregates, including particles of different dimensions. The FESEM image of the PRM and ORM showed a very fine distribution of particles with a size of 64 and 71 nm respectively as shown in Fig. 7. While ORM showed a homogenous distribution of particles than raw red mud with particle size larger than PRM due to formation of organic coating around the particles.

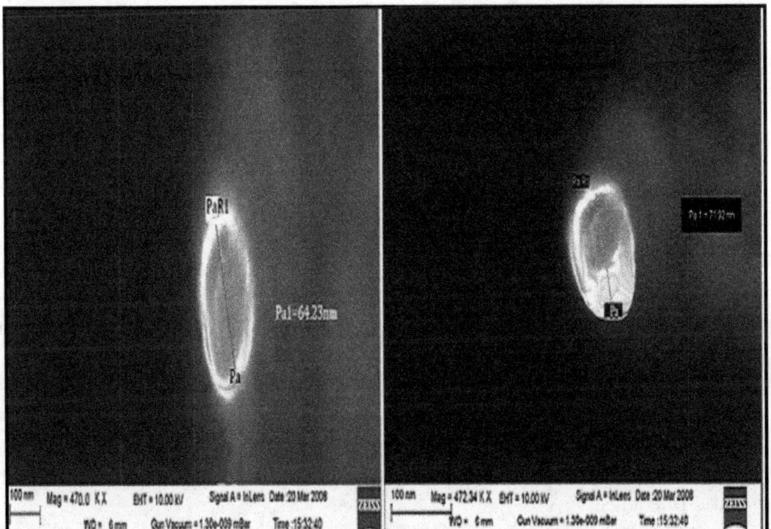

Fig. 25. FESEM images depicting size of a single red mud particle (a) Phosphomolybdic acid modified red mud (PRM), (b) Organically modified red mud (ORM)

4.4 Transmission Electron Microscopy (TEM)

TEM complements XRD by observing a very small section of the material for the possibility of intercalation or exfoliation. It also provides information about the particle size and nanodispersion of particles. It however, supplies information on a very local scale. However, it is a valuable tool because it enables us to see the polymer and the filler on a nanometer scale.

Fig. 8 (a) showed a representative TEM image of the PNC3 with a 3 wt% loading percentage of organically modified red mud in the phenoxy matrix, while fig. 8(b) showed a representative TEM image of PRC3. As seen in fig. 8(a), a rather good dispersion of the organically modified red mud is achieved. It is clear that the area covered by the red mud in fig.8 (a) is larger than in fig. 8(b) indicating a more extended platelet separation and an overall best nanodispersion of the critical loading of 3 wt% ORM in the phenoxy matrix.

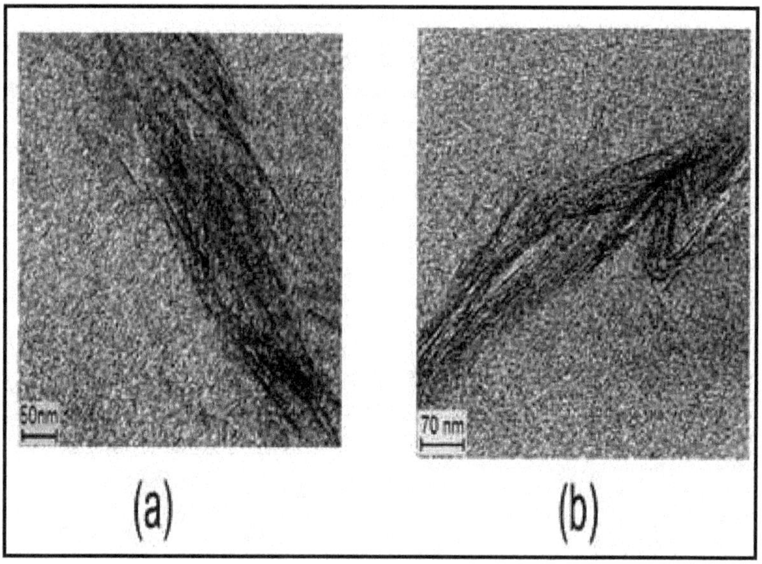

Fig. 26. TEM micrographs of (a) PNC3 (b) PRC3

The average particle size was found as 14 nm in case of PNC3 as shown in fig. 26(a), while the size of the particles in PRC3 was found to be 19 nm as shown in fig. 26(b). In fig. 26(b), the nanoparticles showed certain extent of agglomeration and formation of interconnected aggregates or tactoids of silicates.

There may be several fundamental problems that affect the polymer particle interactions in solution, resulting in disordered nanoparticle aggregates. These problems may arise from competing interactions between solvent, polymer chains and filler particles. The conformation of the polymer chains adhered to the nanoparticles also influences the ordered dispersion of the particles.

The presence of the organically modified red mud and phosphomolybdic acid modified red mud with different loading percentage and dispersing capabilities, may help us to identify the main structural driving force for dispersion in these Phe based PNs. It has been reported that the interaction between the polymer chains and the inorganic surface of the clay (Fornes et al., 2004) is crucial in polar polymers, and therefore it could also exist between the polar Phe and the modified red mud surface. Good dispersion of modified red mud in the polymer matrix will have a significant effect on the properties of the nanocomposites.

4.5 Thermogravimetric properties

The thermal stability of the pure Phe and Phe nanocomposite membranes with different filler content was investigated by thermogravimetric analysis. The TGA thermograms corresponding to the thermal decomposition of pure Phenoxy and Phe- nanocomposite films are shown in fig. 27 and fig. 28.

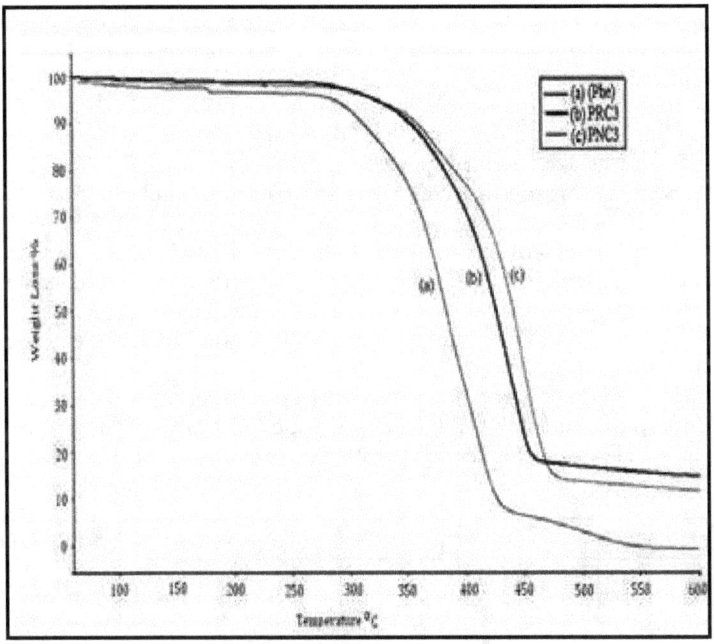

Fig. 27. Thermograms of (a) Phe (b) PRC3 (c) PNC3

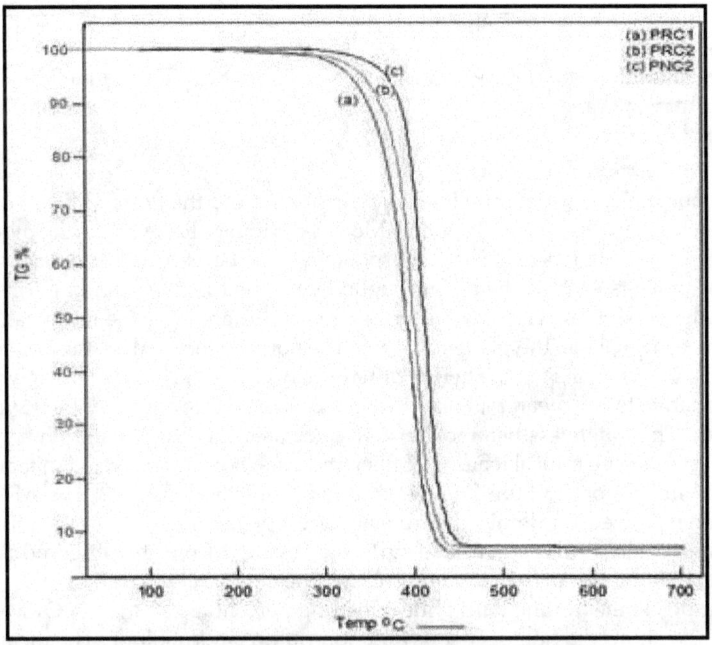

Fig. 28. Shows typical THA thermograms of (a) PRC1 (b) PRC2 (c) PNC2

As can be observed, both pure polymer and nanocomposite materials decompose in a single stage. If we compare the thermal stability of the polymer and the critical loading percentage nanocomposite membranes of PNC3 and PRC3, it is clear that PNC3 is more stable than pure Phenoxy and PRC3. The pristine Phe is thermally stable up to 275 °C. Since, the Phe nanocomposite membranes show enhanced thermal stability, the initial decomposition temperature (T_i) of phenoxy occurred at 276 °C while in case of PNC3 and PRC3, T_i was found at 318 and 298 °C respectively as shown in table 15. Similarly, final decomposition temperature (T_f) of phenoxy, PNC3 and PRC3 were observed at 433, 479 and 458 °C respectively. The enhancement in the thermal stability of PNC3 could be attributed to the better compatibility and intercalation of polymer matrix within the silicate galleries of 3 wt% organically modified red mud. Similarly, from fig.28 and table 15, it was observed that PNC2 showed higher Ti and Tf as compared to the PRC1 and PRC2, thereby showing higher thermal stability comparatively.

In general, major weight losses were observed in the range of 275–500 oC for Phe and polymer nanocomposite films, which may be correspondent to the structural decomposition of the polymer backbones at higher temperatures. After 500oC, all the curves became flat and mainly the inorganic residue (i.e. Al2O3, MgO, and SiO2 etc.) remained.

Sample	T_i (°C)	T_f (°C)	Ash (%)
Pure Phe	276	433	3.2
PRC1	287	439	6.33
PRC2	293	443	7.97
PNC2	312	454	8.23
PNC3	318	479	11.30
PRC3	298	458	14.15

Table 15. Degradation temperature and ash content of Pristine Phe and the Phe-nanocomposite membranes

4.6 Atomic Force Microscopy (AFM)

The surface morphology of the pure Phenoxy membrane and the Ph-modified nanocomposite membranes were analyzed by Tapping Mode - Atomic Force Microscopy (TM-AFM). Quantitatively, the differences in the morphology can be expressed in terms of various roughness parameters such as the mean roughness R_a, the root mean square (rms) of vertical data R_q, and the maximum height R_{max}. Here, the mean roughness is the mean value of surface relative to the center plane, the plane for which the volume enclosed by the image above and below this plane are equal; R_{max} the height difference between the highest and lowest points on the surface relative to the mean plane and R_q is the standard deviation of the Z values within the given area. The roughness parameters were calculated for pure Ph and Ph nanocomposite membrane surfaces with critical loading of filler and have been summarized in tabular format. The TM-AFM images of the Pure Phe membrane is depicted in Fig. 29. The AFM images of Phe membrane showed randomly distributed hard crystalline regions (crystallites) of 50 – 200 nm in dimension on the membrane surface as evident from the observed morphology. The pure Phe membrane surface has a mean roughness of 3.297 nm.

The nanocomposite membrane with critical loading percentage of 3 wt. % ORM (i.e., PNC3) content (Fig. 29) showed a homogeneous and non-porous morphology. The mean roughness for the surface topography of the membrane was found to be 4.045 nm.

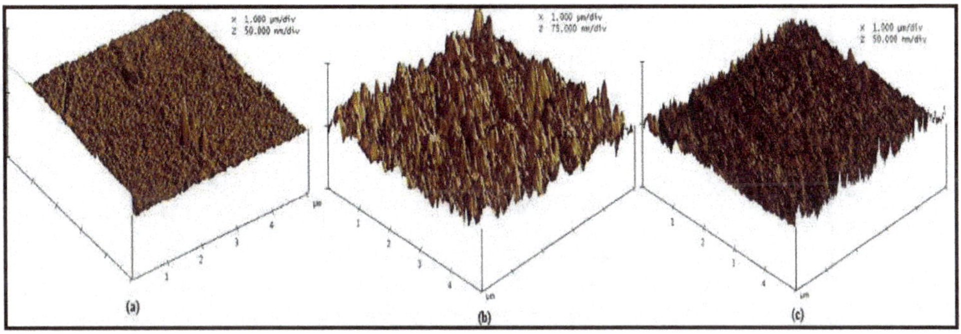

Fig. 29. Tapping mode (3D) AFM images of (a) Phe (b) PRC3 (c) PRC2

This shows that there is a good compatibility between the filler (ORM) and the Ph matrix. This is in consistent with the results of SEM studies. While critical loading surface topography of PRC3 showed some non-uniform regions with non-porous morphology. The mean roughness of PRC3 was observed to be 4.562 nm as shown in table 16. This showed that the phosphomolybdic acid modified red mud with 3 wt% loading percentage tends to form some aggregated particles within the polymer matrix leading to higher roughness values as compared to the PNC3.

Sample	R_a (nm)	R_{max} (nm)	R_q (nm)
Pure Ph	3.297	27.668	4.173
PNC3	4.045	57.648	5.497
PRC3	4.562	116.65	7.003

Table 16. Roughness parameters for Pristine Ph and the Ph nanocomposite membrane surfaces

4.7 Differential Scanning Calorimetry (DSC)

DSC traces of pure Phenoxy and polymer nanocomposite materials are shown in Fig.30. Ph exhibited a shallow endotherm at 61°C corresponding to the glass transition temperature (T_g) of Ph. All the nanocomposite materials with different critical loading percentage of modified red mud were found to have a high T_g as compared to the pristine Phe. This could be attributed to the confinement of the intercalated polymer chains within the red mud galleries that prevents the segmental motions of the polymer chains, thereby enhancing the glass transition temperature of the polymer matrix. The PNC3 nanocomposite membrane showed maximum T_g and melting temperature (T_m) of 66 °C and 220 °C respectively. This depicted that PNC3 is thermally more stable than PRC3 nanocomposite membrane on account of better dispersion and adhesion of the ORM within the polymer matrix. The glass transition temperature (T_g) and melting temperature (T_m) of pure phenoxy and Ph nanocomposite membranes have been tabulated in table 17.

Sample	T_g (°C)	T_m (°C)
Pure Phe	61	206
PNC3	64	211
PRC3	66	220

Table 17. Glass Transition temperature (T_g) and melting temperature of pure Phe and Phe-nanocomposite membranes

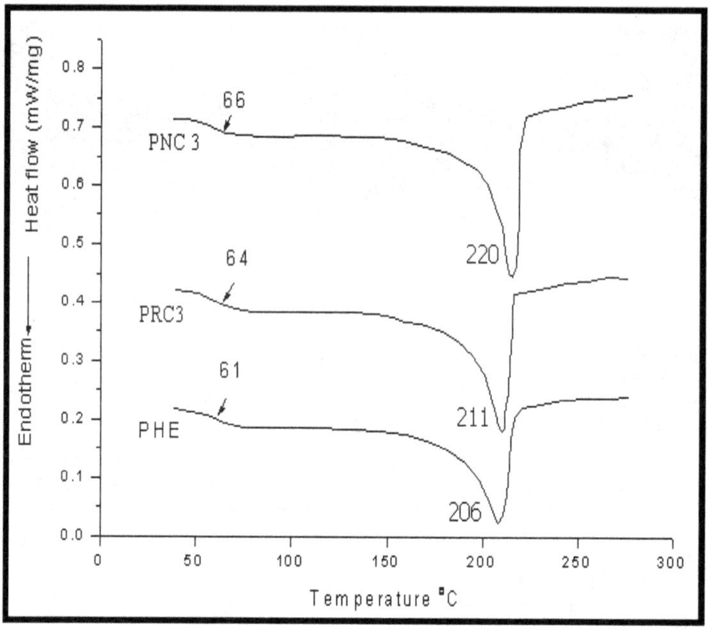

Fig. 30. DSC thermograms of (a) Ph (b) PRC3 (c) PNC3

4.8 Conclusion

- Red mud was successfully modified with the inorganic acids and organic moiety (aniline formaldehyde). The modification ameliorates both the dispersion level and the material properties of the polymer nanocomposites.
- The PVA- boric acid modified red mud nanocomposite films showed better glass transition temperature T_g and thermal stability at a filler concentration of 3.0wt percentage.
- In PVA/ORM nanocomposite systems, 2.0wt% modified red mud (CP4) based nanocomposite film exhibited relatively good dispersion with increase in the thermal stability, glass transition temperature.
- In PVA/PRM nanocomposite systems, 2.5wt% modified red mud (PRM4) based nanocomposite film exhibited relatively good dispersion with increase in the material properties as compared to the pristine poly (vinyl alcohol).
- PVA based nanocomposites showed an increase in the roughness values with the increase in the filler content but the 3.0wt% of boric acid modified red mud nanocomposite membrane showed lower roughness values as compared to the critical loading percentage of ORM and PRM based nanocomposites.
- On comparing the material properties of all the three nanocomposite systems, it was found that the critical loading percentage of BRM based nanocomposite system (i.e., SP5) showed better enhancement as compared to the critical loading percentage of ORM and PRM based nanocomposite systems.
- The TEM images of the PVA-modified red mud nanocomposite membranes showed homogeneous nano phase dispersion of the modified red mud in the PVA polymer matrices at a scale of 8- 23 nm particle size.

- In the poly(hydroxy ether of bisphenol A) (Phe) based nanocomposites as reported in this work, mostly intercalated structures were produced in the acid modified red mud modifications, but modification by organic moiety (ORM) resulted in a mixed intercalated-exfoliated structure.
- Thermal stability of polymer also increases after nanocomposites preparation, because red mud acts as a heat barrier, which enhances the overall thermal stability of the system, as well as assist in the formation of char after thermal decomposition.
- Phe based nanocomposites showed an increase in the roughness values with the increase in the filler content but the 3.0wt% of organically modified red mud nanocomposite membrane showed lower roughness values as compared to the critical loading percentage of PRM based nanocomposites. Thus, PNC3 showed homogenous and non-porous morphology as compared to the coarse morphology of PRC3 type of nanocomposites.
- The average particle size in case of PNC3 is 14 nm while the particle size of PRC3 is 19 nm as revealed by TEM studies.

5. References

Agag, T.; Koga, T.; Takeichi, T. (2001). Studies on thermal and mechanical properties of polyimide–clay nanocomposites, *Polymer*, 42, 3399.

Alexandre, M.; Dubois P. (2000). Polymer-layered silicate nanocomposites: preparation, properties and uses of a new class of materials. *Materials Science & Engineering, R: Reports*, R28 (1-2), 1-63.

Bachtsi, A.R; Kiparissides, C. (1996). Synthesis and release studies of oil containing poly (vinyl alcohol) microcapsules prepared by coacervation, *Journal of Controlled Release*, 38 (1), 49-58.

Bhat, A.H, Banthia, A.K, (2007). Preparation and Characterization of Poly (vinyl alcohol)–Modified red mud composite materials, *Journal of Applied Polymer Science*, 103 (1), pp. 238-243.

Biron, M., (2007). Thermoplastics and Thermoplastic Composites. Elsevier Ltd., Jordan Hill, Oxford.

Chand, N.; Hashmi, S. A. R. (1999). Effect of Red Mud Addition on Abrasive Wear Rate of Isotactic Polypropylene/Low Density Polyethylene Blend under Low Stress Conditions, *JSci Ind Res*, 58, 795.

Chand, N.; Hashmi, S. A. R. (1999). Effect of addition of polycarbonate on sheared flow of red mud-filled isotactic polypropylene *Bull Mater Sci*, 22, 801.

Chen, H.L. Wu, L.G; Tan, J; Zhu, C.L.; (2000). PVA membrane filled β-cyclodextrin for separation of isomeric xylenes by pervaporation. *Chemical Engineering Journal*, 78 (2-3), 159-164.

Dilsiz, N.; Ebert, E.; Weisweiler, W.; Akovali, G. (1995). Effect of plasma Polymerization on Carbon Fibers Used forFiber/Epoxy Composites *JColloid Interface Sci*, 170, 241.

Donnet, J. B.; Bansal, R. C. Carbon Fibers, 2nd ed.; Dekker: New York, 1990.

Fornes TD, Hunter DL, Paul DR. (2004) Nylon - 6 nanocomposites from alkylammonium - modified clay: the role of alkyl tails on exfoliation. *Macromolecules*, 37(5), pp. 1793-1798.

Grjotheim, K.; Welch, B. J. Aluminium Smelter Technology: A Pure and Applied Approach, 2nd ed.; Verlag: Dusseldorf, 1998.

Hajji, P.; David, L.; Gerard, J. F; Pascault, J. P; Vigier, G. (1999). Synthesis, structure, and morphology of polymer-silica hybrid nanocomposites based on hydroxyethyl methacrylate, *Journal of Polymer Science*, Part B: Polymer Physics, 37(22), 3172-3187.

Horii,F; Hu,S; Ito,T; Odani,H; Kitamaru.R, (1992).Cross-polarization/magic-angle-spinning carbon-13 NMR study of solid structure and hydrogen bonding of poly(vinyl alcohol) films with different tacticities, *Polymer*, , 33(11), 2299-2306.

Huang, H H.; Orler, B.; Wilkes, G L. (1987). Structure-property behavior of new hybrid materials incorporating oligomeric species into sol-gel glasses. 3. Effect of acid content,tetraethoxysilane content, and molecular weight of poly (dimethylsiloxane), *Macromolecules*, 20, 1322.

Ishikawa, Y.; Matsumoto, Y. (2001). Electrodeposition of TiO2 photocatalyst into nano - pores of hard alumite *Electrochem Acta*, 46, 2819.

Jang, B. Z. (1992). Control of interfacial adhesion in continuous carbon and kevlar fiber reinforced polymer composites, *Compos Sci Technol*, 44, 333.

Kasliwal, P.; Sai, P. S. T. (1999). Enrichment of titanium dioxide in red mud: a kinetic study *Hydrometallurgy*, 53, 73.

Lee, D. C.; Jang, L. W. (1996). Preparation and characterization of PMMA-clay hybrid composite by emulsion polymerization. *Journal of Applied Polymer Science*, 61(7), pp. 1117-1122.

Novak, B. M; Ellsworth, M. W. (1993). "Inverse" organic-inorganic composite materials: high glass content non- shrinking sol-gel composites, *Materials Science & Engineering*, A: Structural Materials: Properties, Microstructure and Processing, A162(1-2), 257- 64.

Park, S. J.; Cho, M. S. (2000). Effect of anti-oxidative filler on the interfacial mechanical properties of carbon–carbon composites measured at high temperature. *Carbon*, 38, 1053.

Park, S. J.; Kim, M. H. (2000). Effect of acidic anode treatment on carbon fibers for increasing fiber - matrix adhesion and its relationship to interlaminar shear strength of composites *JMater Sci*, 35, 1901.

Park, S. J.; Kim, J. S. (2001).Influence of Plasma Treatment on Microstructures and Acid - Base Surface Energetics of Nanostructured Carbon Blacks: N2 Plasma Environment, *JColloid Interface Sci*, 232, 311.

Peppas, N.A; Merrill, E.W, (1977). Development of semicrystalline poly(vinyl alcohol) hydrogels for biomedical applications, *Journal of Biomedical Materials Research*, 11(3), 423-34.

Pradhan, J.; Das, S. N.; Thakur, R. S. (1999). Adsorption of Hexavalent Chromium from Aqueous Solution by Usin Activated Red Mud, *JColloid Interface Sci*, 217, 137.

Ray S. S and Bousmina M, (2005). Biodegradable polymers and their layered silicate nanocomposites: In *greening the 21st century materials world, Progress in Materials Science*, 50(8), 962-1079.

Shao, C.; Kim, H-Y; Gong, J.; Ding, B.; Lee, D-R.; Park, S-J. (2003). Fiber mats of poly(vinyl alcohol)/silica composite via electrospinning, *Materials Letters*, 57(9-10),1579-1584.

Surivet, F.; Lam, T. M ; Pascault, J. P; Pham, Q. T. (1992). Organic-inorganic hybrid materials. 1. Hydrolysis and condensation mechanisms involved in alkoxysilane-terminated macromonomers, *Macromolecules* , 25(17), 4309-20.

Yuan, L. T.; Shyu, S. S.; Lai, J. Y. (1991). Plasma surface treatments on carbon filbers. II. Mechanical property and interfacial shear strength *JAppl Polym Sci*, 42, 2525.

Zhang H, Ma H, Hongtu Li and Wang (2002). J, Synthesis and characterization of a polymer– metal complex from bisphenol A polyhydroxy ether with sulfur-bearing side-groups, *Polym Int.*, 51, pp.1129–1134.

8

Development of 3D Knitted Fabrics for Advanced Composite Materials

Luminita Ciobanu, Ph.D., Eng.
"Gheorghe Asachi" Technical University of Iasi Romania

1. Introduction

Generally, a composite material is made of distinct materials, that together act in a different way than when considered separately. There are a lot of examples of composite materials, both natural and synthetic, from the human body, to buildings, airplanes and so on.

Most comprehensive definition of the composite materials that characterises their nature is given by P. Mallick. According to Mallick (1997), a composite is a combination of two or more chemically different materials, with an interface between them. The constituent materials maintain their identity in the composite material (at least at macroscopic level), but their combination gives the system properties and characteristics different from those of each component. One material is called matrix and is defined as the continuous phase. The other element is called reinforcement and is added to the matrix in order to improve or modify its properties. The reinforcement represents the discontinuous phase, distributed evenly in the matrix volume.

There are several options for reinforcement and matrix, as illustrated in Fig.1, that are taken into consideration based on the mechanical requirements specific to the application.

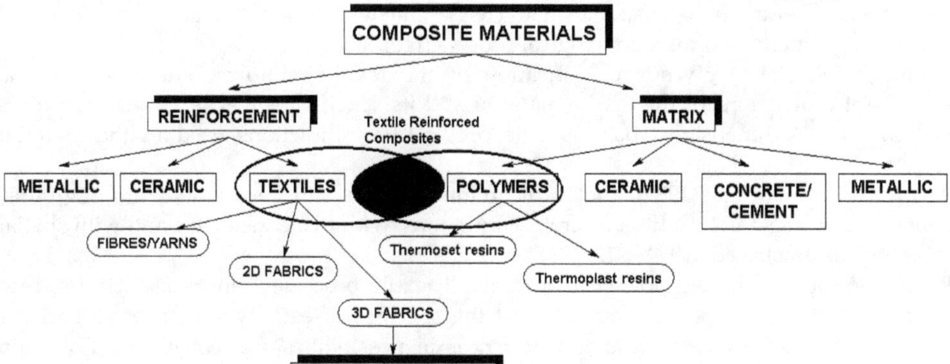

Fig. 1. Structure of composite materials

The development of textile reinforced composites (TRCs) with resin matrix is based on the desire to produce improved materials, with tailored properties. The textile material gives the ensemble strength, while the matrix ensures the composite unity and transmits the strains. The advantages of the textile reinforced composites are:

- controlled anisotropy of the textiles which means that their structure materials can be designed so that the fibres are placed on preferential directions, according to the maximum strain;
- the use of textile reinforcements allows to obtain a better weigh/strength ratio compared with the classic materials, such as steel;
- textile materials maintain their integrity and behaviour under extreme conditions - for example, they do not corrode in a outdoor environment, nor vary their dimensions when there are significant temperature variations, nor are they sensible to electro-magnetic fields;
- TRCs present an improved fatigue life.

The aeronautic industry was the first that used TRCs for airplanes. Currently, there is a high diversity of TRCs applications, with high economic impact (Mouritz, 1999). Composite materials can be found in all fields of technical textiles. Industrial applications of the composites include tanks, storage structures, pipes, hoses, etc. The automotive industry uses TRCs for car frames and other machine parts (manifold, wheels), while in aeronautics the composites developed from 1st level applications to 2nd level that refers to resistance elements in an airplane structure and the future trend is building one exclusively with composites. The composite materials also replaced traditional ones for the rotor blades of helicopters, increasing their life span and their resistance to wear (Mallick, 1997). One field of great interest for textile reinforced composites is the wind energy management - these materials are used to build wind mills. The TRCs are also used to produce sport equipment - tennis rackets, bicycles and motorcycles, etc.

An interesting application is in buildings, where composites (the so called Textile Reinforced Concrete) are used to reinforce walls (cement/concrete matrix), increasing their strength and reducing their thickness and subsequently production costs.

Classification of textile reinforced composites

Two main criteria can be used to characterise the textile reinforcements: the material structure/geometry and the technological process (Hu, 2008).

Fukuta et al. (1984) gives a classification of the textile reinforcement based on the significant dimensions of the textile material and its specific geometry. Fukuta considers not only the 3 dimensions, but also the preset fibres directions used in the material structure.

According to Scardino (1989), the textile reinforcements can be divided into 4 groups, depending on their architecture: discrete, continuous, with plane geometry and with spatial geometry, as illustrated in Table 1.

When considering the technological process, all textile processes can be used to produce reinforcement for composite materials, but the specifics of each type of process and the resultant material geometry lead to differences in possibilities and behaviour. The main processes employed in the production of textile reinforcements are: weaving, braiding, knitting and non-woven. Also there are other processes, such as filament winding and poltrusion, which process filaments. Most used reinforcements are woven fabrics (2D and 3D) and nonwovens (fibre mats), but the knitted fabrics, especially warp knitted structures, present a good development potential.

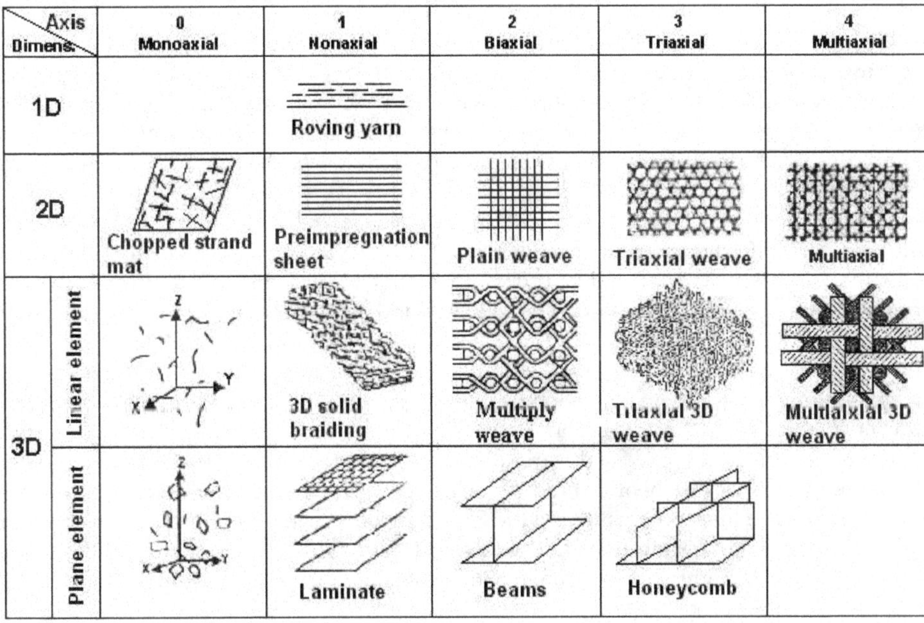

Fig. 2. Textile reinforcement systems, classification according to Fukuta et al.

Level	Reinforcement	Textile construction	Fibre length	Fibre orientation	Fibre entanglement
I	Discreet	Short fibres	Discontinuous	Uncontrolled	none
II	Linear	Filaments	Continuous	Linear	none
III	Plane	2D materials	Continuous	Planar	Planar
IV	Integrated	Advanced materials	Continuous	3D	3D

Table 1. Constructive classification of the textile reinforcements

The selection of a certain process is based on the architectural possibilities, the material characteristics and behaviour (dimensional stability, mechanic strength, drape and formability, etc) and its suitability with regard to the composite processing and its application.

Potential of knitted fabrics for composite reinforcement

The main advantages of knitted fabrics for composite reinforcement are:

- the possibility of producing knitted fabrics with 3D complex shapes
- improvement of fabric handling and matrix injection during composite processing
- acceptable processability of high performance fibres (glass, aramid, PES HT or HM)
- rapid manufacturing of knitted fabrics for reinforcements
- controlled anisotropy (yarn in-laid under preferential angles).

When considered in reference to other types of textile materials, knitted fabrics are not as well developed, mainly due to their lower mechanical properties (Leong et al., 2000). According to Verpoest et al. (1997), knitted fabrics present lower in-plane strength and stiffness in comparison to materials such as woven, braiding, non-crimps. Another problem limiting the use of knitted fabrics for composite reinforcement is the low value for volume fraction, due to the specific geometry of knitted stitches, characterised by areas without yarns.

The reduced mechanical behaviour is determined by the specific bending of fibres in the knitted stitches. Mechanical properties are controlled through fabric structure, structural parameters, yarn characteristics and process parameters. Structure is an effective way of improving properties by the use of float stitches and in-laid straight yarns placed under certain angles. Stitch density also affects the tensile behaviour and fabric stiffness,

Yarns are also important, their properties being transferred to the fabric level. The specifics of the knitting process make bending strength and rigidity the most important characteristics. This situation is essential, considering that high performance fibres are rigid and therefore must be processed carefully. Apart from carbon fibres, all other high performance yarns can be bent around a needle hook and transformed into stitches. The problems related to their processing are the fibre destruction and the modifications brought by the strains during knitting that lead to reduced mechanical characteristics of the fabric. The use of in-laid straight yarns eliminates the problem of fibre damage and also increases the volume fraction.

The multiaxial warp knitted fabrics, presented in part 3 of this chapter, are the most used for the production of composites. They have a laminar structure, with layers of yarns under preset angles, according to the application. The layers are connected and the risk of delamination is reduced.

Another development direction is the production of preforms with complex shapes for advanced composite materials. This is an interesting development direction, considering the complexity of the fabric architecture that can be achieved through knitting. The literature presents a significant amount of references concerning the development, characterisation and mechanical behaviour of these 3D fabrics.

2. Raw materials used for the production of knitted fabrics for advanced composite materials

Textile reinforcements are produced using high performance fibres, like glass, carbon/graphite, Kevlar, PES HM and HT, ceramic fibres, boron and silicon carbide fibres, etc. These yarns have superior mechanical behaviour that can meet the specific demands of composite applications that are illustrated in Table 2. They also have high bending rigidity that affects the knitting process and other characteristics that must be taken into consideration when designing a knitted reinforcement for composite materials (Miller, 1989).

Glass fibres (yarns, rovings) are the most common high performance fibres used to reinforce composite materials. They are characterised by hardness, resistance to chemical agents, stability and inertness, low weight and processability (Muckhopadhyay, 1994). There are more types of glass fibres depending on their chemical composition: E-glass, with good strength and high electrical resistivity, most common in composite materials; S-glass, with high tensile strength, most common in military applications; and C-glass, characterised by chemical stability and corrosion resistance.

	Fibre	Relative density [g/cm³]	Young's Modulus [GPa]	Tensile strength [GPa]
1	Carbon (PAN)	2.0	400	2.0-2.5
2	Boron	2.6	400	3.4
3	E-glass	2.5	70	1.5-2.0
4	S-glass	2.6	84	4.6
5	Kevlar 29	1.44	60	2.7
6	Kevlar 49	1.45	60	2.7

Table 2. Main characteristics for some high performance fibres

Mechanically, the glass fibres are characterised by high strength, low elongation, high bending rigidity and brittleness. Law and Dias (1994) and Savci et al. (2001) showed that the glass fibres can resist when bent around the needle hook and therefore can be processed through knitting. Due to their brittleness and their low resistance to friction, the glass yarns damage easily, thus affecting the knitting process and subsequently the real strength of the reinforcement. Knitting glass fibre therefore requires a preliminary stage to determine the optimum technological conditions that ensure minimum fibre damage while maintaining the fabric quality. The fabric density, essential for the fibre fraction volume of the composite reinforcement gives this quality, together with the amount of fibre damage. High fibre fraction volume is a sine-qua-non requisite for the performance of the composite materials.

Experimental work

The experimental work is based on the direct study of the yarn after knitting, in order to identify the damage degree of the glass fibres inflicted by the knitting process. All models for the mechanical behaviour of the glass knitted fabric are based on the Young's modulus for the glass yarns. During the knitting process the filaments are damaged in a significant proportion, therefore altering the initial value of Young's modulus and altering the fabric properties. No previous study indicated the relation between technological parameters and the final value of the Young's modulus.

Two types of glass fibre were considered for the experiment: EC 11 408 Z28 T6 - Vetrotex and EC 13 136 Z30 P 100. The yarns are knitted using single jersey, as being the simplest possible structure.

The fabrics were produced on a CMS 320 TC (Stoll) flat machine with the following characteristics: gauge 10 E, negative feeding - IRO NOVA (Fig. 3) and holding down sinkers and presser foot (Fig. 4).

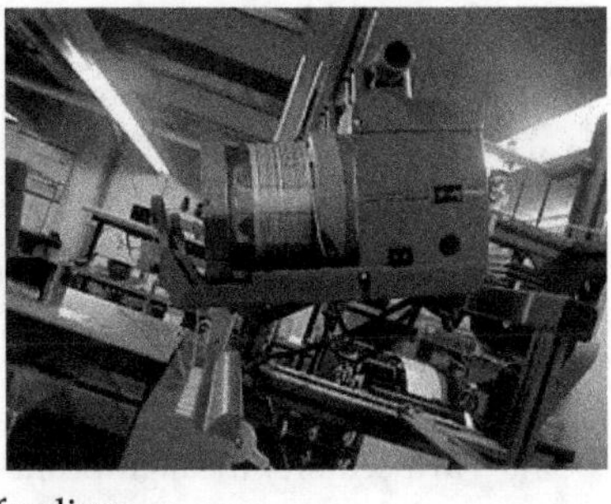

Fig. 3. IRO NOVA negative feeding

Fig. 4. Holding down sinkers and presser foot

In order to adapt to the yarn count, the fabrics are knitted with different values for the stitch quality cam, presented in Table 3. Each sample had 100 wales and 100 courses. The fabrics were relaxed until they presented no dimensional variation. The structural parameters (horizontal and vertical stitch density and stitch length) are illustrated in Table 4.

Yarn count [tex]	Quality stitch cam (NP)					Take down (WM)
	NP 1	NP 2	NP 3	NP 4	NP 5	
408	10.0	10.5	11.0	11.5	12.0	20
136	9.5	10.0	10.5	11.0	11.5	18

Table 3. Technological parameters used for knitting the samples

Yarn count [tex]	NP 2			NP 3			NP 4			NP 5		
	D_w [w/10 cm]	D_r [r/10 cm]	l_s [mm]	D_w [w/10 cm]	D_r [r/10 cm]	l_s [mm]	D_w [w/10 cm]	D_r [r/10 cm]	l_s [mm]	D_w [w/10 cm]	D_r [r/10 cm]	l_s [mm]
408	48	69	7.28	46	64	7.8	48	61	8.25	42	57	8.85
136	56	88	6.00	50	84	6.51	44	78	7.17	40	70	7.52

Table 4. Values for the structural parameters, in relaxed state

After relaxation 10 yarn lengths were drawn from the fabrics in order to determine their tensile properties, avoiding the edges, visibly more damaged then the rest. The tensile strength was tested on a HOUSENFIELD H10K-S (Tinius Olsen), according to ASTM 2256. According to previous studies, the glass yarns break in less than the minimum 20 seconds indicated by the standard. Therefore, the testing speed selected was the minimum value of 50 mm/minute. The data confirmed the breaking of the glass yarns less than 7 seconds.

Experimental results

Knitting conditions

Fig. 5 presents the aspect of a 408 tex glass fibre jersey fabric produced with the quality stitch cams in the limit position, in this case NP = 10.5. The destroyed filaments are placed more at the level of the sinkers loops and not at the level of the needle loops, as Law and Dias pointed out in their study. This situation sustains the idea of other cause for yarn damage than the tension peaks in the knitting point. Furthermore, the significant filament breakage repeats at every two courses, corresponding to the reverse carriage displacement when the needles receive less yarn. If the stitch length is even lower, the yarn gets out of the needle hooks and it can not be knitted. This situation is exemplified in Fig. 6, for a jersey fabric made of 136 tex glass fibres. The filaments appear to be completely destroyed, creating a plush effect.

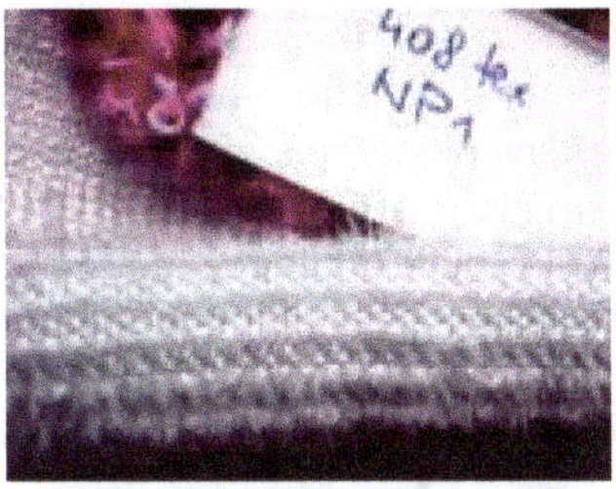

Fig. 5. Aspect of the jersey fabric (408 text) knitted with NP1=10.5

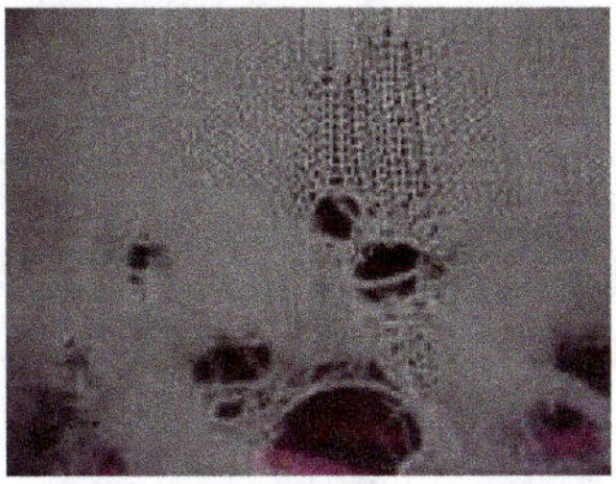

Fig. 6. Aspect of the jersey fabric (136 tex) knitted with NP1=10.5

Test results

The experimental results were calculated based on the raw data from the testing machine, according to the established methodology (ASTM D 2256), to obtain the following values: breaking strength, breaking tenacity, Young's modulus E, breaking elongation, breaking toughness and time to break. The results are centralised in Table 5. Yarn variant called Normal is the witness yarn, while the other variants are yarns knitted with a certain position for the stitch quality cam, defined in Table 3.

Yarn count [tex]	Yarn variant	Breaking force [N]	Breaking tenacity [cN/tex]	E modulus [N/tex]	Breaking elongation [%]	Breaking toughness [J/g]	Time to break [sec]
136	Normal	75.67	55.59	27.37	2.2	6.64	6.89
	NP 3	23.59	17.27	14.79	1.38	1.45	4.14
	NP 4	30.92	22.73	18.26	1.42	1.93	4.24
	NP 5	37.51	27.58	21.24	1.55	2.47	4.66
408	Normal	229.87	56.34	84.65	2.46	6.86	6.76
	NP 2	118.73	29.10	61.18	1.87	3,01	5.60
	NP 3	156.59	38.37	69.52	2.00	4.29	6.01
	NP 4	165.05	40.45	70.08	2.12	4.86	6.38
	NP 5	167.83	41.14	71.66	2.15	4.87	6.44

Table 5. Experimental results for tensile testing

Discussions

Contrary to Law and Dias, the experience of knitting on flat electronic machines showed that the needle could be pulled inside its channel without restrictions for the lowest point. The relation filament breaking - stitch length was found to be opposite to the one presented by Law and Dias. For each yarn count there is an inferior limit for the stitch length, guaranteeing the quality of the fabric. For this limit the degree of filament breaking is so high the yarn is almost completely destroyed and will break when unravelled from the fabric. It is the case of the 136 tex yarn knitted with NP2 stitch cam position. No tensile tests were performed for this variant and therefore it was not included in the experimental data.

The differences in strength and tenacity, compared to the normal values, show that the knitting process has a negative influence on the tensile properties. Furthermore, the decrease in strength is in a direct correlation with the stitch length - the lower the stitch length, the lower the tensile properties. Figs. 7 and 8 present the representative graphics for each type of yarns, illustrating the variation of the breaking force with the elongation.

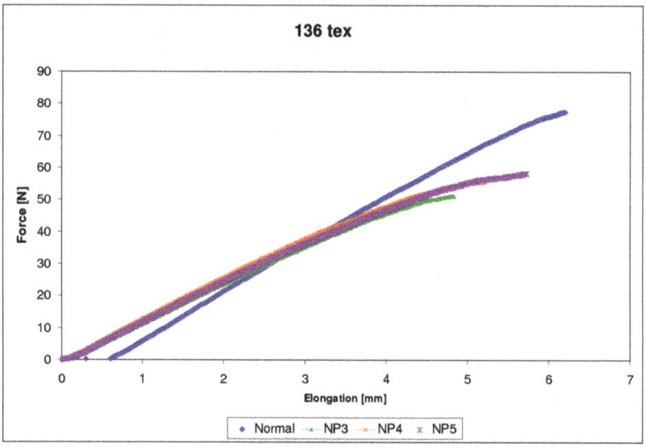

Fig. 7. Force-elongation curves for the EC 13 136 Z30 P 100 glass yarn

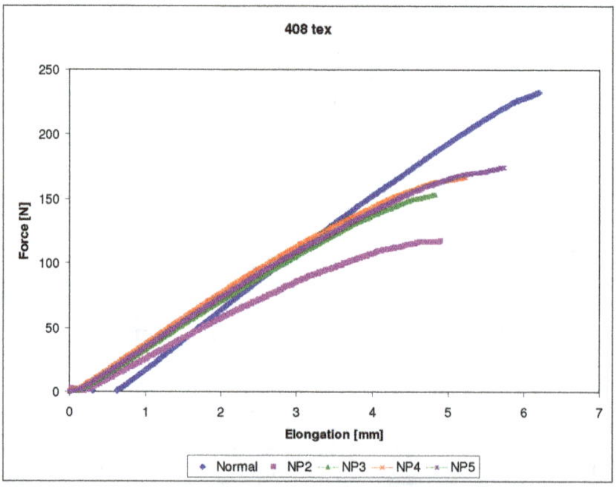

Fig. 8. Force-elongation curves for the EC 11 408 Z28 T6 glass yarn

The experimental data and the graphics are showing a contradiction with Law and Dias, concerning the cause of filament breakage during knitting. If the tension occurring in the looping point is responsible for the filament damage, then higher stitch length should present higher values for the breaking force, but the experimental results contradict this. One answer for this different opinion could be the feeding mechanism. In this case, the yarn was fed using an IRO NOVA feeding device, ensuring the proper quantity of yarn for the process. Without it, knitting proved impossible.

The breaking phenomenon appears due to the friction between the yarns and the knitting elements, especially the knock-over plates that can act like knifes during rob back stage. The longer the stitch length, the smaller the tension in yarns and there is less filament damage.

The differences between the normal values and the ones for the knitted yarns 136 tex are varying from - 71% in case of the smallest stitch length and - 55% for the highest stitch length. In the case the 408 tex yarn, the difference interval is - 48.34% to - 26.99%.

The decrease in breaking toughness is more significant, as illustrated in Fig. 9. The toughness value for the 136 tex NP2 variant was introduced and was considered 0 only for comparison purposes. When compared, the initial toughness values for the two yarns are similar, but the 136 tex yarn shows a much higher decrease in breaking toughness then the other yarn. The decrease interval for 136 tex is extremely high 78-63%, while for 408 tex the decrease is in the interval 56-29%.

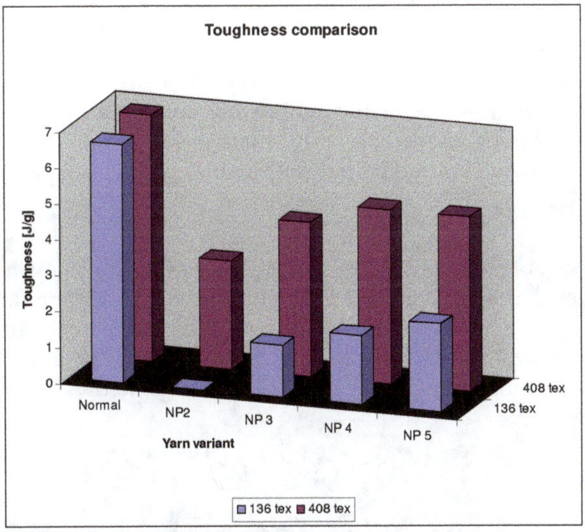

Fig. 9. Influence of yarn count and filament damage on breaking toughness

This situation suggests the influence of the number of filaments in the yarn, and also the fact that the amount of filament damage on the knitting machine is the same, regardless of the yarn count. A superior number of filaments ensure a better knittability, the yarn maintaining better tensile properties.

The breaking force and toughness values determined for the yarns knitted with NP4 and NP5 stitch cam positions are similar indicating that filament damage amount is the same. Even if the best properties are obtained for the highest stitch length - corresponding to NP5 value for the quality cams, this must be balanced with the fabric density, essential for the overall performance of the fabric in the composite material. Therefore, the optimum technological parameters appear to be those for the fourth situation, with NP4 value for the stitch quality cam. In practice, even if the strength level is lower, value NP3 can also be used, due to the higher fabric density.

3. 3D knitted structures

In the case of knitted fabrics, the 3D architecture is facilitated by their high extensibility and formability that allow the production of complex shapes. This is the reason why the knitted fabrics are regarded as a viable option for preforms for advanced composite materials.
The main advantages of the 3D knitted fabrics are:

a. the high formability of the fabrics, especially due to their drape characteristics
b. the high complexity of the shapes that can be produced;
c. the use of existing technology, without major adaptations;
d. knitted fabrics exhibit good impact behaviour.

Knitted three dimensional performs are less studied and used, mostly because the following problems related to their production and their properties:

a. the development of these fabrics is still at laboratory stage;
b. the mechanical characteristics of the resulting composites are at a lower level and require improvement;
c. the specific properties and their prediction are not yet well developed, mainly because of the complexity of the knitted fabrics;
d. the pretension of the perform before its impregnation with resin determines a uneven behaviour for the final composite due to fibre migration in the stitches.

The 3D knitted fabrics can be divided into three main groups: multiaxial fabrics (multilayer), sandwich/spacer fabrics and knitted fabrics with spatial geometry (spatial fashioned).

Multiaxial fabrics

The multiaxial fabrics are characterised by the presence of multiple layers of yarns disposed at preferential angles that are assembled in the knitted fabrics. These fabrics are produced on special warp knitting machines using glass fibre or carbon fibre for the layers. The warp knitting technology is best suited for this kind of structures with in-laid yarns. Multiaxial fabrics are used mostly for the reinforcement of composite materials.

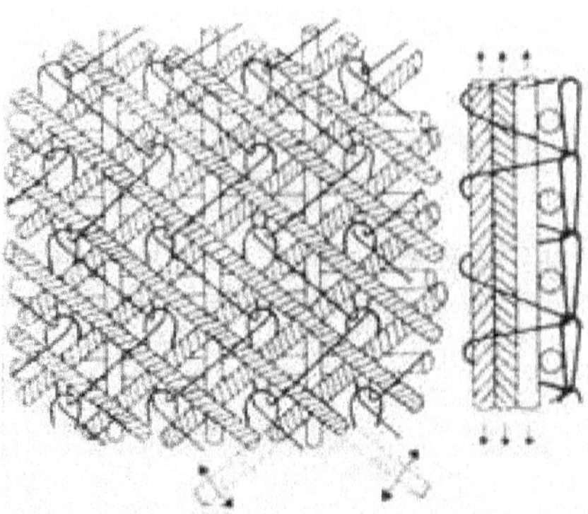

Fig. 10. Structure of the multiaxial knitted fabrics

The different layers of yarns in the multiaxial warp knitted fabrics are independent and the yarns fed under preset angles: 0^0 (weft yarns), 90^0 (warp yarns) and any value between these, like +/- 30^0 and +/- 45^0. The layers are united by the actual knitted fabric, using pillar or tricot stitches, as exemplified in Fig. 10 (Raz, 1989).
The preset angles correspond to the directions requiring higher strength during use and are imposed by the application. Currently, there are two main technologies adapted for the production of multiaxial fabrics: the Karl Mayer technology and the Liba technology.

a. Karl Mayer technology

The Karl Mayer machine uses four yarn systems - one to produce the ground fabric, one to insert the warp yarns, one for the weft yarns and the fourth for the yarns disposed under a preset angle. The inclined yarns are fed with two guide bars that have a rotational movement of 1 pitch every row. The guides are shogged in only one direction, passing from one bar to the other, changing the yarn direction within the fabric structure, as illustrated in Fig. 11.

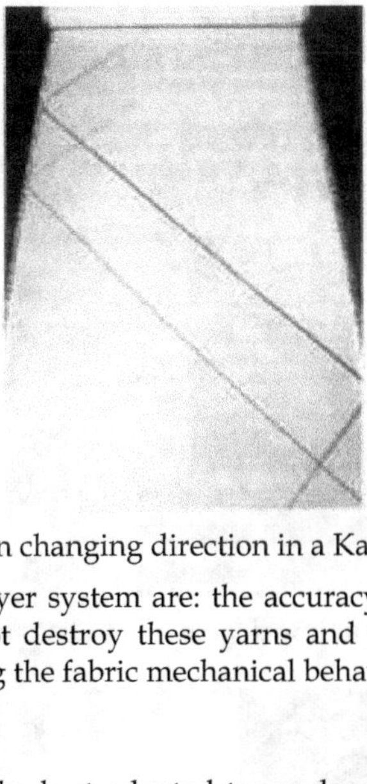

Fig. 11. The modified angle when changing direction in a Karl Mayer multiaxial fabric

The advantages of the Karl Mayer system are: the accuracy of the inclined yarns position, the fact that the needles do not destroy these yarns and the high productivity. Still, the fabrics are less compact affecting the fabric mechanical behaviour.

b. LIBA technology

The Liba technology is by far the best adapted to produce multiaxial fabrics. The MAX 3 CNC machines use a Copcentra machine that is preceded by a number of 3 zones where the layers are formed with the help of carriers feeding the yarns under the desired angle along the table: 0^0 for the weft yarns and 26^0 to 60^0 for the rest. The warp yarns (90^0) are introduced in the knitting zone, using special guides. The layers are brought to the knitting zone and connected using pillar or tricot stitches. A non-woven mat (chopped glass fibres) can also be inserted, if the application requires it.
Fig. 12 illustrates the Liba system (source: www.liba.de). Liba also developed such a system also for carbon fibres (Copcentra MAX 5 CNC), where the carrier course is modified in order to reduce fibre waste.
The system allows obtaining fabrics with different degrees of compactness, made of different raw materials, as exemplified in Fig. 12 (source www.liba.de). The risk of fibre

destruction caused by the needles penetrating the layers is eliminated through the walking needle technique, where the needle bed is moved together with the layers.

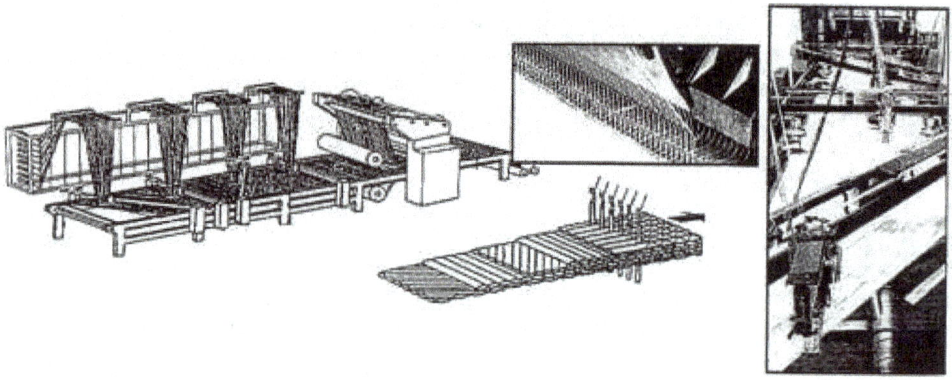

Fig. 12. Liba CNC machine for multiaxial fabrics

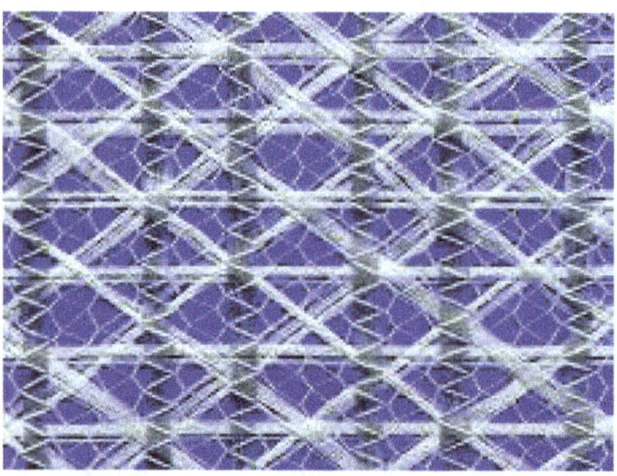

Glass fibre multiaxial fabric with low degree of compactness

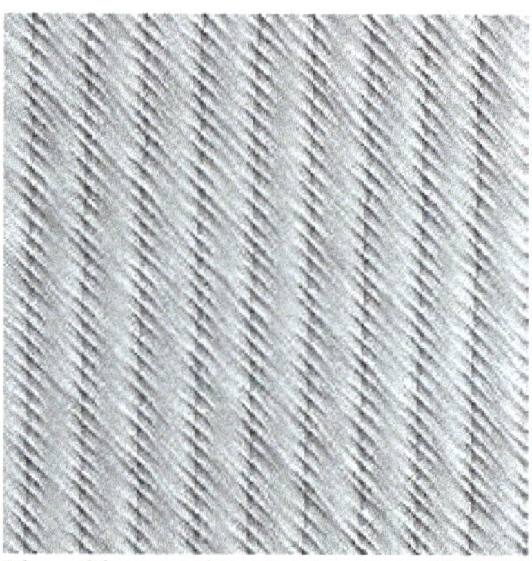

Glass fibre multiaxial fabric with high degree of compactness

Multiaxial fabric made of carbon fibre

Multiaxial fabric made of Kevlar fibre

Fig. 13. Examples of multiaxial fabrics

Spatial fashioned knitted fabrics

The 3D knitted performs started to be developed in the '90, mainly based on the development of electronic flat knitting machines and CAD/CAM systems. Even if there is significant progress in this domain, there are a lot of aspects that need addressed in order for the spatial knitted fabrics to become industrially feasible.

The spatial fashioning of the knitted fabrics is based on the need to produce fabrics with complex shapes that are similar to the shape of the final product. Even if a certain degree of spatial geometry can be obtained by using modules of structures with different patterns or by dynamic stitch length, the technique of spatial fashioning is the only one that has no limitations with regard to the shape complexity and dimensions. This technique (also known as 'flechage') is based on knitting courses on all working needles and courses on a variable number of needles, determining zones with different amount of stitches. The zones with the highest amount of stitches will have in the end a spatial geometry.

A classification of the spatial fashioned fabrics must take into consideration the 3D shape of the product, defined geometrically as a 3D body. These bodies can be divided in solids of revolution, such as tubes (cylinders), spheres and hemispheres, cones and frustum of cones, ellipsoids, hyperboloids, and polyhedrons, such as tetrahedrons, pyramids, parallelepipeds, etc. Apart from these simple bodies, other bodies can be considered: bodies obtained from composing simple bodies or bodies with irregular shape.

The fabric 2D plan is a rectangular area where are positioned fashioning lines that define the final 3D shape of the product. Such a 3D fabric is obtained by placing more fashioning lines with certain characteristics, forming repeating geometric basic forms within the plan or not (as is the case for parallelepiped forms).

The **fashioning lines** can be defined as the zones where the knitting will be carried out on a variable number of needles, these zones generating the spatial geometry. The lines have two components, corresponding to decreasing the number of working needles and the other to increasing them. In the fabric, these two lines become one, the actual fashioning line, as presented in Fig. 14.

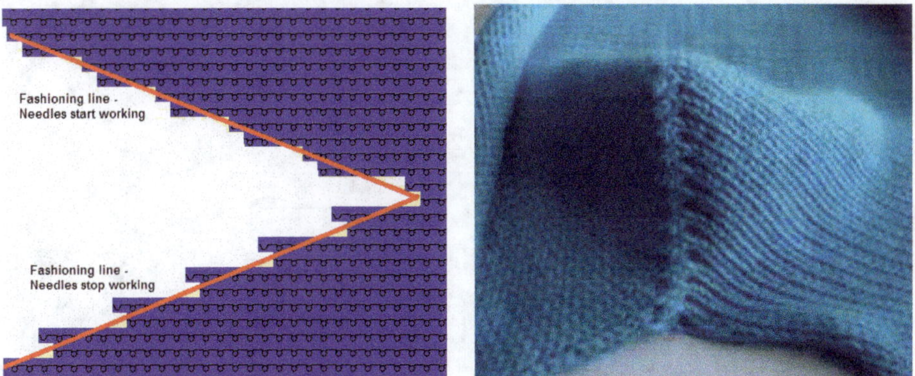

Fig. 14. Fashioning line - in the knitting programme and in the fabric

Fig. 15 illustrates the correlation between the 3D shape of the product and the 2D plan of the fabric, emphasising the most significant elements for the knitting process design. The evolute of the 3D body is obtained using sectioning lines and is the same with the 2D plan of the fabric that contains the fashioning lines. The knitting direction is very important when designing a

3D fabric. Knitting along the transversal or longitudinal direction of the final shape determines the knitting programme, the plan aspect and furthermore the specific behaviour of the fabric in the product. In some cases, only one knitting direction with regard to the product shape is possible, the other option being technologically not feasible. Other limitations concern the positioning of the fashioning lines within the fabric and their dimensions.

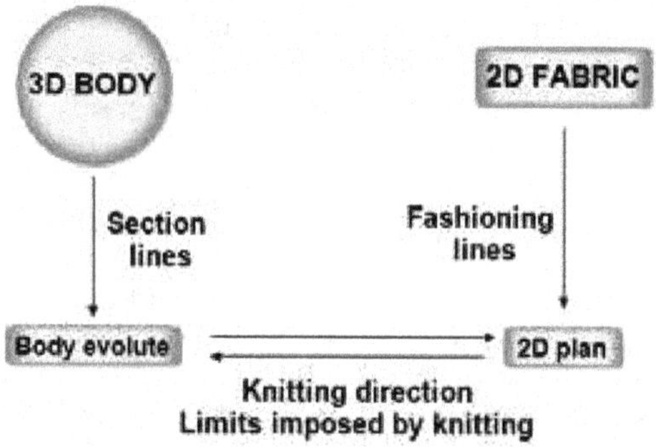

Fig. 15. Correlation between the 3D body and the knitted 2D fabric

Some of the most representative examples are illustrated in Figs. 16 to 20, presenting the 3D shape and the fabric 2D plan.

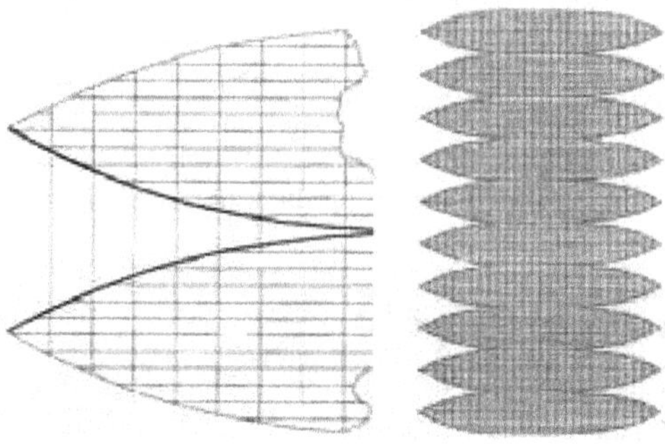

Fig. 16. (Hemi)spherical shape (Cebulla et al., 2000)

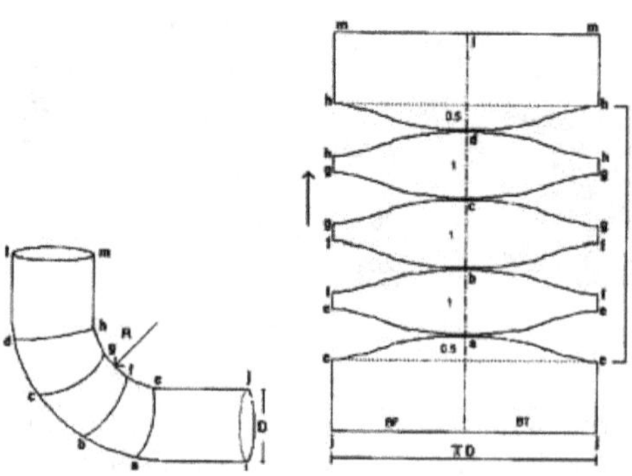

Fig. 17. Bent tubular shape (Song et al.)

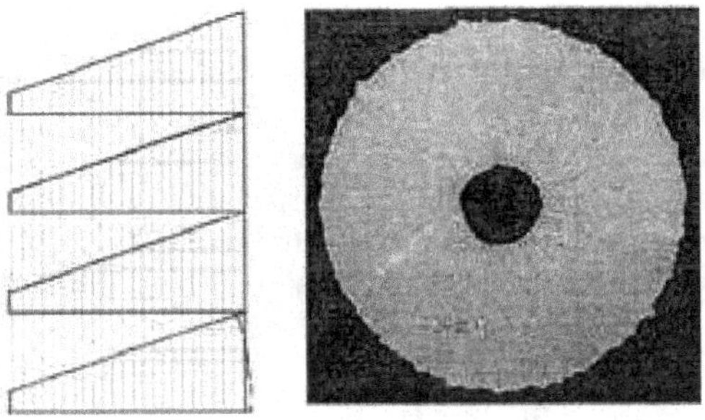

Fig. 18. Discoid shape (Cebulla et al., 2000)

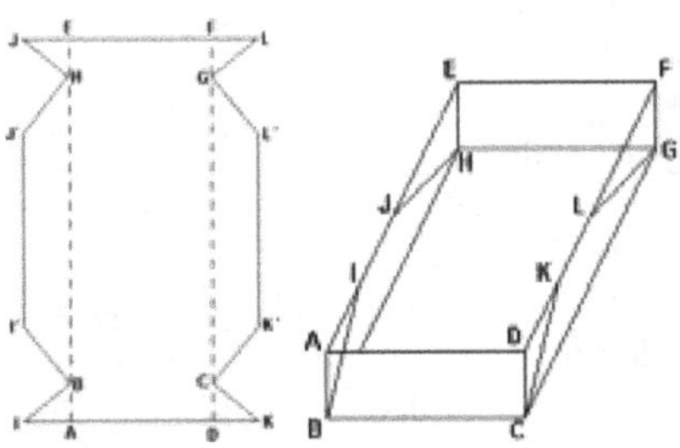

Fig. 19. Parallelepiped shape (Dias, 2000)

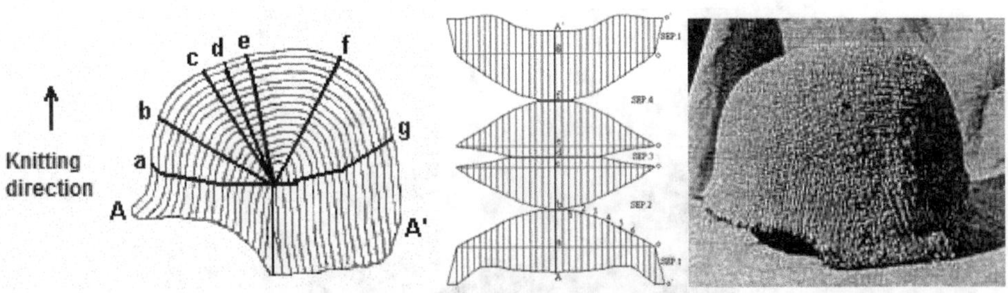

Fig. 20. Helmet made of aramid fibres – theoretical form, 2D fabric and final preform (Araujo et al., 2000)

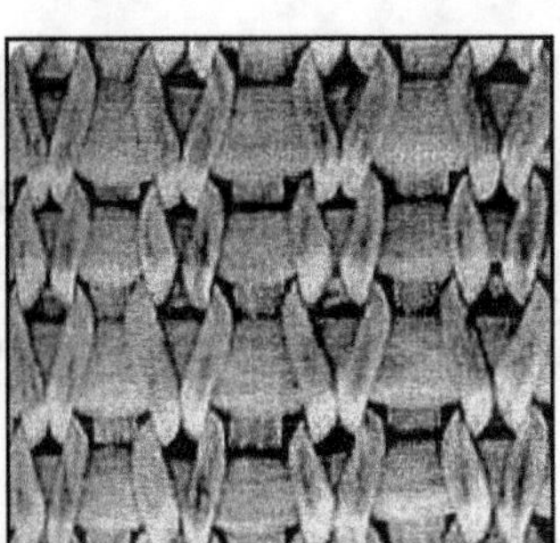

Fig. 21. Jersey fabric reinforced with weft and warp yarns (Cebulla et al., 2000)

There are some issues related to the fashioned fabrics used as preforms. One is the fabric strength, considering that the weft knitted materials exhibit limited mechanical characteristics. The solution to this problem is to insert warp and weft yarns within the structure. Fig. 21 presents a jersey fabric with warp and weft yarns, produced on a flat machine with adapted yarn feeder for the warp yarns (Cebulla et al., 2000). Apart from giving the fabric strength, it also improves the volume fraction of the reinforcement, increasing the quality of the composite materials.

Sandwich/spacer fabrics

A sandwich/spacer fabric is a 3D construction made of two separate fabrics, connected in between by yarns or knitted layers (Ciobanu, 2003). The fabric thickness determined by length of the connecting yarns/layers.

When produced on warp machines, these fabrics are known as spacers. They are obtained on double needle bar machines, with 4 to 6 guide bars - 1 or 2 guide bars produce the independent fabrics by knitting only on one bed and the middle bars create the connection by working on both beds (forming stitches or being in-laid). The fabric thickness depends on the distance between the two beds (spacer distance).

Fig. 22 exemplifies a spacer fabric made of glass fibre. An interesting application for spacer fabrics are the so called textile reinforced concrete that is used in buildings. Liba designed a double needle bar Raschel machine model DG 506-30-2HS that produces spacer fabrics with net structure, as illustrated in Fig. 23. The fabrics present weft and warp in-lays that form the net geometry and give the fabric strength on both directions (source: www.liba.de).

Fig. 22. Spacer fabric made of glass fibre

Fig. 23. Net spacer used for textile reinforced concrete

In the case of weft knitted fabrics, they are known as sandwich fabrics. Even if they can be produced also on circular machines (connection with yarns), they are mainly produced on electronic flat knitting machines that offer the required technical conditions and the development possibilities.
The connection can be generated through yarns fed on both beds or by knitted layers. The first solution is limited with regard to shape complexity and fabric thickness. The second connecting principle requires knitting separately on the two beds and at a certain point to stop and knit the connection layer on selected needles, usually 1x1. These needles can work also for the separate fabrics (if the length of the layer is small enough), or can be used exclusively to produce the connection, if the length and/or the shape complexity require (Fig. 24).

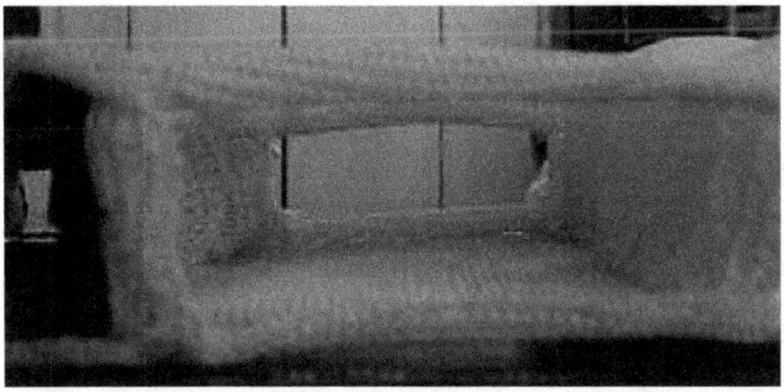

Fig. 24. Sandwich fabric with connecting knitted layers

There are two types of connecting layers (Araujo et al., 2000):

- Single layers (Fig. 25.a) – the layer is produced on one bed (jersey) or on both beds (rib, interlock) and can have a perpendicular or an inclined disposition between the separate fabrics.
- Double layers (Fig. 25.b) - two layers are knitted separately on the beds, connected at a certain point with a rib evolution; if a specified amount of rib courses will be produced also in the exterior fabrics, then the connection will be "X" shaped, with possibilities to extend more the rib dimensions or to alternate the disposition of the two layers.

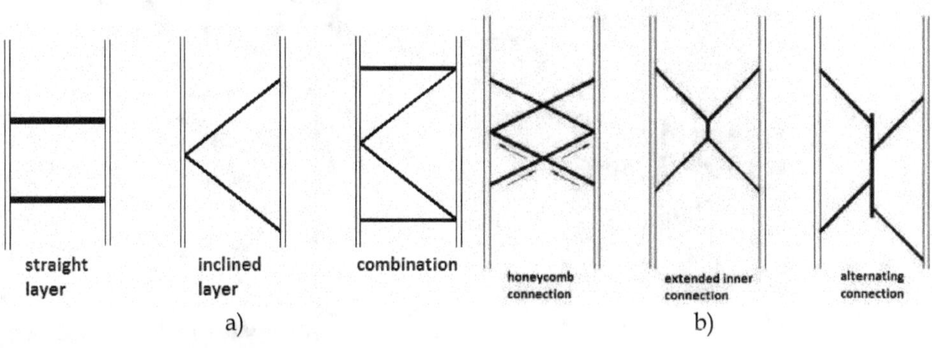

Fig. 25. Types of connecting layers

The sandwich fabrics presented are characterised by constant thickness and rectangular form in cross section. There are three major ways to diversify these fabrics and to obtain structures with complex shape (Ciobanu, 2003):

- The use of connection layers with different length
- The use of connection layers with variable form (the fashioning/flechage technique)
- The use of exterior fashioned separate fabrics

Sandwich fabrics with connecting layers of different length

The simple variation of the layer length will modify the cross-section of the fabric. The shape is created by the specific position of the separate fabrics between the consecutive layers. Two possibilities can be mentioned. One takes into consideration a specified sequence of layers with different length, as illustrated in Fig. 26. The different length of the layers will determine an inclined geometry for the exterior fabrics.

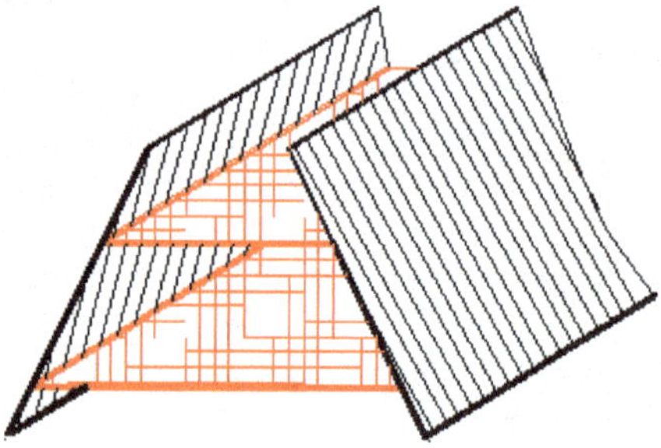

Fig. 26. Sandwich fabric with connecting layers of variable length

The other possibility refers to a sequence of two layers with predetermined (different) length, combined with an according number of courses in the separate fabric(s) that generates a corner effect (position at 90^0). It is the case of fabrics with L and T shapes, presented in Fig. 27 and Fig. 28.

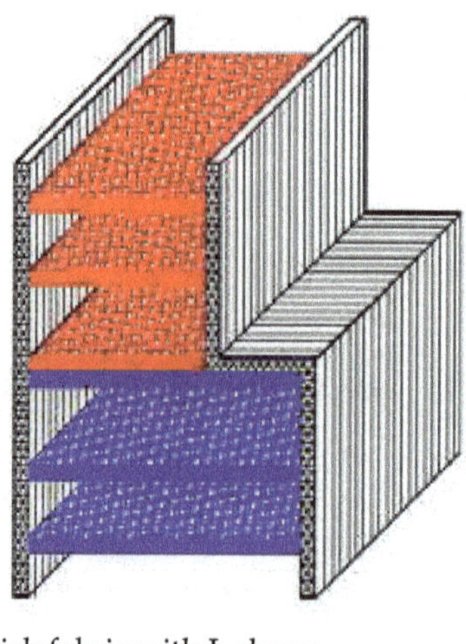

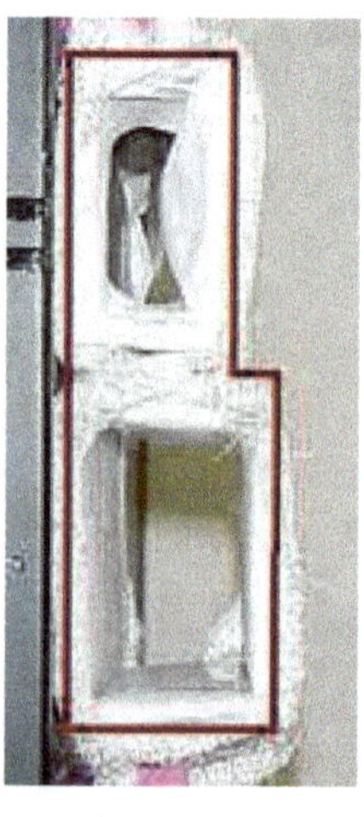

Fig. 27. Sandwich fabric with L shape

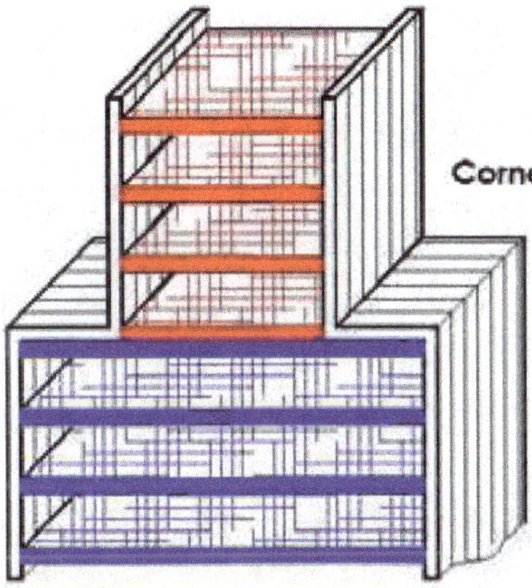

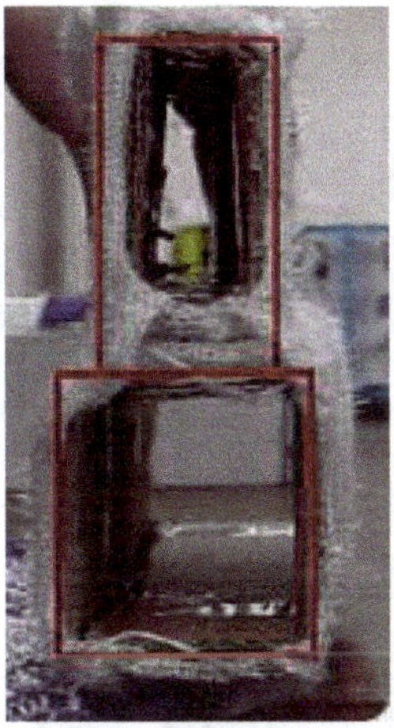

Fig. 28. Sandwich fabric with T shape

Sandwich fabrics with shaped layers (variable courses technique)

In this case the shape is determined directly by the shape of the layers, obtained through the variable courses technique. The technique is based on knitting on only a part of the needles producing the layer, in certain rows, while the others are missing. One example (knitting programme, Sirix) is presented in Fig. 29.

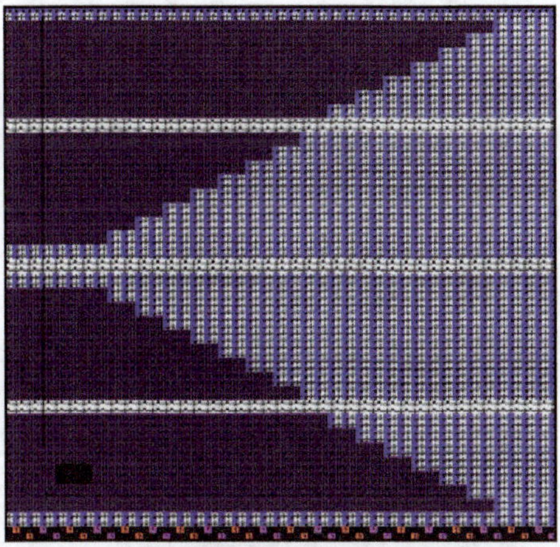

Fig. 29. Knitting sequence for a shaped layer (SIRIX, Stoll)

The fabric cross section given by the connecting layers is controlled through these layers. Different shapes can be therefore produced, as exemplified in Fig. 30. The layers are emphasised in red, while the separate fabrics are lined; the arrows are indicating the knitting direction. Two main types of shaped layers can be identified: with integral shape (the yarn guide is feeding continuously, even if on only a variable number of needles), as in Fig. 30.a. and b. and with divided shape (the evolution is split on distinct groups of needles, requiring separate feeding), as in the case of the fabric presented in Fig. 30.c.

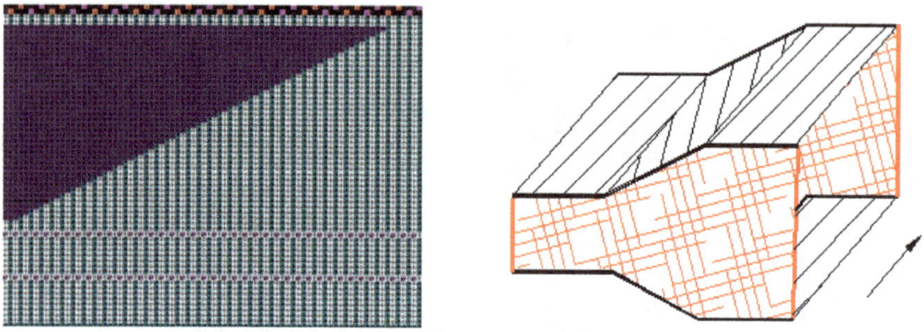

a) Sandwich fabric with trapezoidal cross section

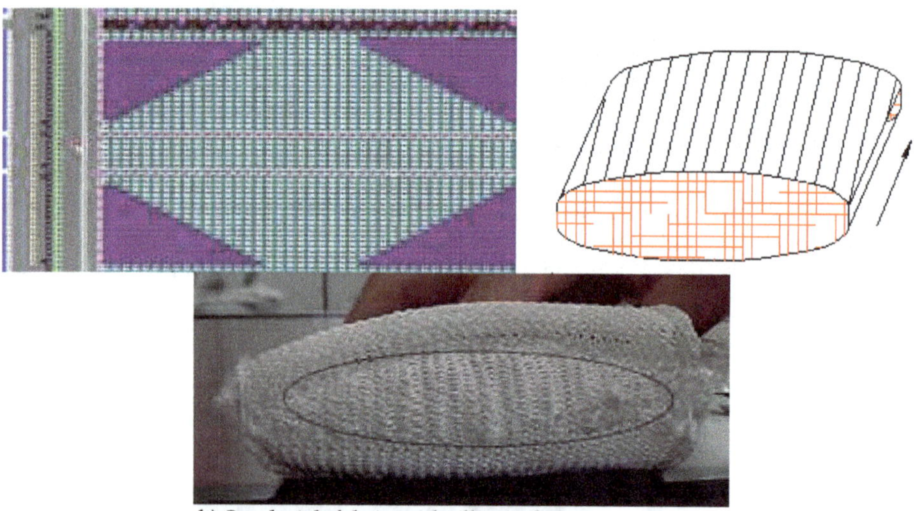

b) Sandwich fabric with elliptical cross section

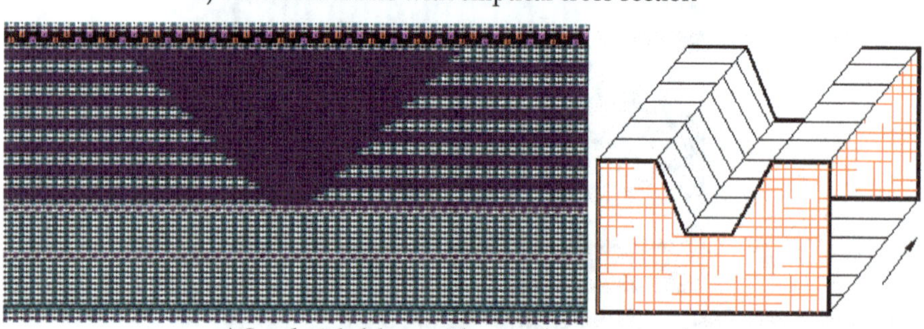

c) Sandwich fabric with complex cross section

Fig. 30. Examples of sandwich fabrics with shaped connecting layers

Sandwich fabrics with fashioned exterior fabric

The fully-fashioning of the separate fabrics is obtained by narrowing and enlargement technique the number of working needles of the fabrics, using stitch transfer and racking. Fig. 31 illustrates one such possibility.

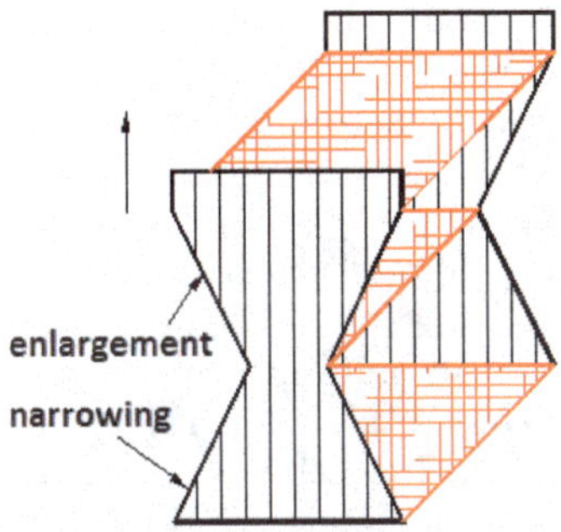

Fig. 31. Sandwich fabric with fashioned exterior fabrics

4. Application example – knitted preform for airplane wing

The application presented in the paper refers to a knitted preform for an airplane wing (glider). It was intended to knit a 3D shape identical to the wing. The preform was used to produce the composite material (through RTM).

Type and geometry of the knitted preform

The airplane wing prototype was defined using NACA 4 digital profile, according to specifications. Fig. 32 presents the geometry and dimensions of the chosen profile.
The wing is characterised by the following aspects:

- A difference in width between the beginning and the end of the wing - that requires successive narrowing;
- The difference between the two extremities is 60 mm, determining the narrowing slope;
- The fabric thickness for the outer layer is 5 mm;
- There are two interior walls with 5 mm thickness and different heights, according to the wing cross section: 79 mm - for the higher wall, and 64 mm - for the smaller wall.

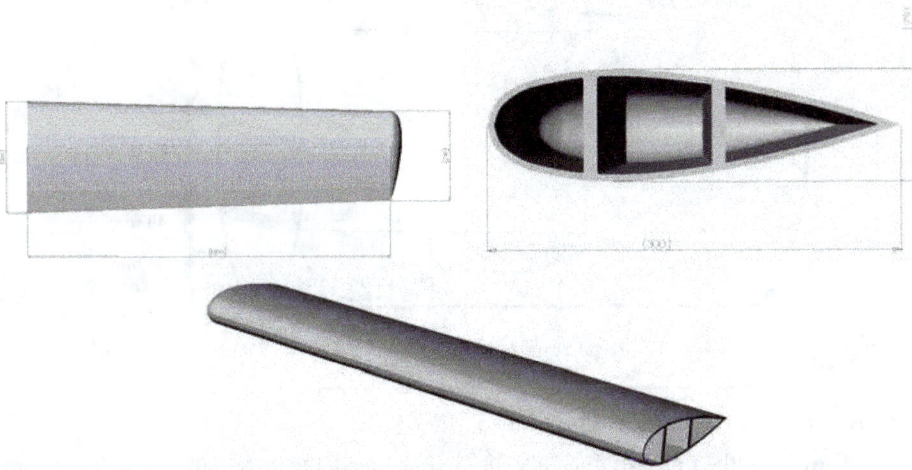

Fig. 32. Wing geometry and dimensions

Constructive, structural and technological characterisation of the preform

The complexity of the wing shape required the production of the knitted preform on a flat weft knitting machine, due to their advantages mentioned before.

The preform was knitted on a CMS 320 TC Stoll machine, using EC 11 408 Z28 T6 (Vetrotex) glass fibre. The yarn was tested on a Housenfield (Tinius Olsen) testing machine model H100 KS in order to determine its mechanical properties. The yarn is characterised in chapter 2 of this work.

The production of a preform that includes the inner walls requires a sandwich structure with connection through single knitted layers, according to the dimensions defined above. A very important aspect is the knitting direction that is determined in this case by the position of the inner walls. In the preform, the inner walls are created by the connecting knitted layers. Choosing another knitting direction, perpendicular to the one considered is not a viable solution due to the fact that this way the inner walls (connecting layers) are impossible to knit. Fig. 33 illustrates the architecture of the knitted preform. The zones marked on the drawing represent:

- border - 1 x 1 rib on selected needles;
- initial zone -1 x 1 rib - this is the part where the rib evolution is produced on all needles and the feeder for the glass yarn is not yet working; it is subsequently removed, together with the border;
- beginning zone - first zone in the preform, corresponding to its inclined inferior extremity;
- zone I - zone made only of the outer fabrics up to the first inner wall (connecting layer);
- zone II - zone of the outer fabrics between the inner walls;
- zone III - zone of the outer fabrics up to the beginning of the inclined superior extremity;
- ending zone - zone corresponding to the superior extremity.

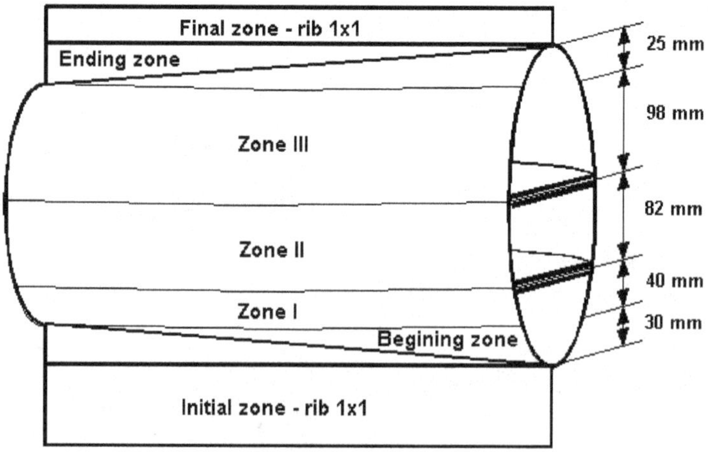

Fig. 33. Architecture of the knitted preform

The lateral edges of the preform must also be characterised, the lower edge being closed and shaped, while the wider edge is open and straight.

In the case of the considered knitted sandwich structure, both the outer fabrics and the inner walls are knitted on selected needles (1x1). A tubular evolution on selected needles with fleece yarns can be used in order to increase the fabric compactness (see Fig. 34). The main problem with last structure is the positioning outside the fabric of the floating fleece yarns that generates an irregular plush aspect, and also increases the snagging. Still, the simple pretension of the fabric, specific to the production of the composite material eliminates this situation.

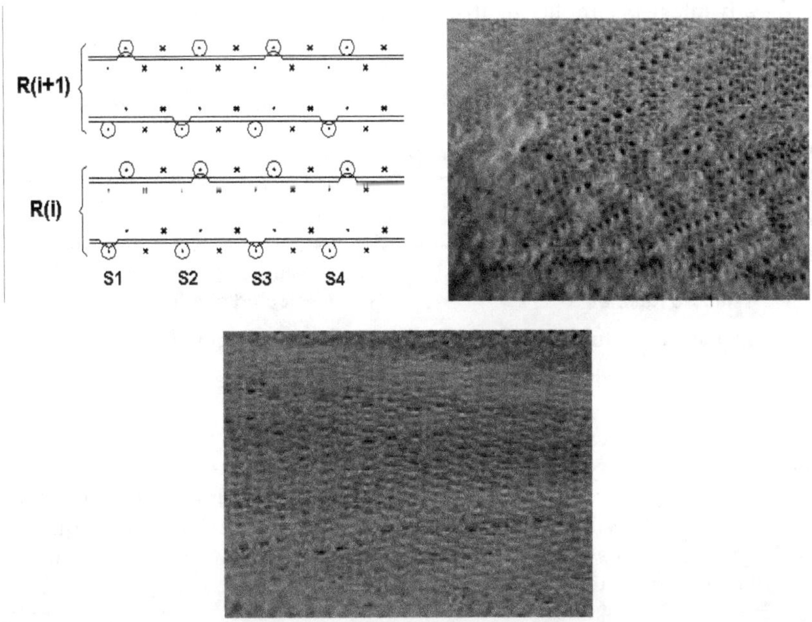

Fig. 34. Tubular evolution on selected needles with fleece in-lay – fabric aspect relaxed and stretched

Table 6 presents the values of the main structural parameters for the two structures (simple tubular on selected needles and tubular on selected needles with fleece yarns).

<table>
<tr><th rowspan="2">Structure</th><th colspan="2">Stitch density</th><th rowspan="2">Stitch pitch A, [mm]</th><th rowspan="2">Stitch height B, [mm]</th><th rowspan="2" colspan="2">Stitch length l_{stitch}, [mm]</th><th rowspan="2">Weight M/m², [g]</th></tr>
<tr><th>D_w [wales/50mm]</th><th>D_r [rows/50mm]</th></tr>
<tr><td>Tubular, on selected needles</td><td>16</td><td>26</td><td>3.12</td><td>1.92</td><td colspan="2">9.6</td><td>842</td></tr>
<tr><td>Tubular, on selected needles, with fleece yarns</td><td>12</td><td>25</td><td>4.17</td><td>1.92</td><td>10.3</td><td>4.7</td><td>950</td></tr>
</table>

Table 6. The values of the main structural parameters for the knitted fabrics

Beginning and ending zones of the preform

The difference in width between the two extremities can be obtained through more knitting sequences. One possibility is to produce some zones that will be subsequently removed. In these zones the evolution is successively changed from 1x1 rib to tubular jersey, specific to the sandwich structure. The use of two feeders for the tubular fabric (row 2) requires 1x1 rib evolutions on selected needles that present the disadvantage of larger floats. To avoid this situation, the 1x1 rib evolutions on selected needles can be maintained only for small zones, the rest of the working needles producing tubular evolutions (see Fig. 35). Such a variant presents the advantage of simplifying the knitting sequence and the process.

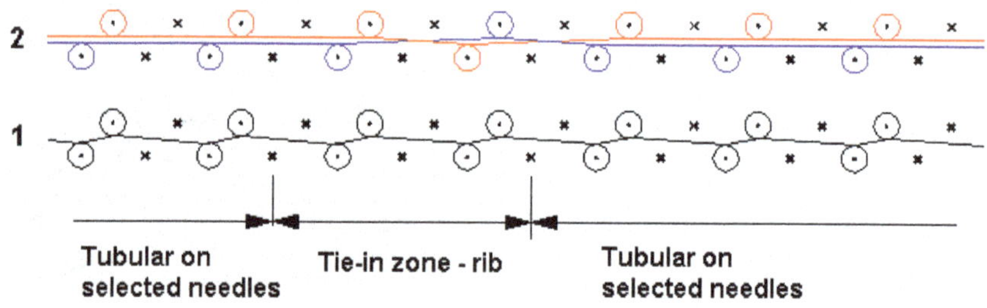

Fig. 35. Tie-in zone – from tubular to rib on selected needles (beginning zone)

Fig. 36 illustrates a fashioning solution for the beginning and ending zones. In this case, a number of incomplete rows are knitted with tubular evolution, using one yarn feeder. The groups of needles are successively introduced to work until all active needles are working. This way, the supplementary zones to be subsequently eliminated are cut out.

The number of incomplete rows will determine the slope of the inferior extremity. The slope of the superior extremity will be determined the same way, but the knitting sequence will be reversed. Regardless of the chosen variant, the number of rows knitted in the beginning zone is calculated with relation (1):

$$N_r = \frac{L}{B} \tag{1}$$

Where:

N_r - number of rows for the beginning and ending zones;

L - height of the beginning or ending zone, mm;

B - stitch height, mm.

The height of the beginning/ending zone is determined in the design stage, according to the specific dimensions of the wing; in this case 30 mm for the beginning zone and 25 mm for the ending zone. The calculated value is used to determine the groups of needles that will pass successively from one evolution to the other, or will successively start to work (2).

$$l_{group} = \frac{N_{needles}}{N_r} \tag{2}$$

Where:

l_{group} - width of a group of needles;

$N_{needles}$ - total number of needles working in each bed;

N_r - number of rows in the beginning/ending zone.

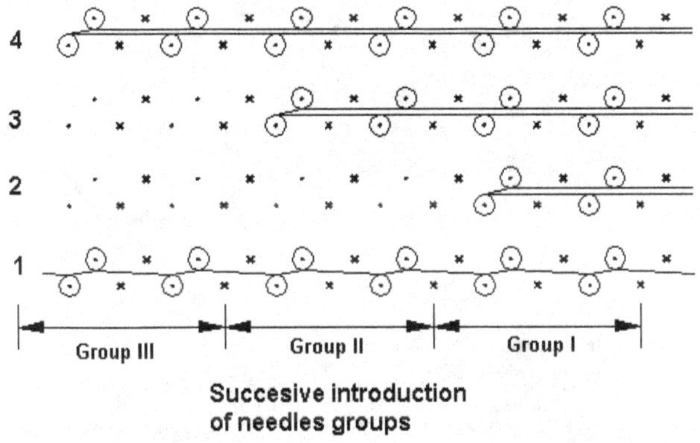

Fig. 36. Knitting sequence for the beginning zone

The inner walls of the preform

The inner walls of the preform are produced as single connecting layers of the sandwich fabric. The layer length is 80 mm, respectively 64 mm. In order to satisfy the required 5 mm thickness for the inner walls, two successive connecting layers with the same length will be knitted. After the introduction of the resin the two neighbouring connecting layers will generate an inner wall. Apart from knitting connecting layers of such high length, the production of two layers one after the other represented another problem.

During the knitting of the single connecting layer, some rows of the outer fabric are produced on the opposite bed in order to reduce the tension generated by the long period in which the needles making the outer independent fabrics miss. These supplementary rows must be taken into consideration when calculating the number of rows between the two connecting layers so that they are horizontal.

Presentation of the preform made of glass fibre

The general aspect of the preform is presented in Fig. 37. The beginning and ending zones of the outer fabrics can be observed, as well as the zones where the connecting layers were produced (see Fig. 38). One connecting zone is encircled in Fig.37.

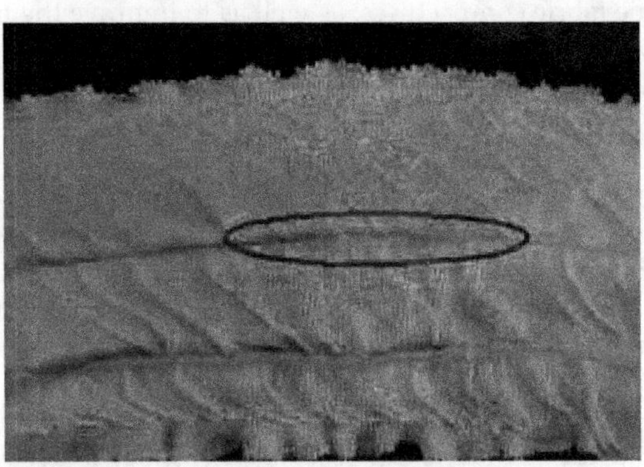

Fig. 37. Preform with jersey evolution on selected needles

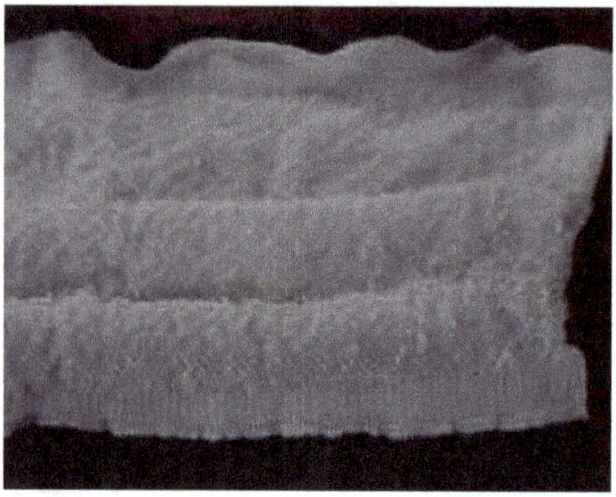

Fig. 38. Preform with jersey evolution on selected needles and fleece in-lays

The successive connecting layers are exemplified in Fig. 39 where the preform is presented in cross section, showing the two connecting layers of 80 mm length.

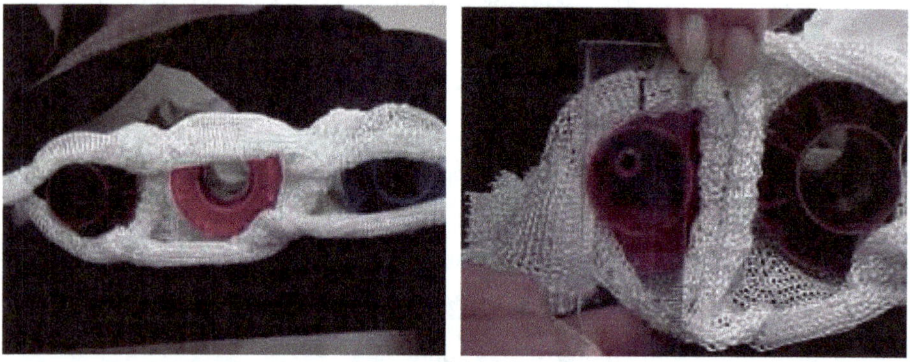

Fig. 39. Connection layers – cross section view and aspect of the first connection layer

Production of a supplementary exterior layer for the wing

In order to increase the fabric compactness, as well as to improve the mechanical behaviour of the preform, a supplementary exterior layer is added, covering the preform. The tubular fabric is knitted perpendicularly in relation to the direction used to knit the preform (see Fig. 40).

The fabric dimensions correspond to the ones presented in Fig. 32. The conical shape is obtained with successive, symmetrical narrowing. The tubular fabric is closed, giving a rounding effect in N_1 and N_2 zones. The closing can be done using the bind-off technique, though it is preferably to avoid its use for fabrics made of glass fabric. The repeated transfers lead to a significant amount of broken filaments, affecting the mechanical behaviour and even destroying the stitches during the introduction of the fabrics in/on the moulds before the RTM process.

Considering that there are no restrictions regarding the structure and its tensioning, the evolution chosen for the tubular fabric can be jersey, 1x1 jersey or even jersey fleece (see Fig. 41).

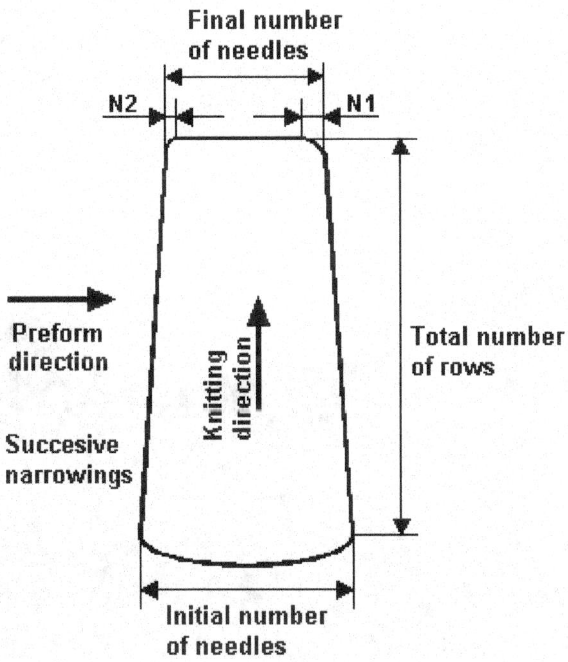

Fig. 40. Production of the supplementary layer

Fig. 41. Front aspect of a jersey and 1x1 miss jersey fabrics made with glass fibre count 408 tex

Production of a glider wing made of composite material using a knitted preform

The composite material glider wing was produced using an injection process RTM (Nicolau et al., 2002). The resin used was unsaturated polyester orthophtalic resin, S226E (Neste), characterised by reduced viscosity, 150 mPa·s. The pre-acceleration was obtained with a 0.15% solution of cobalt 6%. The catalyst used was Trigonox TX 44B. This recipe formula avoids resin curing during the injection process.

Moulds production

The complexity of the preform shape required exterior and interior moulds that would generate the desired shape for the composite. The exterior mould (see Fig. 42) is made of a composite material reinforced with two layers of glass fibre matt. The three interior moulds (Fig. 43) that are to be introduced in the preform, are produced with polyurethane foam and have a protective layer.

Fig. 42. Exterior mould - inferior and superior components

Fig. 43. Internal moulds - front view

Description of the RTM process

The injection process included the following stages (Nicolau et al., 2002):

- mould cleaning and application of the separation liquid;

- introduction of the interior moulds in the preform (Fig. 44) and covering with the supplementary tubular fabric;
- positioning in the exterior mould;
- mould vacuum closing;
- connecting the mould to the devices for resin injection and vacuum;
- vacuum formation inside the mould;
- resin preparation and catalyst introduction inside the bowl under 1 bar pressure;
- starting the injection process (Fig. 45).

Fig. 44. Introduction of the interior moulds in the preform

Fig. 45. Resin injection process

The polyester resin was injected inside the exterior mould at low pressure ($1{,}5 \cdot 10^5$ Pa) using the vacuum device. The injection process took approximately 30 minutes. After injection the composite material was subjected to a thermal treatment in order to complete the curing process. The composite wing is presented in Figure 46.

Fig. 46. Composite wing - aspect before and after resin injection

5. Conclusions

Knitted fabrics are well used in the field of technical textiles, including composite materials with plastic matrix. As composite reinforcement, knitted fabrics have some negative points with regard to their mechanical behaviour, but this can be improved through structure (especially the use of in-laid yarns) and structural parameters.

Other issues concern the knittability of high performance fibres (volume of fibre damage during processing) and there are ways of limiting the filament breaking phenomenon - the modification of knitting elements, lower yarn tension, etc. The study presented in this chapter identifies as main cause of destruction the rob-back stage in the knitting process, when the yarns are pulled over the trick plate under high tension. The best approach to knitting high performance fibres seams to be a case by case approach, when the knitting conditions can be determined in order to obtain good quality fabrics.

Both weft and warp knitting technologies can be used to produce composite reinforcement. Warp knitting is best suited for structures with in-laid straight yarns (multiaxial fabrics), while weft knitting allows obtaining fabrics with three dimensional architecture, used as preforms for advanced composite materials. The examples of 3D knitted fabrics presented in this chapter illustrate the idea that knitting is a viable option when considering the production of complex shapes. Further studies should explore diversification possibilities. Other directions for future development are: the design of knitted fabric in relation to 3D bodies, the simulation of fabric spatial geometry and properties prediction.

6. References

Araujo, M.D., Hong H. & Fangueiro, R. (2000) *Têxteis Técnicos: Materiais do Novo Milénio*, vol. I *Visão Geral*, Williams Lda, ISBN 972 98468 0 4, Braga (in Portuguese)

Araujo, M. D.; Hong, H. & Fangueiro, R. (2000). *Têxteis Técnicos: Materiais do Novo Milénio*, Vol. III, *Aplicações, Novos Processos e Novos produtos.* Williams Lda, ISBN 972 98468 2 0, Braga (in Portuguese)

ASTM (2008). ASTM D2256 Standard Test Method for Tensile Properties of Yarns by the Single-Strand Method, *ASTM Standards on Disc*, section 7 Textiles (I)

Cebulla, H.; Diestel, O. & Offerman, P. (2001). *Modelling* and Production of Fully Fashioned Biaxial Weft Knitted Fabrics, *TECNITEX 2001, AUTEX, Technical Textiles: Designing Textiles for Technical Applications*, vol. I, editor Araujo, M.D., Portugal, pp. 263-269, ISBN 972-98468-3-9, Povoa de Varzim, Portugal, Williams Lda, Braga

Ciobanu, L. (2003). *Contributions to the production and development of technical knitted fabrics*, PhD thesis, „Gheorghe Asachi" Technical University, Iaşi (in Romanian)

Dias, T. et al. (2000). Knitting Seamless Three-Dimensional Shell Structures on Modern Electronic Flat Bed Knitting Machines, *Proceedings of the International Conference, 24&25 August, Bolton, UK*, pp. 36-43, ISBN 1 85573 494 X, Woodhead Publishing, UK

Fukuta et al., in 15th Textile Research Symposium, the Textile Machinery Society of Japan, Osaka, 1984, p. 36-38

Hu, J. (2008). Introduction to three-dimensional fibrous assemblies, in *3-D fibrous assemblies Properties, applications and modelling of three-dimensional textile structures*, Woodhead Publishing, ISBN 978-1-84569-377-0, Cambridge, UK

Lau, K.W. & Dias, T. (1994) *Knittability of High – Modulus Yarns*, JTI, vol. 85, nr. 2, , pp. 173-190, Woodhead Publishing, ISSN 0040 5000, Cambridge, UK

Leong, K.H. et al. (2000). The potential of knitting for engineering composites - a review, *Composites Part A: applied science and manufacturing*, 31, p. 197-220

Mallick, P. (1997). *Composites Engineering Handbook*, Marcel Dekker, USA

Miller, D.M. (1989). Glass Fibers, in *Engineered Material Handbook*, ASM International Handbook Committee, Vol. I - Composites, p. 45-48, ISBN 0 87170 279 7

Mouritz, A.P., et al. (2000). Review of applications for advanced three-dimensional fibre textile composites, *Composites Part A: applied science and manufacturing*, vol. 30, , issue 3, p. 1445-1461, ISSN 135-835X

Muckhopadhyay, S.K. (1994). High-performance Fibres, *Textile Progress*, vol. 25, nr. 3/4, Textile Institute, ISSN 0040 5167, ISBN 1870812662

Nicolau, A. et al. (2002). Structural Composite Parts Production from Textile Preforms, Materiais 2001, 1st International Materials Symposium, Pos M5, Porto, Portugal

Raz, S. (1989). *Warp Knitting Production*, Verlag Melliand Textilberichte, Heidelberg, ISBN 3-87529-022-4

Savci, S.; Curiskis, J.I. & Pailthorpe, M.T. (2001). Knittability of Glass Fiber Weft - Knitted Preforms for Composites, *TRJ* vol. 71, 1, pp. 15-21, ISSN 0040-5175

Scardino, F. (1989). An Introduction to Textile Structures and their Behaviour, in *Textile Structural Composites*, Composite Materials Series, vol. 3, editors Tsu-Wei Chou, Ko, F.K., p. 1-24, ISBN 0-444-42992-1, Elsevier Science Publishers B.V., Amsterdam

Song, G.; Wu, J. & Wei, Y. Computer Aided-Design of Three Dimensional Knitwear, JOTI, vol. 97, no. 6, pp. 549-552

Verpoest, I. et al. (1997). The potential of knitted fabrics as a reinforcement for composites, *Proceedings Volume of the 11th International Conference on Composite Materials*, Gold Coast, Australia, 14-18 July, 1997, 1:108-33

Permissions

All chapters in this book were first published in ACMEA, by InTech Open; hereby published with permission under the Creative Commons Attribution License or equivalent. Every chapter published in this book has been scrutinized by our experts. Their significance has been extensively debated. The topics covered herein carry significant findings which will fuel the growth of the discipline. They may even be implemented as practical applications or may be referred to as a beginning point for another development.

The contributors of this book come from diverse backgrounds, making this book a truly international effort. This book will bring forth new frontiers with its revolutionizing research information and detailed analysis of the nascent developments around the world.

We would like to thank all the contributing authors for lending their expertise to make the book truly unique. They have played a crucial role in the development of this book. Without their invaluable contributions this book wouldn't have been possible. They have made vital efforts to compile up to date information on the varied aspects of this subject to make this book a valuable addition to the collection of many professionals and students.

This book was conceptualized with the vision of imparting up-to-date information and advanced data in this field. To ensure the same, a matchless editorial board was set up. Every individual on the board went through rigorous rounds of assessment to prove their worth. After which they invested a large part of their time researching and compiling the most relevant data for our readers.

The editorial board has been involved in producing this book since its inception. They have spent rigorous hours researching and exploring the diverse topics which have resulted in the successful publishing of this book. They have passed on their knowledge of decades through this book. To expedite this challenging task, the publisher supported the team at every step. A small team of assistant editors was also appointed to further simplify the editing procedure and attain best results for the readers.

Apart from the editorial board, the designing team has also invested a significant amount of their time in understanding the subject and creating the most relevant covers. They scrutinized every image to scout for the most suitable representation of the subject and create an appropriate cover for the book.

The publishing team has been an ardent support to the editorial, designing and production team. Their endless efforts to recruit the best for this project, has resulted in the accomplishment of this book. They are a veteran in the field of academics and their pool of knowledge is as vast as their experience in printing. Their expertise and guidance has proved useful at every step. Their uncompromising quality standards have made this book an exceptional effort. Their encouragement from time to time has been an inspiration for everyone.

The publisher and the editorial board hope that this book will prove to be a valuable piece of knowledge for researchers, students, practitioners and scholars across the globe.

List of Contributors

Enzo Martinelli, Antonio Bilotta, Ciro Faella and Emidio Nigro
Department of Civil Engineering, University of Salerno, Fisciano (SA) ,Italy
Department of Structural Engineering, University of Naples "Federico II", Naples, Italy

Wilma Polini
Dipartimento di Meccanica, Strutture, Ambiente e Territorio, Italy
Università degli Studi di Cassino, Cassino, Italy

Wojciech Błażejewski, Paweł Gąsior and Jerzy Kaleta
Wrocław University of Technology, Poland

Boqin Gu, Ye Chen and Jianfeng Zhou
Nanjing University of Technology, China

Denis Mihaela Panaitescu, Adriana Nicoleta Frone, Marius Ghiurea, Catalin Ilie Spataru, Constantin Radovici and Michaela Doina Iorga
National Institute of Research and Development in Chemistry and Petrochemistry, Romania

Elena-Felicia Beznea and Ionel Chirica
University Dunarea de Jos of Galati, Romania

A.H. Bhat and H.P.S. Abdul Khalil
School of Industrial Technology, Universiti Sains Malaysia, Penang-11800, Malaysia

A.K. Banthia
Materials Science Centre, Indian Institute of Technology Kharagpur-721302, India

Luminita Ciobanu, Ph.D., Eng.
"Gheorghe Asachi" Technical University of Iasi, Romania

Index